# Food 88

# 边伐木边造林

Cutting Wood while Growing Forests

Gunter Pauli

冈特·鲍利 著

凯瑟琳娜·巴赫 绘
李欢欢 牛玲娟 译

## 丛书编委会

主　任：贾　峰
副主任：何家振　郑立明
委　员：牛玲娟　李原原　李曙东　吴建民　彭　勇
　　　　冯　缨　靳增江

## 丛书出版委员会

主　任：段学俭
副主任：匡志强　张　蓉
成　员：叶　刚　李晓梅　魏　来　徐雅清　田振军
　　　　蔡雩奇　程　洋

特别感谢以下热心人士对译稿润色工作的支持：

姜竹青　韩　笑　贾　芳　刘　晓　张黎立　刘之杰
高　青　周依奇　彭　江　于函玉　于　哲　单　威
姚爱静　刘　洋　高　艳　孙笑非　郑莉霞　周　蕊

# 目录

| | |
|---|---|
| 边伐木边造林 | 4 |
| 你知道吗？ | 22 |
| 想一想 | 26 |
| 自己动手！ | 27 |
| 学科知识 | 28 |
| 情感智慧 | 29 |
| 艺术 | 29 |
| 思维拓展 | 30 |
| 动手能力 | 30 |
| 故事灵感来自 | 31 |

# Contents

| | |
|---|---|
| Cutting Wood while Growing Forests | 4 |
| Did you know? | 22 |
| Think about it | 26 |
| Do it yourself! | 27 |
| Academic Knowledge | 28 |
| Emotional Intelligence | 29 |
| The Arts | 29 |
| Systems: Making the Connections | 30 |
| Capacity to Implement | 30 |
| This fable is inspired by | 31 |

猩猩忧心忡忡,加里曼丹岛的森林正不断被砍伐。也许很快他就没有可以悬挂的树枝了。这时他看见一只侏儒象正沿着河岸试图逃跑。
"你为什么这么匆忙呢?"猩猩问道。

The orangutang is worried. Forests are being cut down in Kalimantan. Soon he may not have a branch to hang from. He spots a pygmy elephant trying to escape along the river.
"Why are you in such a hurry?" asks the orangutang.

你为什么这么匆忙呢?

Why are you in such a hurry?

# 我们的家园快要消失了

Our home is about to disappear

"我们的家园快要消失了,这片物种丰富的茂密森林被棕榈树取代了,没人愿意住这里了。"侏儒象尖叫着说。

"我不清楚人类是否知道野生生物无法生活在非洲棕榈树林里,但是棕榈树适合制造不会污染欧洲河流的肥皂。"

"Our home is about to disappear, the thick and diverse forest is replaced with oil palm trees where no one wants to live," squeals the elephant.

"I wonder if people know that wildlife cannot live in African oil palm plantations. On the other hand, the palm oil is good for making soap that do not pollute rivers in Europe."

"让欧洲的河流保持干净是很不错的想法，但要以破坏我们的家园为代价吗？这不合理。"

"唉，以前欧洲的河流被泡沫覆盖。后来，有人发现棕榈油制成的肥皂清洁度高，可以让泡沫迅速消失，河面就不再有泡沫了。"

"这给了企业家在亚洲种植非洲棕榈树，破坏我们栖息地的特权吗？"

"Cleaning up rivers in Europe is a great idea, but at the cost of destroying our home? That can never be justified."

"Well, Europe used to have rivers which were covered in foam. Then they discovered that palm oil-based soap cleans well, disappears fast, and keeps the rivers foam free."

"Does that give these entrepreneurs license to plant an African tree in Asia and to destroy our habitat?"

棕榈油制成的肥皂清洁度高

Palm oil-based soap cleans well

他们最初并不知道

But at the beginning they didn't know

"不是,他们最初并不知道。"

"你是说,他们不知道种植非洲的树种会破坏我们家族世世代代生活的森林?"

"我相信他们不知道。但由于肥皂十分好用,越来越多的森林被破坏了,由棕榈树取而代之。"

"No, but at the beginning they didn't know."

"You mean they didn't know that African trees were going to be planted on destroyed forest land where my family has been roaming for ages?".

"I trust they didn't know. But as the soap was a success, more and more forests needed to be destroyed to be replaced by plantations."

"他们意识到这点时,做了什么吗?"
"他们继续种植棕榈树。"
"什么?这是犯罪!"
"他们中有些人意识到了自己所犯的错误,为此感到难过,希望创造出可持续的棕榈林。"

"Once they had figured it out, what did they do?"
"They continued to plant the palm trees."
"What? That is criminal!"
"Some of them realised their error, felt bad, and wanted to create sustainable palm oil plantations."

创造出可持续的棕榈林

To create sustainable palm oil plantations

改正过去的错误

To reverse the mistakes of the past

"破坏土地来进行种植,怎么能称为可持续?更糟糕的是,这些棕榈林完全是空荡荡的,它们已经被破坏了,不会再有生命存在了!人类没有意识到吗?"

"巨大的破坏已经造成,但人类已经学会了如何改正过去的错误。"

"怎么做呢?"

"How can you call anything sustainable when it is planted on destroyed land? And what's ever worse is that these palm plantations are as good as empty forests. This damage is done and there will be no life here ever again! Don't people realise that?"

"Great damage has been done, but we have learned how to reverse the mistakes of the past."

"How?"

"人类在荒芜的大草原上种植了成千上万的松树,那里过去是一大片森林。"
"只种植松树,真是个坏主意。"
"松树只是个开始,它是先锋植物。"
"那最后会发生什么?"

"People planted millions of pines in the barren savannah that was once a great forest."
"Planting only pines is a bad idea."
"The pine is only the beginning. It is the pioneer plant."
"And what will happen in the end?"

在大草原上种植了成千上万的松树

Planted millions of pines in the savannah

出现了一片幼小的森林

A young forest can emerge

"也许你无法预测到最终结果,但你可以决定发展方向。"

"通往什么方向?我可以和我的朋友一起走一趟吗?"

"由于松树的绿荫遮挡,土壤温度很低。雨水渗过土壤,形成水源,帮助休眠种子生长,所以出现了一片幼小的森林。"

"You can never predict the end, but you can decide on the direction."

"What direction is this road leading to? Can I walk it with my friends?"

"Thanks to the shade of the pine trees, the soil is now cooler and rainwater filters through it, producing drinking water, and helping all the dormant seeds grow so that a young forest can emerge."

"太神奇了！"

"人们从每十株松树中移走八株，这样当地自然生长的植物能够茁壮成长。于是当地的物种更丰富了。"

"这意味着我们可以边伐木边造林。我们都应该这么做！"

……这仅仅是开始！……

"Fantastic!"

"The community removes eight out of every ten young pine trees so that native plants that grow there naturally can thrive. This improves the biodiversity of the area."

"That means we can cut wood while growing new forests. We should all do that!"

... AND IT HAS ONLY JUST BEGUN!...

……这仅仅是开始！……

... AND IT HAS ONLY JUST BEGUN! ...

# Did You Know?

## 你知道吗？

In three countries alone (Indonesia, Malaysia, and Papua New Guinea) 3.5 million hectares of forestland were lost to palm plantations between 1990 and 2010.

1990 年至 2010 年之间，印度尼西亚、马来西亚和巴布亚新几内亚三个国家就消失了 350 万公顷森林，用于种植棕榈树。

The loss of rainforests threatens the survival of the last remaining Sumatran tigers, the orangutang, and the pigmy elephant, as well as the world's largest biodiversity of vascular plants.

热带雨林的减少威胁着濒临灭绝的苏门答腊虎、猩猩和侏儒象的生存，以及世界上最多样化的维管束植物。

When deforestation occurs on peat soils, enormous amounts of CO2 is emitted. It is not necessary to deforest so much land as 12.5 million hectares of degraded land is already available for tree plantations.

在泥炭土上滥伐森林，会释放大量的二氧化碳。有1250万公顷的退化土地可以用来造林，没有必要如此滥伐。

Palm oil is the most used vegetable oil in the world. 42 million tonnes of palm oil are exported each year, representing 65% of all internationally traded oil. It is used in everything from chocolate to toothpaste, to lipstick, and cookies.

棕榈油是世界上使用最多的植物油，每年出口4200万吨，占全球油量交易的65%。它广泛用于制作巧克力、牙膏、唇膏和小甜点。

Deforested land in the Vichada Department of Colombia regenerated its rainforest more than 200 years after its destruction. Las Gaviotas demonstrated that the combination of science and human commitment could correct the errors of the past.

在热带雨林遭破坏的两百年后,哥伦比亚比查达省在原来的土地上重建了一个雨林。拉斯卡维塔斯的实践证明,科学加上人类的努力可以改正以前的错误。

The regeneration of the rainforests in Las Gaviotas includes blending oil palm trees into the new forest. The oil is processed locally and made available for local consumption. The excess wood is used for energy and to make fuel.

拉斯卡维塔斯雨林的重建包括在新森林里混合种植棕榈树。棕榈油在当地生产,供当地消费。木材用于能源和燃料生产。

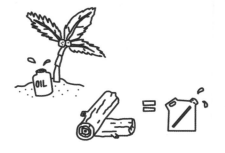

The denuded land in Colombia had 17 plants, including 11 non-native grasses. 25 years after the initiation of the project, the regenerated forests have 256 plant species.

哥伦比亚这片被砍伐过森林的土地曾经只有17种植物，包括11种外来物种。项目启动25年后，重建的森林中已经有了256种植物。

Pollen studies enables the identification of all the plants that once thrived in this region for over six thousand years. Small bio-reserves along the rivers provided the natural source of biota carried by birds, bees, and the wind to "fill" the forest.

通过花粉研究，确定了过去六千年曾在这片土地生长的所有植物。沿河的小型生物保护区是这片区域的生物源头，鸟类、蜜蜂和风把这些生物带入森林。

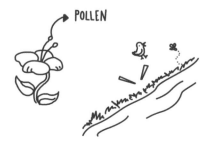

# Think About It
# 想一想

Would you like to eat chocolate knowing that the oil it is made of is responsible for destroying rainforests and the habitat of tigers, elephants, and primates?

喜欢吃巧克力吗？你知道用来做巧克力的棕榈油是毁坏热带雨林，破坏老虎、大象和猩猩栖息地的原因吗？

如果你不知道自己的行为有不良后果，那就不是故意的，可以被原谅。但是一旦知道了后果，你会怎么做？

If you did not know your actions had negative consequences, they were unintended and forgivable, but what do you do once you find out the truth?

Do you think that planting one tree a year is enough? How many trees are needed to grow a forest?

你认为一年种一棵树够吗？需要种植多少棵树才能形成一片森林？

如果你有办法能消除已经造成的伤害，你还会对自己之前犯的错误感到难过吗？

Would you still feel bad about your errors of the past if you knew there was a way to reverse the damage done?

# Do It Yourself!
## 自己动手!

Check all the food products in your house. Which ones contain palm oil? Most of the time the label only states "vegetable oil", without any further details. Put all the products you find together on a table and take a photograph of it. Then you will realise that palm oil has become a big part of our lives.

检查家里所有的食品,哪些含有棕榈油?大部分标签只标有"植物油",没有更详细的说明。把你找到的所有食品放在一张桌上,拍一张照片,然后你会发现棕榈油已成为我们生活中的重要组成部分。

# TEACHER AND PARENT GUIDE

## 学科知识
### Academic Knowledge

| | |
|---|---|
| 生物学 | 非洲象、亚洲象和侏儒象的区别；濒临灭绝的苏门答腊虎；具有丰富生物多样性的沿河森林走廊的重要性；森林重建时树木很快就开始锁住土壤，生长过程中，多达80%的树会被移除，以便让最强壮的树生长；重建森林过程中，热带松树在贫瘠草原恶劣环境中的生存能力。 |
| 化 学 | 棕榈油的饱和脂肪酸含量低，具有抗氧化性，含有脂肪酸、甘油和胡萝卜素；棕榈油用于生产甲酯和加氢脱氧生物柴油。 |
| 物 理 | 植物遮挡阳光，树荫降低了土壤温度，使其低于雨水温度，能够更好地吸收水分，保护种子和幼苗免受大量紫外线照射。 |
| 工程学 | 棕榈油生产需要研磨、分馏精炼、结晶、固体分离、熔化、脱胶、过滤和漂白；红棕榈油冷榨装瓶，用于烹饪；制油产生的有机废物能制成燃料。 |
| 经济学 | 一个标有"有机"或"可持续"的产品能轻易取代同类或稍有不同的产品；许可经营一项业务似乎还包含允许破坏和污染。 |
| 伦理学 | 棕榈油曾是非洲和亚洲穷人的主要食品，使用于成千上万种产品，并转换成燃料，这使棕榈油价格提升，引起了棕榈油究竟是用于燃料还是食品的争论；如果是无意导致的后果，那么有责任改正以前的错误。 |
| 历 史 | 棕榈油的使用最远可追溯到5000年前的埃及；19世纪70年代，棕榈油是加纳和尼日利亚的主要贸易产品；英国工业革命期间，棕榈油是优选润滑剂。 |
| 地 理 | 加里曼丹地区隶属于印度尼西亚；棕榈树生长在全球的热带地区；雨水丰富的森林一旦消失，就会出现稀树草原，后者之后会变成沙漠。 |
| 数 学 | 每公顷棕榈树每年生产3.7吨棕榈油，是大豆产量的10倍、油菜籽产量的5倍；全球种植了将近1亿公顷大豆、1000万公顷棕榈树，但两者都被少数企业垄断。 |
| 生活方式 | 消费者还未意识到时，棕榈油就已经取代了多种当地油产品；消费者只看见产品及其功能，不会意识到自己的生活会影响到那些遥远的国家。 |
| 社会学 | 西方文化的线性时间轴观念：那些已发生的事无法改变；东方文化的循环时间轴观念：丢失的机会会回来，因此已发生的错误能改正。 |
| 心理学 | "格式塔"心理学强调看到整体以及在对某事形成看法前了解事情的来龙去脉；当我们意识到能挽回损失，过去的负面影响将来能被正面影响取代时，会觉得安心；如何把愤怒转变成兴奋的能量。 |
| 系统论 | 单一种植（如在一片土地上只种棕榈树、玉米、小麦或西红柿）常常会导致很多意想不到的问题，企业很少考虑这些问题，因为他们只注重核心业务和利益，忽略了潜在的外部影响。 |

# 教师与家长指南

## 情感智慧
## Emotional Intelligence

### 猩猩

猩猩关心森林里的朋友，表现了同情心。他仔细思考并意识到使用棕榈油的利弊，认为快速生物降解的棕榈油能清洁欧洲河流。猩猩认为企业没有意识到自己造成的破坏，而且在发现事实后也没有采取任何措施。猩猩肯定了那些对所犯错误感到难过，进而想要解决问题并促进可持续种植的人，他坚定地相信过去的负面影响能消除。在侏儒象的敦促下，猩猩给出了一个积极解决方法的例子：从松树开始种植一片新森林。他解释了已经做的尝试，也承认现在还没有明确结果。但他向侏儒象保证，这一发展方向是前进之路。

### 侏儒象

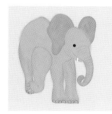

侏儒象容易激动，说话语调高昂。他担心失去自己的家园，基于道德和正义的标准，表明毁坏家园永远都没有正当理由。他说人类不应该为了达到利益而去做一些坏事。他质疑颁发企业运营许可证的系统，起初那些企业没有意识到会造成破坏，但发现事实后还继续之前的行为，侏儒象为此感到很生气。他作出肯定的判断，认为这是犯罪，并对人们执意认为在退化的土地上进行新型种植是可持续种植提出问题。当猩猩指出以松树为先锋植物重建森林的方法时，侏儒象起初强烈反对，但猩猩讲清了过程后，他同意了这个方法，由愤怒变成了兴奋。

## 艺术
## The Arts

我们看到了森林被砍伐后土地被焚烧、树木被移走的图片，也看到了美丽的热带雨林图片。为什么不画两幅图，一个展现翠绿，另一个展现褐色和淡绿。讨论一下这种改变的原因，讲述热带雨林被毁变成单一种植的棕榈林的情况，以及改正错误的可能性：有了决心和耐心，光秃秃的土地也能重新变成雨林。

# TEACHER AND PARENT GUIDE

## 思维拓展
### Systems: Making the Connections

在寻找高产作物的过程中，人类发现了起源于非洲西部的棕榈树，将它纳入农业产业化的一部分。这种棕榈树的产油量是大豆的10倍，且相当稳定。于是许多动物油和植物油被棕榈油取代，衍生出了一个大型产业，集中于马来西亚和印度尼西亚。不幸的是，为了生产更多标准化的油，人类扩大棕榈树种植，导致大量本地物种死亡。人类面临的主要挑战是没有足够的土地来种植棕榈树，因此只能牺牲雨林，牺牲老虎、猩猩和大象的栖息地。如今，大量的破坏仍在继续。在热带地区，单一种植棕榈树导致真菌病害，从内部破坏棕榈树。而大量使用杀真菌剂伤害了生物网里的所有其他成员，造成了光秃或贫瘠的森林。过去被烧掉的木材，现在越来越多地用作燃料和家具制造，但问题依然存在。在马来西亚，最初的棕榈园越来越多地变成了房地产项目。快速的城市化进程和住房的高需求导致森林有了第二个被砍伐的理由。吉隆坡的城市化模式证实了农田变房产，土地价值会增长1000倍。亚洲还将要建立至少1000个百万人口城市，土地会像水、能源和公路一样，变得非常值钱。然而，也还有另外一个选择，即重建森林。我们不能继续在世界各地砍伐森林，改变大气的化学成分了。因为过去是我们砍伐了森林，利用了土壤，所以现在重要的是，我们应保持积极的态度，并承担重建森林的所有责任。修复表层土壤，让重生的生态系统创造价值，提供食物、水、工作和收入——这些都是社会的迫切需求。

## 动手能力
### Capacity to Implement

无论你生活在世界何处，都一定会与某种植物油或者动物油接触。让我们研究一下可选择的本地油，而不是全球销售的油。它没有被标准化，但会让我们发现哪些油在批量生产中被遗弃。你能列出本地的油吗？

# 教师与家长指南

## 故事灵感来自

## 更家悠介
## Yusuke Saraya

更家悠介出生于日本大阪，获得微生物学学位，大半生就职于更家有限公司，现任总裁。悠介是日本青年商会主席，负责开拓国际关系。作为肥皂的主要生厂商，他知道棕榈油对环境的挑战，决定参加可持续棕榈油圆桌会议，并成立了一个基金会，救助加里曼丹侏儒象。他还带领自己公司的研发团队找到棕榈油的一系列替代品。他所投资的新产品，不仅清洁度高，而且确保人类和地球的健康。他也是联合国大学第十届和第二十届零排放国际大会主办人。

### 更多资讯

http://www.theguardian.com/sustainable-business/ng-interactive/2014/nov/10/palm-oil-rainforest-cupboard-interactive

## 图书在版编目（CIP）数据

边伐木边造林：汉英对照 /（比）冈特·鲍利著；
（哥伦）凯瑟琳娜·巴赫绘；李欢欢，牛玲娟译. -- 上
海：学林出版社，2016.6
（冈特生态童书. 第三辑）
ISBN 978-7-5486-1052-6

Ⅰ. ①边… Ⅱ. ①冈… ②凯… ③李… ④牛… Ⅲ.
①生态环境－环境保护－儿童读物－汉、英 Ⅳ.
① X171.1-49

中国版本图书馆 CIP 数据核字 (2016) 第 125754 号

---

© 2015 Gunter Pauli
著作权合同登记号 图字 09-2016-309 号

## 冈特生态童书
### 边伐木边造林

| | |
|---|---|
| 作　　者—— | 冈特·鲍利 |
| 译　　者—— | 李欢欢　牛玲娟 |
| 策　　划—— | 匡志强 |
| 责任编辑—— | 程　洋 |
| 装帧设计—— | 魏　来 |
| 出　　版—— | 上海世纪出版股份有限公司 学林出版社 |
| | 地　址：上海钦州南路81号　电话/传真：021-64515005 |
| | 网址：www.xuelinpress.com |
| 发　　行—— | 上海世纪出版股份有限公司发行中心 |
| | （上海福建中路193号 网址：www.ewen.co） |
| 印　　刷—— | 上海丽佳制版印刷有限公司 |
| 开　　本—— | 710×1020　1/16 |
| 印　　张—— | 2 |
| 字　　数—— | 5万 |
| 版　　次—— | 2016年6月第1版 |
| | 2016年6月第1次印刷 |
| 书　　号—— | ISBN 978-7-5486-1052-6/G·387 |
| 定　　价—— | 10.00元 |

（如发生印刷、装订质量问题，读者可向工厂调换）

# Food
## 97

# 欢迎每个人

Everyone is Welcome

Gunter Pauli

冈特·鲍利 著

凯瑟琳娜·巴赫 绘
李欢欢　牛玲娟 译

## 丛书编委会

主　　任：贾　峰

副主任：何家振　郑立明

委　　员：牛玲娟　李原原　李曙东　吴建民　彭　勇
　　　　　冯　缨　靳增江

## 丛书出版委员会

主　　任：段学俭

副主任：匡志强　张　蓉

成　　员：叶　刚　李晓梅　魏　来　徐雅清　田振军
　　　　　蔡雩奇　程　洋

特别感谢以下热心人士对译稿润色工作的支持：

姜竹青　韩　笑　贾　芳　刘　晓　张黎立　刘之杰
高　青　周依奇　彭　江　于函玉　于　哲　单　威
姚爱静　刘　洋　高　艳　孙笑非　郑莉霞　周　蕊

# 目录

| | |
|---|---|
| 欢迎每个人 | 4 |
| 你知道吗？ | 22 |
| 想一想 | 26 |
| 自己动手！ | 27 |
| 学科知识 | 28 |
| 情感智慧 | 29 |
| 艺术 | 29 |
| 思维拓展 | 30 |
| 动手能力 | 30 |
| 故事灵感来自 | 31 |

# Contents

| | |
|---|---|
| Everyone is Welcome | 4 |
| Did you know? | 22 |
| Think about it | 26 |
| Do it yourself! | 27 |
| Academic Knowledge | 28 |
| Emotional Intelligence | 29 |
| The Arts | 29 |
| Systems: Making the Connections | 30 |
| Capacity to Implement | 30 |
| This fable is inspired by | 31 |

早晨，小猪正在农场上散步，他停下来和他的鸡朋友们聊天。

"早上好，昨晚睡得好吗？"小猪问。

"嗯，挺好的，谢谢！"一只母鸡说，"你把这里收拾得很温馨舒适！整个房间暖融融的！和你住在同一个屋檐下，真好！"

A pig is taking a morning stroll around the farm and stops to talk to his friends.

"Good morning. Did you sleep well last night?" asks the pig.

"Oh yes, thank you!" says one of the hens. "You've made it so cosy for us here! You warmed up the whole room. How good it is to live under the same roof."

住在同一个屋檐下，真好！

How good it is to live under one roof!

你们可能会把流感传染给我

you could give me the flu

"很高兴为大家服务。不过,我很多朋友认为我疯了才和一群鸡住一起。他们说你们可能会把流感传染给我。"

"嗯,我们不用挤在小笼子里,呼吸自己的粪便浮尘,而是睡在你上面,享受你身体散发的热量。同时你保护我们的安全,防止狐狸进来。多放松啊。"

"My pleasure. However, most of my friends think I'm crazy to sleep in the same place as chickens. They say you could give me the flu."

"Well, we are not packed tightly into small cages, breathing the dust of our own manure. Instead, we sleep on a perch above you to enjoy your body heat while you keep us safe by keeping the foxes at bay. It is so relaxing."

"你们吃掉了所有烦人的虫子和苍蝇,帮助我保持清洁和健康。你不介意帮我捉身上的寄生虫,对吧?那感觉就像温柔的按摩。"

"和你住一起,真好!你瞧,你吃的东西,我们不吃。你吃剩的残渣养活了土里的生物,这些生物让我们更强壮。"母鸡说。

"And you help me stay clean and healthy by eating all the worms and flies that bother me. You don't mind picking a few parasites off my skin, do you? It feels like a gentle massage."

"It is so great to have you in our community. You see, what you eat, we don't eat, and the droppings you leave behind feeds creatures in the soil that make us so much stronger," says the hen.

让我们更强壮

Make us so much stronger

有我们生存所需的所有东西

We have all we need to live

"的确是这样。这周围有森林、灌木丛和草地,有我们生存所需的所有东西。"
"但我们中有些鸡并没能活很长时间。"母鸡突然伤心地说。

"That's true. Around here, with the forest, the bush, and the meadow we have all we need to live."
"But some of us do not get a chance to live long," says the hen, suddenly sad.

"谁想杀害你们？记住了，我会赶走所有的狐狸！"

"不，不是的。我不担心那些捕食者。我是担心那些人类，他们总是想要更快地得到更多的食物。"

"哦，我明白了！没错，人类压力很大，他们觉得有很多穷人会死于饥饿，除非企业能生产更多更便宜的食物。"

"Who wants to kill you? Remember, I will keep all foxes away!"

"No, no, it's not our predators that I'm worried about. I'm worried about those people who always want more and more food faster and faster."

"Oh, I see. Yes, people are stressed, as they believe that many poor people may die of hunger unless businesses produce more and cheaper food."

谁想杀害你们?

Who wants to kill you?

是因为你的儿子们不能下蛋吗?

Is that because your boys do not lay eggs?

"专家声称为了能更高效地生产更多食物,一旦我的儿子们孵化出来,全都得被杀死。"
"是因为你的儿子们不能下蛋吗?我猜想主营鸡蛋销售的人认为养公鸡没用。"

"These experts claim that to make more food more efficiently, you need to kill all my boys as soon as they hatch."

"Is that because your boys do not lay eggs? I suppose that people whose main business is selling eggs think that there is no use in keeping the males."

"那正是养鸡企业不再需要我的兄弟们，想要全部杀了他们的原因。"

"嗯，欢迎你所有的兄弟们来这里，他们在这里可以过着幸福健康的生活。"

"非常感激你的提议。但你知道吗？同样的事情也发生在我女儿们的身上。"母鸡说。

"That's exactly why the chicken industry does not want our brothers anymore and want to kill them all."

"Well, all your brothers are welcome here where they can lead a happy and a healthy life."

"We are so grateful for that. But did you know the same happens to my girls?" asks the hen.

同样的事情也发生在我女儿们的身上

The same happens to my girls

不能长出和公鸡一样多的鸡肉

Not able to produce as much meat as the boys

"我以为他们会留着母鸡下蛋呢？"

"唉，他们会留下一些品种中的所有母鸡来下蛋。但另一种用来产肉的鸡就不一样了。一旦那些鸡孵出来了，人们会把母鸡和公鸡分开，杀死所有的母鸡，因为她们不能长出和公鸡一样多的鸡肉。"

"I thought they keep the girls to lay eggs?"

"Well, for some kinds of chickens they do keep all the hens for laying eggs. But it is not the same for another kind of chicken, those reared to produce meat. Once those chicks hatch, people separate the boys from the girls. All the girls are then killed, as they are not able to produce as much meat as the boys."

"所有这些都是以提高产量为名义吗?让母鸡们也都来这里和我们住在一起吧。"

"谢谢你如此关心并邀请别人分享你的住所。我们确实生活在富足的土地上,在这里大自然永远都为我们提供充足的食物。"

……这仅仅是开始!……

"And all this is going on in the name of productivity? Let the girl chicks come live here with us as well."

"Thank you for being so caring and inviting others to share your space. We certainly live in the land of plenty where nature's table is always laden with food."

... AND IT HAS ONLY JUST BEGUN!...

……这仅仅是开始!……

...AND IT HAS ONLY JUST BEGUN! ...

# Did You Know?

## 你知道吗？

For every female, egg-laying chicken that hatches, a day-old male chicken is killed. For every male chicken that will produce meat that hatches, a day-old female chicken is killed.

每孵化一只下蛋的母鸡，一只刚孵化的公鸡就会被杀害；每孵化一只产肉的公鸡，一只刚孵化的母鸡就会被杀害。

It takes only four weeks to fatten a rooster with soy, corn, and power food, including hormones and antibiotics. A chicken eating grass, seeds, and insects would need four months to reach maturity.

用大豆、玉米和含有激素和抗生素的强力饲料养肥一只公鸡只需四个星期，一只吃草、种子和虫子的小鸡需要四个月才能长成熟。

A "super" hen lays around 330 eggs per year and will survive for a maximum of 18 months. A "natural", domesticated hen would lay 220 eggs per year.

一只"超级"母鸡每年产出约 330 枚鸡蛋，最多只能存活 18 个月。一只自然家养的母鸡每年能产出约 220 枚鸡蛋。

All specialised egg-laying and meat-producing chickens are inbred to the point that they are genetically degenerate and cannot procreate anymore. All 50 billion chickens that are produced industrially each year are the product of artificial insemination.

专门产蛋和专门产肉的鸡是近亲繁殖，基因退化，无法再生育。每年批量孵化的 500 亿只小鸡是人工授精的结果。

Pigs are living in perfect symbiosis with the forest. They get food from the forest without destroying it, while making the forest fire resistant by keeping the undergrowth in check.

猪和森林是完美的共生关系。猪不破坏森林，从森林获得食物，同时控制灌木生长以防森林着火。

One thousand hectares of forest, fields, and bush can host 10 000 chickens, one thousand pigs and one thousand sheep without any need to provide them with commercial feed.

一千公顷森林、田地和灌木可以养活一万只鸡、一千头猪和一千只羊，无需提供任何人工饲料。

Pigs were domesticated from wild boars in Turkey 10 000 years ago. They were originally tuber-eating forest dwellers. Pigs do not sweat and therefore cannot cool their bodies down. That is why they live under the canopy of trees and enjoy rolling in mud.

猪是由土耳其人在一万年前驯养野猪而来，猪最初是食用块茎的森林居民。猪不流汗，因此无法让身体降温。这也是它们住在树底下，喜欢在泥里打滚的原因。

Pigs are very smart and are ranked more intelligent than dogs. Every year two billion piglets are fattened to 200-kg hogs in only six months.

猪很聪明，比狗还机智。每年，仅在六个月内就有二十亿只小猪被养到 200 公斤。

Do you agree that as the world needs more food all female chickens should be removed from meat production and all male chickens from the fertilisation process?

因为这个世界需要更多的食物，就要在肉类生产中杀害所有母鸡，在鸡蛋生产中杀害所有公鸡，你赞同吗？

你认为生活在森林里的动物们需要外界的食物吗？

Do you think that animals living in a forest will need any imported food?

Which system will produce more and healthier food and replenish topsoil: the one where only chickens are raised or the one many animals are raised together, alongside a variety of plants?

只养殖鸡的系统和同时养殖多种动植物的系统，哪种能产出更多更健康的食物并补充表层土？

生产更多的相同产品是最好的解决方法吗？或者，我们能同时生产多种食物，而不是只生产越来越多的鸡蛋吗？

Is producing more of the same always the best solution? Or, instead of producing only more and more eggs, could we have several sources of food produced at the same time?

Have you ever visited a chicken farm? (We are often kept from visiting modern battery chicken farms because of the risk of contamination, as you may become infected and get sick.) Calculate how many eggs you eat per day and then per week. Also look at the labels of any packaged and processed foods you have in your house as many of these contain eggs. Extend you research to also calculating how many eggs other members of your family consume. How many chickens in total are needed to supply your household with eggs? Now have a look at how much kitchen waste and leftover food there is available every day. Do you think that it would be enough to feed chickens of your own?

参观过养鸡场吗？（我们通常不参观现代化养鸡场，因为有染风险，也许会感染生病。）计算一下你每天吃几个鸡蛋，每吃几个。再看看你家里食品包装袋的产品信息，有多少食物含鸡蛋。家里其他成员消耗多少鸡蛋？共需要多少只鸡才能满足里的鸡蛋需求？现在看看每天有多少厨房垃圾和残羹剩饭，你为这足以养活你家需要的鸡吗？

# TEACHER AND PARENT GUIDE

## 学科知识
### Academic Knowledge

| | |
|---|---|
| 生物学 | 植物蛋白相对鸡肉、猪肉和牛肉动物蛋白的转化率；有草原、灌木和森林的生态系统的再生能力；人类起源于热带，如今我们模仿最初的住处来设计保温的房子；共生关系，动植物如何相互协作创造高效系统；鸟类控制自然虫害。 |
| 化 学 | 鸡的粪便富含氮(4%)、磷(2%)和钾(1%)；家禽粪便的PH值随着年龄和食物变化，维持在6.5至8之间。 |
| 物 理 | 热空气会上升，因此，猪散发的热量上升至鸡笼的地板，给鸡提供温暖，这是生物气候学的房屋设计基础。 |
| 工程学 | 猪有逃跑的倾向，因此需要智能门系统把猪圈起来；猪、鸡、鸭和鱼共生场地的设计。 |
| 经济学 | 把鸡养在笼子里以提高产量；每增加1千克体重，猪需要吃掉2.5千克食物；人口增长和食物产量息息相关，需要提高产量。 |
| 伦理学 | 健康食物和食物量之间的平衡；生命和产量之间的选择。 |
| 历 史 | 2500年前，赫拉克利特已讨论生命的和谐，指出所有事物都受制于养分、物质和能量之间的流动；大约9000年前，中国和印度已驯养小鸡。 |
| 地 理 | 土耳其位于欧洲和亚洲的交界处，第一次世界大战后，庞大的奥斯曼土耳其帝国瓦解，现在的土耳其是其剩余的一部分。 |
| 数 学 | 土地面积固定，人口增长，所以食物产量必须提高；每平方米需要8至400条蚯蚓才能保证土壤健康，这就意味着每英亩地需要超过一万条蚯蚓。 |
| 生活方式 | 我们的饮食越来越依赖于动物产品，似乎没有人意识到如果我们摄入更多的素食，能更快地解决世界饥饿问题；给流浪者、孤儿和难民提供蔽护的文化。 |
| 社会学 | 多元文化社会的生活与相同阶级、相同种族和相同文化人群一起生活的对比；语言形成了雄性、雌性和幼崽的特定称呼。 |
| 心理学 | 拥有足够的安全感，能睡得更香甜；放松程度影响脑海中的信息。 |
| 系统论 | 共生的重要性：共同协作制造更多养分、物质和能量。 |

# 教师与家长指南

## 情感智慧
### Emotional Intelligence

**小猪**

小猪对自己身边的每一个人都怀有深深的同情心,关心他人的幸福。即使存在禽流感的威胁,他还是希望和鸡住在一起。小猪感到庆幸,优雅地解释他如何给鸡提供温暖,保护他们远离捕食者。他意识到应对世界饥饿问题需要更多食物,那就需要多元化生态系统。为了产量,杀害公鸡或母鸡,他认为这不合理,并非常同情他们,愿意给所有不被接受的小鸡提供住所,甚至欢迎两倍数量的鸡(公鸡和母鸡)来这里。

**母鸡**

母鸡对小猪给予她的关心很高兴。对于小猪提供的温暖和保护,母鸡感到幸福快乐并表达了自己的感激之情。她坦率地讨论起伤心的话题,说出了自己对其他鸡的担忧,他们没有机会像她一样感受生活。母鸡明确地回答了小猪的问题和疑虑,揭开了问题背后的真实原因。当小猪说欢迎其他人来这片富足的土地上时,母鸡感到兴奋。

## 艺术
### The Arts

你在鸡蛋上画过画吗?拿出三个煮熟的鸡蛋,在一个鸡蛋上画棵树,在另一个鸡蛋上画只鸡,在第三个鸡蛋上画一只猪,试着顺着鸡蛋的曲线画出他们的脸。记住可以用白色或棕色作为鸡蛋的底色。如果喜欢,你可以画更多的鸡蛋!

## TEACHER AND PARENT GUIDE

## 思维拓展
### Systems: Making the Connections

　　社会不停地寻找提高食物产量的方法，一方面希望解决世界饥饿问题，另一方面希望获取更多的利润。我们似乎完全认识到了食物生产对动物生存质量的巨大影响。稀缺资源的有效管理带来了复杂的喂养系统，为了达到更高的生产水平，喂养过程中的每一步都经过详细研究。让我们想想人类食物生产行为如何影响和我们一起生活在地球上的其他物种的生存质量，这些行为看似没有人性。因为性别原因无法达到生产的最高水平，所以雄性蛋鸡和雌性肉鸡被杀害了。这种杀害不仅限于鸡类，还有牛类，因为商人只关注牛奶的产量。动物们更愿意和谐地一起生活在食物富足的地方，这样会带来多样化：多种动物有多种养分来源，提供了强大的共生关系基础。比起单纯地关心环境，这还需要改变我们的计算方式和商业模式。通过做出改变，我们能不断地创造、支持和维护共生关系，补充土壤，加强生态系统，以便后代能收获超乎我们想象的大量优质产品。所有这一切使我们能够在地球有限的资源下可持续地生存。

## 动手能力
### Capacity to Implement

　　我们首先要意识到食用肉类或蛋类是个人选择。我们作为消费者，要求高品质食品很重要。只有消费者需求，高品质食品才会被生产，政府才会制定食品生产法则。食品生产商通常只会关注降低成本，赚取利润。我们的购买力很大程度上决定了能否实现改变。如果成千上万的消费者要求高品质鸡蛋，那么就会生产出这样的鸡蛋。如果现有的企业不能生产这样的鸡蛋，那就给创业者一个机会满足市场。和家人朋友探讨一下，告诉他们你会为杀死公鸡和母鸡的现状感到伤心。看看周围有多少人支持你，坚信鸡应该被养大，改变杀害幼鸡的现状也能创造足够的食物。

# 教师与家长指南

## 故事灵感来自

## 格奥尔格·施魏斯富特
### Georg Schweisfurth

格奥尔格·施魏斯富特在德国长大，生活在欧洲最大的腊肠加工厂之一的附近，那里每天加工牛的数量多达 5000 头。他是一个屠夫，在慕尼黑和弗莱堡学习商科。比起他父亲的食品生产方式，他坚信应该有更好的方式。《互联思维艺术》作者弗雷德尔克·韦斯特、永续生活设计的共同创立者比尔·莫里森、《一根稻草的革命》作者福冈正信，启发他 1988 年在德国慕尼黑东南的格隆镇成立了 Hermannsdorfer Landwerkstätte。之后，他还创立了 EPOS（一切皆有可能）组织 Gut Sonnenhausen 生态酒店和 Basic AG 连锁生物超市。他和家人致力于给竞争激烈的农业生产和肉类加工制定新的质量标准。他著有《清醒的安德斯》一书。

更多资讯

https://de.wikipedia.org/wiki/Herrmannsdorfer_Landwerkst%C3%A4tten

图书在版编目（CIP）数据

欢迎每个人：汉英对照 /（比）冈特·鲍利著；
（哥伦）凯瑟琳娜·巴赫绘；李欢欢，牛玲娟译. —— 上
海：学林出版社，2016.6
（冈特生态童书. 第三辑）
ISBN 978-7-5486-1053-3

Ⅰ. ①欢… Ⅱ. ①冈… ②凯… ③李… ④牛… Ⅲ.
①生态环境－环境保护－儿童读物－汉、英 Ⅳ.
① X171.1-49

中国版本图书馆 CIP 数据核字 (2016) 第 125760 号
————————————————————————
© 2015 Gunter Pauli
著作权合同登记号 图字 09-2016-309 号

## 冈特生态童书

### 欢迎每个人

| | |
|---|---|
| 作　　者 —— | 冈特·鲍利 |
| 译　　者 —— | 李欢欢 牛玲娟 |
| 策　　划 —— | 匡志强 |
| 责任编辑 —— | 程　洋 |
| 装帧设计 —— | 魏　来 |
| 出　　版 —— | 上海世纪出版股份有限公司 学林出版社 |
| | 地　址：上海钦州南路 81 号　电　话／传真：021-64515005 |
| | 网　址：www.xuelinpress.com |
| 发　　行 —— | 上海世纪出版股份有限公司发行中心 |
| | （上海福建中路 193 号　网址：www.ewen.co） |
| 印　　刷 —— | 上海丽佳制版印刷有限公司 |
| 开　　本 —— | 710×1020　1/16 |
| 印　　张 —— | 2 |
| 字　　数 —— | 5 万 |
| 版　　次 —— | 2016 年 6 月第 1 版 |
| | 2016 年 6 月第 1 次印刷 |
| 书　　号 —— | ISBN 978-7-5486-1053-3/G·388 |
| 定　　价 —— | 10.00 元 |

（如发生印刷、装订质量问题，读者可向工厂调换）

# Food
## 75

# 螃蟹大餐

## Crabs for Dinner

### Gunter Pauli

冈特·鲍利 著

凯瑟琳娜·巴赫 绘

李欢欢 牛玲娟 译

## 丛书编委会

主　　任：贾　峰
副 主 任：何家振　郑立明
委　　员：牛玲娟　李原原　李曙东　吴建民　彭　勇
　　　　　冯　缨　靳增江

## 丛书出版委员会

主　　任：段学俭
副 主 任：匡志强　张　蓉
成　　员：叶　刚　李晓梅　魏　来　徐雅清　田振军
　　　　　蔡雩奇　程　洋

特别感谢以下热心人士对译稿润色工作的支持：

姜竹青　韩　笑　贾　芳　刘　晓　张黎立　刘之杰
高　青　周依奇　彭　江　于函玉　于　哲　单　威
姚爱静　刘　洋　高　艳　孙笑非　郑莉霞　周　蕊

# 目录

| | |
|---|---|
| 螃蟹大餐 | 4 |
| 你知道吗？ | 22 |
| 想一想 | 26 |
| 自己动手！ | 27 |
| 学科知识 | 28 |
| 情感智慧 | 29 |
| 艺术 | 29 |
| 思维拓展 | 30 |
| 动手能力 | 30 |
| 故事灵感来自 | 31 |

# Contents

| | |
|---|---|
| Crabs for Dinner | 4 |
| Did you know? | 22 |
| Think about it | 26 |
| Do it yourself! | 27 |
| Academic Knowledge | 28 |
| Emotional Intelligence | 29 |
| The Arts | 29 |
| Systems: Making the Connections | 30 |
| Capacity to Implement | 30 |
| This fable is inspired by | 31 |

小猪和蘑菇正在讨论为什么农场池塘里的鱼在逐渐消失。

"一定是有鸟在晚上偷了我们的鱼。"蘑菇这样认为。

"不,那是不可能的,我们晚上可是很警惕的,"小猪反驳说,"一定会注意到的。我想知道在鱼类家族里有没有吃鱼的种群。"

The pig and the mushroom are debating why the fish are vanishing from the pond on their farm.

"It must be the birds who come to steal our fish at night," argues the mushroom.

"No, that's not possible; we are very vigilant," defends the pig. "We would have noticed. I wonder if one family of fish isn't eating the other."

为什么池塘里的鱼在逐渐消失？

Why are the fish vanishing from the pond?

没有鱼对吃其他鱼感兴趣

No fish is interested in eating the other

"看，"蘑菇说，"在3米深的池塘中有7种鱼，每个食物链层次都有一种。没有鱼对吃其他鱼感兴趣，我们可以不用专门为鱼买饲料。那些鱼可能生病了，正在池塘的底部奄奄一息。"

关于食物链级别，你说得对，顶层的食草鱼类从不吃食底层鱼类。由于有藻类，水里有许多浮游生物和水蜗牛，这也意味着池塘里所有的鱼都有充足的食物。我实在不明白，过去池塘里鱼的数量是现在的5倍多，发生什么了？"

"Look," answers the mushroom, "we have seven types of fish here – one for each food level of our three metre deep pond. No fish is interested in eating the other. It is possible for us to farm fish without having to buy feed for them. Perhaps they got sick and are dying at the bottom of the pond."

"You're right about the levels. The grass eaters at the top will never bother the bottom feeders. Thanks to the algae, there is a lot of plankton and many tiny snails in the water, which means that there's enough food to feed all the fish. I don't understand it. We used to have five times more fish than we have now, what is going on here?"

"我真的不知道，你觉得是不是池塘里的水出问题了？"蘑菇问道，"我知道你的猪舍里有很多固体流出来，但这些不是应该被沼气池的过滤器捕获了吗？"

"是的，这些固体确实是被沼气池捕获了，池塘里的水也来自猪舍。猪粪流入沼气池中，细菌会将水净化，之后才会进入外界。你知道沼气池也能生产沼气吗？"

"I really don't know. Do you think there's a problem with the water?" wonders the mushroom. "I know that too many solids flowing out of your sty, but aren't these caught in the digester's filter?"

"Yes, the solids are indeed left behind in the digester. The water for this pond comes from our pigsty. The manure flows into a digester where bacteria purify the water. And the clean water is then returned to nature. Did you know that the digester also creates biogas?"

# 沼气池也生产沼气

The digester also creates biogas

真有趣!

That's interesting!

"真有趣！"蘑菇说道。

"但如果不是水的问题，那是什么问题呢？"小猪问，"农场里一直都有充足的食物来源。我们也提供了这么多食物，甚至周边曾被认为贫瘠的土壤现在也已经肥沃得超乎人们的期待了。"

"有些人认为贫瘠的土壤会一直贫瘠，这种想法是不正确的。"蘑菇回应道，"有些人认为我们池塘里的水被污染了，因为里面有太多的食物。但池塘流出的水可以神奇地使土壤变得肥沃！"

"That's interesting!" the mushroom replies.

"But if it is not the water, what could the problem be then?" the pig asks. "We have always had plenty of food sources on this farm. We are providing so much food; even the soil around us, which was considered poor for farming, started to flourish beyond everyone's expectations."

"Those who think poor soil will always remain that way, don't think properly," the mushroom replies. "Some people thought the water in our pond was polluted, as there was too much food in it. But the water from the pond turned out to enrich the soil wonderfully!"

"但是，我们还是需要为防止我们的鱼减少做些什么。"小猪说道。

"所以我们要想一想，池塘里过剩的水会流到哪里？"蘑菇问。

"流到大海里。"

"水流向的海岸带会有什么生物生存呢？"蘑菇想知道。

"But we still need to do something about our fish disappearing," says the pig.

"So let's think. Where is the excess water from the pond flowing to?" the mushroom asks.

"To the sea."

"And what lives in the coastal zone of the sea where our water flows to?" the mushroom wants to know.

我们要想一想

Let's think

红树林湿地是虾的天堂……

Mangrove swamps are a paradise for shrimp...

"红树林。"

"是的,红树林湿地是虾、螃蟹、海草和海藻的天堂。"

"那你认为其中哪些生物可以从海洋里移到岸上呢?"小猪问道。

"Mangroves."
"Yes, the mangrove swamps are a paradise for shrimp, crabs, seaweed, and algae."
"So which one of these creatures do you think can move up from the sea onto the land?" asks the pig.

"海草和海藻不行,虾也很难。"蘑菇回答道。
"那么,唯一能进入红树林沼泽的就是螃蟹了!"

"Seaweed and algae have no chance, and shrimp can hardly walk," replies the mushroom.
"Well, then the only ones that can walk up into the mangrove swamp, are the crabs!"

唯一能进入红树林沼泽的就是螃蟹了！

The only ones that can walk up are the crabs!

螃蟹吃了我们的鱼？

Crabs are feasting on our fish?

"你的意思是这些走起路来很搞笑的螃蟹吃了我们的鱼？"蘑菇问道。

"我是这样想的。你知道螃蟹是唯一的腿在身体两边而不是在下面的动物吗？"

"You mean those funny walking crabs are feasting on our fish?" the mushroom asks.

"I think so. Do you know that crabs are the only animals who have their legs on the sides of their bodies instead of under their bodies?"

"显然没法阻止他们进入我们的池塘。"蘑菇说道。"他们可能现在就藏在泥土里，随时准备吃掉我们的鱼！"

"让我们把池塘里的水抽干，抓住这些螃蟹吧！当地的农民会非常高兴的，他们可以把螃蟹卖掉赚很多钱，然后供孩子上学。"

……这仅仅是开始！……

"Clearly that has not prevented them from invading our pond," says the mushroom. "They are probably hiding in the mud right now, ready to feast on our fish!"

"Let's dry out the pond and catch all these crabs! It will make our farmer very happy. He will make a lot more money by selling the crabs. And then he will have enough money to pay his kids' school fees."

... AND IT HAS ONLY JUST BEGUN!...

……这仅仅是开始！……

... AND IT HAS ONLY JUST BEGUN! ...

# Did You Know?

## 你知道吗？

Integrated farming produces fish, agricultural crops, and livestock in such a way that the byproducts of one subsystem becomes a valuable input for another, generating more available food than if each had been cultivated separately.

综合农业生产鱼类、农作物和牲畜，其方法是让某类生物的代谢副产品成为另一种生物的营养源，这比单独培养能够产生更多的食物。

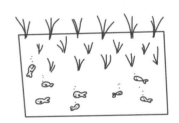

Farming fish in combination with crops and animals means that feed for the fish can be provided naturally and that there is no need to buy fish feed. By doing this, the farmer can save money.

养殖鱼类与种植农作物、饲养动物结合起来，可以自然地为鱼类提供食物而不用专门购买鱼饲料。这样做，农民还可以省钱。

Animals excreta create problems for the environment when not used. Excreta can be used as a source of nutrients and energy for bacteria when they digest the biomass.

动物粪便如果不处理，会带来环境问题。粪便被消化分解时可以作为细菌的营养物来源。

Fruit and plants could create problems when they rot and become waste. This waste is, however, rich in fibres and an ideal substrate for mushrooms. The waste of mushrooms, the used substrate, is rich in amino acids and therefore a great feed for animals.

水果和植物在腐烂变成废弃物时会产生问题。然而，这些废弃物富含纤维素，对于蘑菇来说是理想的生长基床。蘑菇的废弃物，即它们用过的基床富含氨基酸，是非常好的动物饲料。

There is a lot of food in water and each fish species eats a specific kind of food. Some species like algae, others benthos, and others like grass or diatoms. If there is only one kind of fish in a pond, all the other types of food goes to waste.

水里有许多食物，每种鱼吃特定的食物。有的喜欢海藻，有的爱吃底栖生物，还有的喜欢海草和硅藻。池塘里如果只有一种鱼的话，其他的食物或许就会成为废物。

When several species of fish live in a pond, their excrement creates too much food. The excess biomass can be removed by converting the waste into nutrients for other species. This can be achieved by placing floating flower or rice gardens into the pond.

当不同种类的鱼生活在池塘里时，其粪便会成为更多其他生物的食物。将废弃物转化为其他生物的营养物质，可以去除多余的生物质。这可以通过在池塘里放置水上花卉或种植水稻实现。

ℬecause crabs spend most of their life buried in sand, they have developed flat bodies with long legs on both sides. Some species can walk only sideways because of the way their knees bend.

因为螃蟹大部分时间隐藏在沙子里，所以它们的腿长在扁平的身子两边。一些螃蟹品种膝盖是弯曲的，所以只能横着走路。

𝒸rabs use their pincers to communicate with each other. During mating season, they find comfortable places for the females to release eggs. They work together to provide food and protect their families. Crabs also like to congregate in large numbers to socialize.

螃蟹通过钳子和同伴相互联系。在交配季节，雌性螃蟹会找到舒适的地方进行产卵。螃蟹喜欢聚集在一起生活，一起工作，为家人提供食物并保护它们。

# Think About It
## 想一想

If fish can find their own food in nature, why do people have to feed farmed fish?

如果鱼在自然界中能够找到食物，为什么人们还要养殖鱼呢？

生活在河流和湖泊里的鱼类通常不止一种。为什么人们只在鱼缸或池塘里养殖一种鱼呢？

There is always more than one species of fish in a river or a lake. Why do fish farmers keep only one type of fish per tank or pond?

If nature cultivates many fish without providing extra feed for them, why do crabs not have the right to feast on it as well?

如果自然界里有很多鱼，但又没有足够的饲料，为什么螃蟹不能捕食鱼呢？

你觉得农民应该是为了生存多挣一些钱，还是也有权利为了娱乐而挣更多的钱呢？

Do you think the farmer should earn enough money only to live, or does he have the right to earn enough to afford entertainment as well?

Let us see who eats what. Start with mushrooms, followed by pigs, and then bacteria, track the nutrition of algae, map the different forms of fish feed, and then figure out who will eat what food in the pond. Can you map the five kingdoms of nature? Once you have figured out all five, try to calculate how much food will eventually be produced if you start with 100 kg of mushroom substrate. This will give you an idea of how productive nature really is.

一起来看看谁都吃些什么。先从蘑菇开始，然后是小猪、细菌，跟踪藻类的营养流动，绘制不同类型鱼饲料的食物链图谱，然后想想它们都吃池塘里的什么食物。你能绘制出大自然的五个生物王国吗？在你找到所有的五类生物后，试着计算一下，以100千克蘑菇生长基床算起，最终能生产出多少食物。这会使你认识到自然界的创造力究竟有多大。

# TEACHER AND PARENT GUIDE

## 学科知识
### Academic Knowledge

| | |
|---|---|
| 生物学 | 食物链中生物体的营养级别；浮游生物和海藻是初级生产者；鲤鱼和其他鱼种有不同的饮食结构，在鱼塘里一起养殖是互补的；不同种类的螃蟹有不同的饮食结构，所以生活在不同深度的水里；综合生态系统确保了养殖鱼类不需要额外提供饲料——一种生物会为另一种生物提供食物；红树林具有独特的生态系统。 |
| 化学 | 脂类与聚合物的区别；ω3脂肪酸是含有酯基官能团的多元不饱和脂肪酸，人体自身不能合成；甲烷由产甲烷细菌在厌氧条件下生成；厌氧细菌不能降解木质素；沼气的成分有$CH_4$、$CO_2$、$N_2$、$H_2$和$H_2S$。 |
| 物理 | 综合生态系统利用了重力原理：水从池塘的高处流到低处。 |
| 工程学 | 沼气池的设计可以是单室或双室，沼气池污泥含有高浓度的生化需氧量（BOD）和化学需氧量（COD）。 |
| 经济学 | 养殖户把大部分钱花在资本投资和饲料上，单一养殖鱼的利润很小；在中国，沼气用户量已经有3000万，沼气池将废弃物转化为清洁能源，增加可支配收入。 |
| 伦理学 | 人们或许认为某些土壤是贫瘠的，但却没想过贫瘠的土壤可以转变成肥沃的土壤。因此，我们限制了自己的思维，常常忽略了我们面前的宝贵资源。 |
| 历史 | 伏打在1776年发现了厌氧消化；第一个沼气厂1859年在印度孟买建成；1895年生产了第一个通过化粪池沼气发电的灯泡。 |
| 地理 | 河流和湖泊的沉积物会产生可燃气体。 |
| 数学 | 线性过程和计算的区别，复杂多变量过程与反馈型非线性计算的区别。 |
| 生活方式 | 许多人不了解食物是怎样生产的，认为土豆来自超市。 |
| 社会学 | 欠发达地区通常以农业为主，进步依赖于就业比重逐渐从农业和工业转变到服务业。 |
| 心理学 | 由于不知道某个东西为何不能运作带来的压力；我们需要分析一个系统的整体环境，以便找出哪个部分不合适，这个系统也被称作"格式塔"（有组织的整体）。 |
| 系统论 | 在食物链中，属于高营养级别的大型食肉鱼类减少后，渔业会使食物网中鱼类减少；厌氧沼气池的设计力里含有污水、污泥和粪便的处理过程。 |

# 教师与家长指南

## 情感智慧
### Emotional Intelligence

小　猪

　　小猪具有保护意识，他很伤心鱼的数量在减少，于是开始寻找凶手。小猪明确了答案，通过一步步逻辑分析找出原因。他指出生态系统的优势，并疑惑鱼类在并不缺少食物的前提下数量为什么会减少。蘑菇一连串的问题激发他给出了快速并有针对性的答案，展示了他的洞察力。他对螃蟹身体结构的了解让我们吃惊于他的知识量。小猪认为螃蟹是凶手，对怎样处理这件事提出了可行性建议，这让大家感到很满意。

蘑　菇

　　蘑菇原本想到了一个凶手，但很快就选择了进行详细的分析，找出问题所有可能的原因。他追溯鱼类的食物链，排除了同类相食的可能，开始探讨鱼类减少是不是因为健康问题。蘑菇也提到了水质的问题，他知道一些人会把富营养化的水视为污染物，而这些水对其他生物其实具有高营养价值。蘑菇反驳贫瘠的土壤就一直贫瘠的简单思维。他清晰而果断地表达了自己的观点。蘑菇决定找到答案，问了一系列问题。这帮助他和小猪缩小了可能的凶手范围，得出最终的结论。

## 艺术
### The Arts

　　你知道鲤鱼有很多种类吗？不同品种可能看起来很像，但是它们有着自己的食物。为什么不画出不同种类的鲤鱼呢：（1）银鲤（2）鳙鱼（3）草鱼（4）黑鱼（5）普通鲤鱼（6）鲫鱼（7）鲮鱼。然后画出一个池塘，以及每一种鱼吃的食物。任何两种鱼的饮食组合都不一样。这些鲤鱼的典型食物分别是：（1）浮游植物（2）浮游动物（3）植物（4）底栖生物（5）昆虫和幼虾（6）硅藻、蓝藻和植物种子（7）硅藻、绿藻、植物碎屑和水底腐殖质。你会觉得在池塘里有丰富的食物，每种鱼类都会生活得很好。探索不同种类的鲤鱼及其食物，会增加你对生物学的理解。

## TEACHER AND PARENT GUIDE

## 思维拓展
### Systems: Making the Connections

综合生态系统按照自然界中的方式模拟营养物质和能量的产生。它展示了大自然中五个生物王国之间的相互联系与不同生物体间营养物质和能量的转移，显示了资源的有效性与生态系统里营养级别的多样性。生态系统越多，就会产生越多的营养物质。一个营养丰富的水体可以被视为一个生态系统。一定区域（如池塘）里多余的营养成分可以供给其他物质（如土壤）。这种生态系统方法与目前饲养业中使用的方法是非常不一样的。目前的方法是，为农业中的每种生物体分别购买饲料，并对每种生物产生的废物分开管理，这增加了经济和环境成本。现代农业注重遗传品质改进和成本控制，综合生态系统保证了所有的营养物质被利用，使现有资源产生更多价值，不需要运输或再处理。这种综合生态系统不仅增加了营养物质的产出，也通过平衡生产和消费，对生物多样性和生态系统健康产生了积极的影响。

## 动手能力
### Capacity to Implement

试想你将经营一个企业，销售罗非鱼鱼种、鱼饲料和养鱼所需的设备。把所有的数字加起来，计算每个人能创造多大价值。以综合农业系统为例，从蘑菇开始，到喂养动物、生产沼气、培育藻类，最终养殖7种鱼类，这些鱼能提供大量富含营养的水，使原本贫瘠的土壤变得肥沃。比较一下：不同类型的两种企业哪一家会有更多收入？哪一家会生产更多的食物？

# 教师与家长指南

故事灵感来自

## 陈绍礼教授
## Prof. George Chan

陈绍礼在毛里求斯岛出生与长大。二战期间他在英国军队服役，之后有机会在英国学习，获得了伦敦帝国理工学院的卫生工程学位。完成学业后，他曾在毛里求斯首都路易港市做城市工程师，后来为驻南太平洋的美国环境保护署工作。1983年，他来到了祖先的故土中国，并花了五年的时间，作为"赤脚工程师"进行综合农业的试验工作。他周游世界，设计了各种各样的农业综合系统方案。他最大的成功在于依靠他的智慧，在巴西把80个养猪场转化为能源供应商，并提供肥沃的土地。他参与了40多个工程，不仅包括所有的工程设计，还有繁重的体力劳动。他在80高龄时仍在参与这些项目，直到后来因为身体原因不得不停止工作，在家乡毛里求斯退休。

更多资讯

http://www.fao.org/docrep/field/003/ac264e/ac264e00.htm#fo1

https://sites.google.com/site/edenspaceproject/terms-definitions/appropriate-tech/george-chan

图书在版编目（CIP）数据

螃蟹大餐：汉英对照／（比）冈特·鲍利著；
（哥伦）凯瑟琳娜·巴赫绘；李欢欢，牛玲娟译．－－上
海：学林出版社，2016.6
（冈特生态童书．第三辑）
ISBN 978-7-5486-1049-6

Ⅰ．①螃… Ⅱ．①冈… ②凯… ③李… ④牛… Ⅲ．
①生态环境－环境保护－儿童读物－汉、英 Ⅳ．
① X171.1-49

中国版本图书馆 CIP 数据核字（2016）第 125755 号

——————————————————————————

© 2015 Gunter Pauli
著作权合同登记号 图字 09-2016-309 号

## 冈特生态童书
### 螃蟹大餐

| | |
|---|---|
| 作　　者—— | 冈特·鲍利 |
| 译　　者—— | 李欢欢　牛玲娟 |
| 策　　划—— | 匡志强 |
| 责任编辑—— | 程　洋 |
| 装帧设计—— | 魏　来 |
| 出　　版—— | 上海世纪出版股份有限公司 学林出版社 |
| | 地　址：上海钦州南路81号　电　话／传真：021-64515005 |
| | 网址：www.xuelinpress.com |
| 发　　行—— | 上海世纪出版股份有限公司发行中心 |
| | （上海福建中路193号 网址：www.ewen.co） |
| 印　　刷—— | 上海丽佳制版印刷有限公司 |
| 开　　本—— | 710×1020　1/16 |
| 印　　张—— | 2 |
| 字　　数—— | 5万 |
| 版　　次—— | 2016年6月第1版 |
| | 2016年6月第1次印刷 |
| 书　　号—— | ISBN 978-7-5486-1049-6/G·384 |
| 定　　价—— | 10.00元 |

（如发生印刷、装订质量问题，读者可向工厂调换）

# Food 83

# 慢 餐

## Slow Food

### Gunter Pauli

冈特·鲍利 著

凯瑟琳娜·巴赫 绘
李欢欢 牛玲娟 译

## 丛书编委会

主　任：贾　峰

副主任：何家振　郑立明

委　员：牛玲娟　李原原　李曙东　吴建民　彭　勇
　　　　冯　缨　靳增江

## 丛书出版委员会

主　任：段学俭

副主任：匡志强　张　蓉

成　员：叶　刚　李晓梅　魏　来　徐雅清　田振军
　　　　蔡雩奇　程　洋

特别感谢以下热心人士对译稿润色工作的支持：

姜竹青　韩　笑　贾　芳　刘　晓　张黎立　刘之杰
高　青　周依奇　彭　江　于函玉　于　哲　单　威
姚爱静　刘　洋　高　艳　孙笑非　郑莉霞　周　蕊

# 目录

| | |
|---|---|
| 慢餐 | 4 |
| 你知道吗？ | 22 |
| 想一想 | 26 |
| 自己动手！ | 27 |
| 学科知识 | 28 |
| 情感智慧 | 29 |
| 艺术 | 29 |
| 思维拓展 | 30 |
| 动手能力 | 30 |
| 故事灵感来自 | 31 |

# Contents

| | |
|---|---|
| Slow Food | 4 |
| Did you know? | 22 |
| Think about it | 26 |
| Do it yourself! | 27 |
| Academic Knowledge | 28 |
| Emotional Intelligence | 29 |
| The Arts | 29 |
| Systems: Making the Connections | 30 |
| Capacity to Implement | 30 |
| This fable is inspired by | 31 |

兔妈妈有十二个嗷嗷待哺的孩子,但家里已经没有食物了。小兔子们饿了,她得赶紧采取行动。

"我们去吃快餐吧。"最小的兔子提出建议。

Mother rabbit has a dozen kids to feed and there is no food left. The little ones are hungry and she needs to act quickly.

"Let's get some fast food," proposes the youngest of the litter.

我们去吃快餐吧

Let's get some fast food

你是说寿司吗?

Are you talking about sushi?

"你是说寿司吗?"兔妈妈问道,"那是唯一的我同意你们吃的快餐。"

"妈妈,我们知道你喜欢和食,也知道是日本人发明了快餐。但我们对饭团、紫菜和鱼不感兴趣,我们想吃胡萝卜和各种胡萝卜糖果,而且我们现在就想吃!"

"Are you talking about sushi?" responds his mother. "Because that's the only fast food I'm willing to let you have."

"We know you love Japanese food, Mum, and that the Japanese invented fast food. But we are not interested in rice, seaweed and fish. We want carrots and all the goodies that come along with it, and we want it now!"

"从一粒小种子长成一颗胡萝卜,你知道需要多长时间吗?"

"不知道,但肯定需要很长的时间,没法很快就让我不饿着肚子的!"

"在凉爽的沙土地里胡萝卜需要生长三个月。如果太热了,它们就根本长不好了。"

"Do you know how long it takes to grow a carrot from a tiny seed?"

"No idea, but certainly too long to still my hunger any time soon!"

"It takes three months in cool, sandy soil. If it is too hot, they will never look right."

如果太热了，它们就根本长不好了

If it is too hot, they will never look right

为什么胡萝卜有益健康呢?

Why are carrots healthy?

"我喜欢它们鲜艳的橘黄色,吃起来一定味道好极了!"

"无论什么样子的胡萝卜都很好吃。我喜欢胡萝卜,它们是很好的幼儿食物。"

"为什么胡萝卜有益健康呢?"小兔子很好奇。

"I love their bright orange colour; you can see that they must taste great!"

"Carrots always taste good, no matter how they look or what shape they are. I love carrots. They are excellent baby food."

"Why are carrots healthy?" wonders the little one.

"胡萝卜的β胡萝卜素含量是其他蔬菜的两倍,而且维生素含量丰富,甜中带点微苦。"

"但是我不喜欢苦的食物。"

"苦味有益健康,这是我从你们外婆那里学到的。"

"They have double the beta-carotene than other veggies, are full of vitamins, and have a sweet taste with a hint of bitterness."

"But I don't like bitter food."

"Bitter means healthy; that is what I learnt from my mother."

苦味有益健康

Bitter means healthy

每千克大约有一百万粒种子

A million seeds in one kilo of seeds

"妈妈,胡萝卜是水果吗?"另一只小兔子问。
"不是,亲爱的。胡萝卜是蔬菜,两年开一次花,每次结出成千上万的小种子,每千克大约有一百万粒。"
"哇噢!真的很多。那胡萝卜独自生长吗?"

"Mum, are carrots fruit?" asks another of the little ones.
"No, my dear. They are vegetables and bear flowers every other year, producing thousands of tiny seeds. There are more or less a million seeds in one kilo of seeds."
"Wow! That's a lot. Do carrots grow alone?"

"没有人喜欢独自生长。胡萝卜喜欢和香葱、迷迭香、鼠尾草生长在一起。鼠尾草赶跑了所有黏着胡萝卜的苍蝇。"

"因此,尽管胡萝卜需要数月才能成熟,但只要一撒下种子,很快就能让世界不再饥饿了。"

"是的!甚至不用煮熟,我们就可以吃。"

"No one likes to grow alone. Carrots prefer to grow alongside chives, rosemary, and sage. Sage keeps all the carrot flies away."

"So, even though carrots need months and months to grow, once you have seeds blowing around, you could quickly feed the world."

"Yes! And they don't even have to be cooked for us to enjoy them."

很快就能让世界不再饥饿了

you could quickly feed the world

完全煮熟的胡萝卜最有营养

The most nutritious carrots are boiled whole

"不需要煮熟？现在我们谈论的是真正非常健康的快餐，就像兔笼里给我们提供的大部分水果和蔬菜，连皮都不用削。"

"嗯，很简单。没有切开的完全煮熟的胡萝卜最有营养。"

"煮熟需要消耗更多的精力和耐心。妈妈，我认为我们没有足够的耐心。除非你加些糖，把它做得很甜，不然我们不会等的。"

"No need to cook them? Now we are talking about really super healthy fast food! That's like most of the fruit and vegetables that we are served in our rabbit hutch. You don't even need to peel them."

"Well, it is not that much work. The most nutritious carrots are the ones boiled whole, not cut."

"Boiling means using more energy and needing more patience. I don't think we have that, Mum. Unless you of course add some sugar and make it really sweet. Then we will wait."

"加糖？胡萝卜本身就含有很多糖分了，我们也吃了很多像胡萝卜一样的含糖食物。你们外婆教我用丁香煮胡萝卜，只要你慢慢煮，胡萝卜就会神奇地变得更甜。我喜欢当地的胡萝卜，价格公平。但是我真正喜欢的是新鲜的未加工的胡萝卜，或者慢慢煮熟的、味道香甜的胡萝卜。"

"妈妈，你是说我们应该吃慢餐吗？"小兔问。

"是的，亲爱的，这比快餐健康多了。"

……这仅仅是开始！……

"Add sugar? Carrots are rich in sugar already, and we eat far too much sugary food as it is. Grandma taught me to cook carrots with cloves. That miraculously makes them sweeter, as long as you cook them slowly. I like it it that carrots will be available locally, and be sold at a fair price. But what I really like is that we can eat food fresh and raw or cook it slowly for taste and sweetness."

"Mummy, do you mean we should all rather eat slow food?" asks the little one. "

"Yes dear, it is much healthier than fast food!"

… AND IT HAS ONLY HAS JUST BEGUN!…

……这仅仅是开始！……

...AND IT HAS ONLY JUST BEGUN!...

# Did You Know? 你知道吗?

Carrots are root vegetables and even though we associate them with the colour orange, varieties exist that are purple, red, white, and yellow. When carrots were first farmed, they were grown for their aromatic leaves and seeds, rather than for their roots.

胡萝卜是根茎类蔬菜，尽管我们认为胡萝卜都是橘黄色的，但也有紫色、红色、白色和黄色的胡萝卜。人类最初种植胡萝卜是为了收获芳香的叶子和种子，而不是根茎。

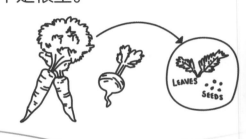

Carrots can be stored for a long time, provided they have never been washed. Carrots are 88% water and nearly 5% sugar. Carrots contain beta-carotene, vitamin K, and vitamin $B_6$.

如果未经水洗，胡萝卜可以存放很长时间。胡萝卜含有88%的水分和接近5%的糖分，还含有β胡萝卜素、维生素K和维生素$B_6$。

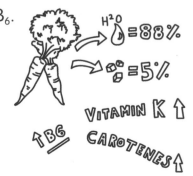

Carrots have been used to sweeten cakes since the Middle Ages. Carrot cake was once voted as the favourite cake in the United Kingdom. Today, carrot cakes are typically made from odd-shaped carrots that supermarkets will not buy.

中世纪以来，胡萝卜就被用来增加蛋糕的甜味。胡萝卜蛋糕曾被选为英国最受欢迎的蛋糕。现在，胡萝卜蛋糕主要由超市不愿收购的奇形怪状的胡萝卜制成。

1 g of carrot seeds contains more than 500 seeds. That is 500 000 seeds in 1 kg. Since every seed can turn into a carrot, it is one of the most prolific vegetables.

1克胡萝卜种子有500多粒，1千克有500 000粒。每粒种子都能长成一根胡萝卜，因此成为产量最多的蔬菜之一。

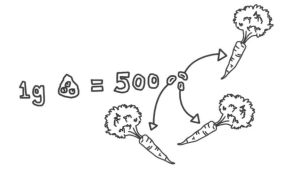

If you give a rabbit a carrot with a green top, it will eat the green and disregard the carrot. If you try to feed a rabbit nothing but carrots, it will die. It is like giving kids only candy to eat.

如果给兔子喂一根带绿叶的胡萝卜，兔子会吃了绿叶而留下胡萝卜。如果只给兔子喂胡萝卜，兔子会死亡，就和只给孩子吃糖一样。

Carrots need to pass through a prolonged cold spell to develop from the juvenile edible carrot root stage to reproductive maturity, during which the plant sprouts flowers. The seeds are embedded in a mericarp filled with oil that inhibits seed germination.

胡萝卜需要经历一段较长的寒冷期才能从幼嫩长至成熟，在这其间会开花。种子嵌在含油分果中，受到抑制而无法发芽。

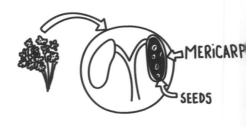

Nearly all carrots you buy in the shop are hybrids and thus sterile. Even if you have the patience to wait for two years to let the flowers grow and harvest the seeds, they are unlikely to grow.

在商店购买的大部分胡萝卜是杂交产品，没有繁殖能力。即使你有耐心等两年，它们也不可能开花结果。

It is easier and requires less energy for the body to digest cooked food. Cooked carrots supply more antioxidants and beta-carotene. The downside is that cooking reduces vitamin levels.

对人体来说，消化熟食更容易，需要的能量更少。熟胡萝卜产生更多的抗氧化物和β胡萝卜素，但不足是维生素含量降低了。

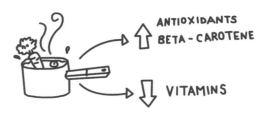

# Think About It

## 想一想

Do you enjoy taking time to have a meal with the family or do you prefer to have fast food and to finish your meal quickly?

你喜欢花时间和家人一起吃饭，还是更喜欢节省时间吃快餐？

马上吃胡萝卜，或在两年时间里用种子种植大量的胡萝卜，三年时间甚至会有更多的胡萝卜。如果必须二选一……你会选择哪个？

If you have to choose between eating a carrot now or to grow plenty of carrots from their seeds within two years, and even more carrots within three years … what would you choose?

Would you plant carrots alone or would you plant something to grow alongside them?

你是会只种胡萝卜，还是会把胡萝卜和其他植物一起种？

西红柿是水果还是蔬菜？

Is a tomato a fruit or a vegetable?

# Do It Yourself!
## 自己动手！

Let's bake a carrot cake. Ask your mother or grandmother for her favourite recipe or find one on the Internet. To make gluten free carrot cake you can substitute wheat flour with almond and walnut flour. To make your cake extra tasty, decorate it with icing made with coconut cream.

我们烘焙一个胡萝卜蛋糕吧！问问你妈妈或奶奶喜欢的食谱配方，也可以上网找一个。用杏仁粉和核桃粉代替面粉，制作无麸胡萝卜蛋糕。用椰乳味的糖霜点缀蛋糕，让蛋糕更美味。

# TEACHER AND PARENT GUIDE

## 学科知识
### Academic Knowledge

| | |
|---|---|
| 生物学 | 胡萝卜是有益的伴生植物，开花时会引来黄蜂，消灭许多园林害虫；萝卜是二年生植物，发芽需要经历两至三周，第二年才能开出伞状花并结出种子；杂交种是由两种或两种以上品种人工杂交而成；水果是物的果实部分，蔬菜是植物的其他部分，如根、叶和茎。 |
| 化 学 | 如果过度授粉，根茎类蔬菜可能含有大量的硝酸钠，危害婴幼儿健康；类胡萝卜素和高铁酸具有抗氧化作用；丁香有生物活性成分能驱赶蚊，并且曾用作牙疼止痛药。 |
| 物 理 | 光谱上，橘黄色处在黄色和红色之间。 |
| 工程学 | 快餐需要严格的物流运输和供应管理。 |
| 经济学 | 单份快餐的利润很低，但成千上万的单份交易提高了利润，这是以规经济和降低边际成本为基础的：销售量越多，额外成本越低；慢餐运倡议食用代表本地气候和生物多样性的本地食物。 |
| 伦理学 | 企业说他们会检查硝酸盐含量，但是，检查并没有根除过量的硝酸盐也没有控制不断过量使用的硝酸盐，而过量的硝酸盐最终会进入地水；超市拒绝销售某些胡萝卜的原因仅是胡萝卜的外形不好。 |
| 历 史 | 公元前3000年到前2000年，瑞士和德国南部已经耕种胡萝卜；十四世纪，国才开始种植胡萝卜；拿破仑使胡萝卜成为第一种罐装蔬菜。 |
| 地 理 | 荷兰推销橘黄色的胡萝卜，因为橘黄色是荷兰皇室奥兰治家族的颜色是荷兰的国色；慢餐运动始于意大利北部的皮埃蒙特。 |
| 数 学 | 斐波那契的兔子：如果第一个月有一对兔子，12个月后将会繁殖出144对。 |
| 生活方式 | 快餐文化：不花时间准备食物或饮食不合理会导致消化问题；和日本细烹饪的文化相比，寿司作为快餐而发明；对于什么是水果，什么是菜，厨师和生物学家持有不同的看法。 |
| 社会学 | 我们创作的动物故事并不准确，例如，兔子只吃胡萝卜将无法存活，但动画让我们相信兔子只吃胡萝卜；橘黄色与欢乐、外向、火焰和危险相关。 |
| 心理学 | 胡萝卜和大棒：奖励和惩罚并用以引导行为。 |
| 系统论 | 喂养孩子不仅仅和食物量有关，也和食物品质、准备过程和食用方式关；正确饮食能带来更健康的生活 |

# 教师与家长指南

## 情感智慧
### Emotional Intelligence

**小兔子**

小兔子们有些迫不及待。他们理解妈妈，知道妈妈喜欢什么。他们也了解自己，知道自己想要什么喜欢什么。关于如何快速得到更多食物，他们想知道更多方法。他们选择食物基于颜色和形状，而健康因素是次要的。他们拒绝食用苦的东西，想要深入了解胡萝卜，兴奋地发现胡萝卜能有那么多种子，能帮助解决世界上的饥饿问题。他们权衡了选择：如果能耐心地等待更长时间，他们能吃到甜甜的东西；但如果想马上吃到东西，食物也许会有点苦味。

**兔妈妈**

对于养育孩子的方式，兔妈妈很坚定。她知道孩子能吃什么，不能吃什么。她愿意让孩子们思考，并讲清楚了她的决定和标准的来龙去脉。她鼓励孩子们不要光看表面，如食物的颜色。在合适的时间，她说出了孩子们无法想象的惊人事实。兔妈妈分享了代代相传的智慧，用胡萝卜作隐喻，提倡促进多样性和朋友间的友谊。除了加糖，兔妈妈提供了其他选择，希望能给孩子们最好的。

## 艺术
### The Arts

该是画一些疯狂胡萝卜的时间了！不是漫画和动画片里那些无趣的胡萝卜，而是大自然里我们见到的奇形怪状的胡萝卜。它们看似来自另一个世界，但只要你去找一些网络图片和书本插画，这样疯狂的胡萝卜可能真能在大自然中找到。

# TEACHER AND PARENT GUIDE

## 思维拓展
### Systems: Making the Connections

几千年来，胡萝卜已经成为我们饮食的一部分。蔬菜经销商希望把生产标准化，因此，挑选胡萝卜不仅基于颜色，还有形状。口感好的胡萝卜不被接受，因为他们的形状不标准。无论什么吃法，胡萝卜都是营养丰富的。新鲜的生胡萝卜富含维生素，煮熟后的胡萝卜富含番茄红素。除了健康营养的食物，胡萝卜还有很多价值，提供了控制害虫的方式，最惊人的特点是丰富的种子产量和相对较快的生育期，从种子发芽到成熟，只要两年。胡萝卜可以生吃或煮熟，由于富含5%糖分，不必加糖。胡萝卜有苦的有甜的，就像大多数水果、蔬菜和谷物一样。不管生吃还是煮熟，下咽前的充分咀嚼很重要，适当的胡萝卜准备时间能保证最佳营养价值且易于消化。土地生产什么就吃什么，采用保护地球的方式进行种植，给农民支付合理的工资，这样既有美味的食物，又是我们当地文化的一部分。总之，这样能有更高的生活质量。

## 动手能力
### Capacity to Implement

公元1202年，斐波那契第一次讲述关于兔子的数学，计算出一对兔子一年繁殖了144对，那是很快的繁殖速度。但是现在，让我们比较一下兔子和胡萝卜的繁殖能力。一根胡萝卜结出种子需要两年，搞清楚每次能收获多少胡萝卜。从一对兔子和一根胡萝卜开始，假设在无限的土地上，五年和十年能生产多少吨胡萝卜？多少只兔子？如果我们能合理利用，你认为世界上还会有饥饿问题吗？如果你能算出数量，建立你的论点论据并与其他人分享吧。

# 教师与家长指南

## 故事灵感来自

## 卡洛·彼得里尼
### Carlo Petrini

　　卡洛·彼得里尼坚信食物应该是优良、干净、公平的，即食物应该有益健康、优质、无污染，以公平价格销售。他以共产主义运动政治活动家身份开始自己的职业生涯。他反对在罗马著名的"西班牙广场"开设第一家速食店，因此迅速成名。他发起了国际慢餐运动，在四大洲经营项目和开展活动。2004年，在他的出生地意大利皮埃蒙特区的库内格省布拉城，他成立了世界第一家烹饪大学，并被评为2004年度《时代周刊》的杰出人物之一。

更多资讯

　　http://www.maths.surrey.ac.uk/hosted-sites/R.Knott/Fibonacci/fibnat.html#Rabbits

　　http://www.carrotmuseum.co.uk/trivia.html

## 图书在版编目（CIP）数据

慢餐：汉英对照／（比）冈特·鲍利著；（哥伦）凯瑟琳娜·巴赫绘；李欢欢，牛玲娟译．－－上海：学林出版社，2016.6
（冈特生态童书．第三辑）
ISBN 978-7-5486-1051-9

Ⅰ．①慢… Ⅱ．①冈… ②凯… ③李… ④牛… Ⅲ．①生态环境－环境保护－儿童读物－汉、英 Ⅳ．① X171.1-49

中国版本图书馆 CIP 数据核字 (2016) 第 125753 号

—————————————————————————

© 2015 Gunter Pauli
著作权合同登记号 图字 09-2016-309 号

### 冈特生态童书
#### 慢餐

| | |
|---|---|
| 作　　者—— | 冈特·鲍利 |
| 译　　者—— | 李欢欢　牛玲娟 |
| 策　　划—— | 匡志强 |
| 责任编辑—— | 程　洋 |
| 装帧设计—— | 魏　来 |
| 出　　版—— | 上海世纪出版股份有限公司 学林出版社 |
| | 地　址：上海钦州南路81号　电话／传真：021-64515005 |
| | 网址：www.xuelinpress.com |
| 发　　行—— | 上海世纪出版股份有限公司发行中心 |
| | （上海福建中路193号　网址：www.ewen.co） |
| 印　　刷—— | 上海丽佳制版印刷有限公司 |
| 开　　本—— | 710×1020　1/16 |
| 印　　张—— | 2 |
| 字　　数—— | 5万 |
| 版　　次—— | 2016年6月第1版 |
| | 2016年6月第1次印刷 |
| 书　　号—— | ISBN 978-7-5486-1051-9/G·386 |
| 定　　价—— | 10.00元 |

（如发生印刷、装订质量问题，读者可向工厂调换）

Food
81

# 疯狂的蟋蟀

Crazy Crickets

Gunter Pauli

冈特·鲍利 著

凯瑟琳娜·巴赫 绘
李欢欢 牛玲娟 译

学林出版社
www.xuelinpress.com

## 丛书编委会

主　任：贾　峰

副主任：何家振　郑立明

委　员：牛玲娟　李原原　李曙东　吴建民　彭　勇
　　　　冯　缨　靳增江

## 丛书出版委员会

主　任：段学俭

副主任：匡志强　张　蓉

成　员：叶　刚　李晓梅　魏　来　徐雅清　田振军
　　　　蔡雩奇　程　洋

特别感谢以下热心人士对译稿润色工作的支持：

姜竹青　韩　笑　贾　芳　刘　晓　张黎立　刘之杰
高　青　周依奇　彭　江　于函玉　于　哲　单　威
姚爱静　刘　洋　高　艳　孙笑非　郑莉霞　周　蕊

# 目录

| | |
|---|---|
| 疯狂的蟋蟀 | 4 |
| 你知道吗？ | 22 |
| 想一想 | 26 |
| 自己动手！ | 27 |
| 学科知识 | 28 |
| 情感智慧 | 29 |
| 艺术 | 29 |
| 思维拓展 | 30 |
| 动手能力 | 30 |
| 故事灵感来自 | 31 |

# Contents

| | |
|---|---|
| Crazy Crickets | 4 |
| Did you know? | 22 |
| Think about it | 26 |
| Do it yourself! | 27 |
| Academic Knowledge | 28 |
| Emotional Intelligence | 29 |
| The Arts | 29 |
| Systems: Making the Connections | 30 |
| Capacity to Implement | 30 |
| This fable is inspired by | 31 |

一只蟋蟀上蹦下跳的,看起来很不安。一只黑猩猩很想吃蟋蟀这种高蛋白的食物,并且很想知道蟋蟀被抓住后会想些什么,他害怕成为灵长类动物的食物吗?"我亲爱的蟋蟀,"猩猩调侃道,"你知道吗?你可是我营养美味的大餐噢。"

A cricket is running up and down looking very worried. A chimpanzee, keen on eating this concentrated form of protein, wonders what idea the cricket has got hold of now. Could it be that it is scared to form part of a primate's diet? "So, my dear cricket," teases the chimp, "are you aware that you are a really rich, nutritious dinner for me?"

你可是营养美味的大餐噢

you are a rich nutritious dinner

# 大家都应该吃昆虫

Everyone should be eating insects

"我知道,似乎到处都在流传说,大家都应该用吃昆虫代替吃肉。"

"人类现在才知道这一点,真令人惊讶。我们早就知道你富含蛋白质了。"

"是的,但至少两条腿行走的人类现在已经清楚认识到,他们不能继续通过吃动物来填饱肚子了。"

"I know, it seems that the buzz has gone around – that instead of eating meat, everyone should be eating insects."

"It is surprising that people only figured this one out now. We have known for years that you are a protein bomb."

"Yes, but at least those walking on two legs are now well aware that they can't continue to eat animals to fill their empty stomachs."

"蟋蟀是最富含蛋白质的食物来源之一。如果你和你的兄弟姐妹同意被吃掉，可以挽救许多动物的生命。"

"这不公平。难道只有付出我的生命，才能挽救他们的生命吗？"

"嗯，你和你的家族成员繁殖很快，还吃别人不吃的食物，所以你们是食物链的关键部分。如果你们能一直存在，每个人都可以有足够的食物来源。"

"Crickets are one of the richest sources of protein. If you, and your brothers and sisters, will agree to be eaten, it could save the lives of many animals."

"That is not fair. I have to pay with my life so they can have theirs?"

"Well, you and your families can reproduce so fast and eat what no one else will eat, so you are a key part of the food chain. If you play along, everyone could finally have enough to live on."

你繁殖很快

you can reproduce so fast

我愿意成为大家的点心

I will agree to be turned into a snack

"你说得对,我们确实繁殖得非常快,如果能让我的孩子们生活得更幸福,我愿意成为大家的点心。"

"我们很钦佩你能有这种生命循环的想法!"

"You are right, we do reproduce very fast, and I will agree to be turned into a snack if it means my kids would live more happily."

"We appreciate that you have such a circular vision of life!"

"我想知道，为什么人们要从原来四条腿行走变成两条腿站立行走？用四条腿跑得更快啊！"

"是的，四条腿行走会更稳一些，还有助于提高在地面的移动速度。但如果你用两条腿行走，一个优势是你可以腾出两只胳膊携带食物了。"

"I wonder why people, who could once run on four legs, decided to stand up and walk on two. You can run faster on four!"

"Yes, running with four legs is more stable, and gives better ground speed. But there is one advantage: if you run on two legs then you have two arms free to carry food."

如果用两条腿行走你可以携带食物

If you run on two legs you can carry food

可以和家人一起分享食物

You can share food with your family

"携带食物有什么用?你还是要吃掉啊!"

"是的,但是如果你用四条腿行走的话,就不能边走边吃了。"

"有意思!所以你现在可以边行走边携带食物,会怎样呢?"

"想想啊,这样你就可以回家和家人一起分享食物了!如果你是一个好猎手,能找到更多的食物,你就可以给家人带回去更多。"

"What is the use of carrying food? You have to eat it!"

"True, but if you need four legs to get around, you can't run and eat at the same time."

"Interesting! So now that you can walk and carry food, what then?"

"Come on, now you can get home and share food with your family! If you are a good hunter and find more food, then you can bring a lot of it with you."

"你的意思是,人类——好斗的智人——会愿意分享吗?"

"嗯,不仅是合作与分享,而且是想活得更好。"

"你确定吗?人类是好斗的,会强烈捍卫他们的领土——只存在于脑海中的一个假想边界。"

"You mean that man, this aggressive Homo sapiens, is prepared to share?"

"Ah, not only cooperate and share, he is better now prepared to survive."

"Are you sure? Human beings can be aggressive, and fiercely defend their territory, based on the illusion of a border that only exists in their mind."

# 人类乐于分享

Homo sapiens is willing to share

分享创造了爱与关心!

Sharing leads to loving and caring!

"好吧,你说得没错,我们都有过一些不好的经历。但我们都知道,当你带着食物回家时,家人会喜欢你。如果你天天都带着食物回家,他们会与你更加亲密。"

"你是说正因为做到了分享,从而得到了爱与关心这样的回馈?"

"Well, you are right and we all have lived through a few bad experiences, but you and I both know that when you come home with food, family do like you. And, if you come home with food every day, they will bond with you."

"You mean that there is in fact a reward, such as loving and caring, because you are sharing?"

"事实的确如此!如果你多与家人以及周围的人分享,就可以形成一个群体,即使在困难的时刻也相互关心。"

"我很快就会被吃掉了,这岂不是很遗憾吗?"

……这仅仅是开始!……

"In fact, there is! And, if you share more with your family and others around you, then you can build a community where people care for each other – even when the going gets tough."

"Isn't it a pity that I am going to be eaten soon?"

... AND IT HAS ONLY JUST BEGUN!...

……这仅仅是开始!……

...AND IT HAS ONLY JUST BEGUN!...

# Did You Know? 你知道吗?

The cradle of humankind is in Africa. Biologically speaking there are no races. We are all Africans. It is here where people learned about sharing and cooperation millions of years ago.

人类起源于非洲,从生物学角度讲没有种族之分,我们都是非洲人。数百万年前人类在这里学会了分享与合作。

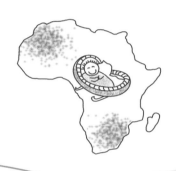

An adult brain is a forest of neurons that contains 100 billion nerve cells that have more than 100 trillion connections. Unfortunately we only use 10% of its potential.

成人的大脑包含1000亿个神经细胞和100万亿个神经突触,它们相互连接。可是,人类只开发了其中10%的潜能。

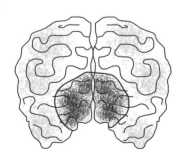

Two billion people worldwide eat insects on a regular basis: in Africa, Asia, Latin America and Australia, but not in Europe or America.

全世界有 20 亿人定期吃昆虫，他们分布在非洲、亚洲、拉丁美洲和澳大利亚，但是在欧洲和美国没有分布。

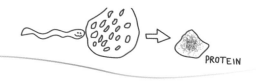

Insects eat two kg of feed to produce one kg of protein, and this is richer in iron than beef. Even better, insects require no land; they eat leftovers and emit fewer greenhouse gases.

昆虫吃掉 2 千克的饲料能生成 1 千克的蛋白质，这种蛋白质含有的铁元素比牛肉更丰富。更好的是，昆虫不需要专门的土地，它们吃剩下的食物，减少了温室气体的排放。

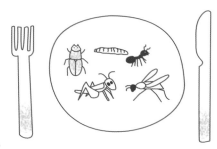

烤蠕虫有坚果的味道，蠕虫吃的黄粉虫影响了味道。所以，如果蠕虫一直吃苹果，它们的味道将会尝起来像苹果。

Roasted worms have a nutty flavour, while mealworms take on the taste of their last meal. So, if they have been eating apples, they will taste like apples.

昆虫蛋白可以缓解水产养殖和家畜养殖所需饲料的压力，也可以防止过度捕捞，一吨的昆虫蛋白可以替代一吨鱼类的蛋白。

Insect protein relieves the pressure on making fish meal for aquaculture and to feed livestock. This can reduce over-fishing. One ton of insect protein saves one ton of fish.

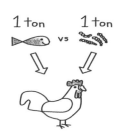

There are 1 900 popular insects eaten by humans and the favourite ones are beetles, caterpillars, wasps, ants, grasshoppers, locusts and crickets.

有1900种常见昆虫被人类食用，最受欢迎的种类有甲虫、毛毛虫、黄蜂、蚂蚁、蚱蜢、蝗虫和蟋蟀。

One kg of cricket flour contains 18 000 crickets, who grew to maturity in just 6 weeks.

一千克的蟋蟀粉由18 000只蟋蟀制成，蟋蟀在短短的6周时间内就能发育成熟。

# Think About It
## 想一想

How do you feel about having insects for breakfast?

你觉得早餐吃昆虫怎么样?

如果有人能劝你吃不健康的肉,其他人还能说服你吃健康的昆虫吗?

If someone convinced you to eat unhealthy meat, could someone else convince you to eat healthy insects?

Are you happy when someone in your family prepares food for everyone when they come home for meals?

回家吃饭时,已经有人为大家准备好了食物,你高兴吗?

你喜欢人类起源于非洲这个说法吗? 为什么?

Do you like the idea that we are all from African origin? Why or why not?

Try to run on all fours, using your legs and arms. It does not work well, does it? We have no rhythm and we lack the right body posture and even feel uncomfortable. Still, give it another try. The experience gives you a sense of the transformation humans went through over thousands of years. Now, try to carry something while you are on all fours. The only place where you could perhaps keep anything is in your mouth. You will experience the power and the logic of walking and carrying something at the same time. It was not only a smart thing to do, it was necessary.

尝试四肢着地用腿和胳膊移动，好像不是很协调，是不是？我们这样节奏不对，身体姿势不对，甚至会感到不舒服。不过，再做一次尝试，这会让你感受到人类几千年的变化。试着趴在地上携带一些东西，这时身上唯一能携带东西的部位就是嘴了，你将会感受到边行走边携带东西的力量。这不仅是一种聪明的行为，而且是一种有必要的行为。

# TEACHER AND PARENT GUIDE

## 学科知识
### Academic Knowledge

| | |
|---|---|
| 生物学 | 食用昆虫包括甲虫、毛毛虫、蜜蜂、黄蜂、蚂蚁、蚱蜢、蝗虫、蟋蟀、蝉、飞虱、白蚁、蜻蜓和苍蝇；人类大脑进化了六百万年后停了下来；人脑由大脑、小脑和脑干三部分组成；动脉为大脑输送体内25%的血液，当你思考问题时大脑会消耗体内50%的氧气。 |
| 化 学 | 大脑会被神经毒素损伤；一些物质如铅（添加到汽油和油漆里的）和乙醇（添加到燃料和饮品里的）会破坏神经组织；科伊桑岩石艺术所用的颜料成分是人的尿液、血液和赭石、木炭、鸵鸟蛋，在一万年后仍会保持亮度。 |
| 物 理 | 记忆信号以微电流形式穿过神经细胞，当电荷到达神经突触（细胞连接处）时触发化学物质（神经递质）的释放。 |
| 工程学 | 在地面依靠四肢行走或直立行走；四条腿行走可以保证更快的速度和更高的稳定性，两条腿行走可以发挥手的作用，并把头抬到地面以上呼吸凉爽的空气。 |
| 经济学 | 价格由供需平衡和人们的期待值决定；昆虫粉的价格和小麦、玉米粉相近，但是它的蛋白质含量比更贵的鱼肉还要高。 |
| 伦理学 | 利他主义认为从道德出发，个体有义务帮助他人。 |
| 历 史 | 非洲南部遗传变异显示，现代人类是非洲南部早期人类混合的结果；洞穴壁画表明科伊桑人生活在2万至3万年以前。 |
| 地 理 | "人类摇篮"遗址位于约翰内斯堡附近，是联合国教科文组织世界遗产之一，这里发现的人类祖先化石占全球的三分之一；南非是拥有石器时代雕塑、绘画和雕刻收藏品最多的国家。 |
| 数 学 | 智商高低与物种的脑体积有关；绝对数与相对数的区别。 |
| 生活方式 | 如果你付出的代价是你无法承受的，那是一种牺牲；父母为自己的孩子们做出牺牲，通常是为了让自己的孩子们实现自己未完成的梦想。 |
| 社会学 | 科伊桑人具有非暴力和愈合文化，他们相信自己生活在"愈合的土地上"，他们对着星星唱歌，几乎不发动战争；循环或线性的时间概念决定着人们是否相信轮回转世；社会神经学显示，人们对与他人保持联系的需求比对食物和住所的需求更迫切。 |
| 心理学 | 人们相互分享是因为他们彼此关心，希望能保持联系；礼物是利他主义的爱、赞赏和感激的普遍表达；理解别人的想法，明白别人的希望、恐惧和动机，能让我们生活得更和谐，却也约束自身的更好发展；为什么一些人可以很快从挫折中恢复而另一些人却不能，这与语境敏感性、社会直觉、应变能力和自我认知相关。 |
| 系统论 | 人类进化的方式与分享经历相关；人类社区随着时间推移而关系加深，这使我们看到完整的生活体系。 |

# 教师与家长指南

## 情感智慧
### Emotional Intelligence

蟋蟀

蟋蟀感受到压力,表现出了很多焦虑,不知道在人类对昆虫蛋白的需求中该如何存活下来,但他仍然保持清醒的头脑,看到不仅是他,而且是整个世界都处于危险当中。蟋蟀反对这样的观点——为解救动物生命以及保存足够多的食物,自己必须死去。蟋蟀为了孩子们的未来,最终接受了自己死亡的归宿,认为这是自己在改造世界中发挥的作用。在经历了这个艰难的选择后,蟋蟀开始提出关于人类进化的基本问题。蟋蟀的问题很有逻辑性,他根本不信任人类,不认为人类会与他人分享任何利益。但在听了黑猩猩的解释后,他看到了建立强大生物群体的可能性。但接着又要直面现实问题:很快就会被吃掉了吗?

黑猩猩

黑猩猩很自信,用黑色幽默的方式调侃蟋蟀,提醒蟋蟀可能会成为他的食物,同时也对蟋蟀富含营养表示感激。黑猩猩正确地理解蟋蟀的牺牲:拯救了其他动物的生命,为自然界提供了更充足的食物来源。黑猩猩与蟋蟀一起互动,解释了人类随着时间是如何进化以及人类直立行走的原因。另外,黑猩猩深入探讨群体的形成,何种分享是可以得到回报的。当蟋蟀存有疑虑时,黑猩猩提供了清晰的思路,让蟋蟀知道何时需要做出牺牲。

## 艺术
### The Arts

看一看科伊桑岩画,它们至少有1万年的历史了。岩画本身不是关键,颜料为什么能保持这么长时间才是关键。你想画这样的画吗?查找颜料成分,尝试自己去画。如果科伊桑人上万年前都可以做到,你也同样可以做到。不要惊讶颜料的成分,它们可能是一些你意想不到的东西!

# TEACHER AND PARENT GUIDE

## 思维拓展
### Systems: Making the Connections

人类出现在地球上后,数量不断增加,超出了自然界的承载能力。而且,随着人类饮食结构的转变,问题日益凸显。人类更想吃动物蛋白的原因是复杂的,实际上,以谷类、豆类和鱼类为食物来源的鸡、猪和牛,可以直接提供给人类营养物质。利用食物生产更多的食物,是食物链效率低,以致不能满足人类对营养物质的基本需求的主要原因。人类人口增加了数十亿,这使得形势越来越严峻。虽然科研人员在试图寻找食物增产的方法,但现在应该是时候改变策略了:在可持续生产上做一些事情。如果人类消耗的食物,不仅仅可再生,也更好、更健康,这样我们可以有更好的生活。许多反对者可能会不喜欢"其他食物",或者是"习惯"很难改变。然而,在过去几十年中我们周围的食物已经改变。如果我们有那么多人愿意吃转基因玉米,或者接受吃含激素的汉堡包,那么能不能想象一下,把昆虫作为地球上70亿人口的食物呢?

## 动手能力
### Capacity to Implement

列出全球最受欢迎的十种昆虫,问问自己哪种是可以吃的。然后,查看所有可以添加昆虫粉的食谱,选出自己喜欢的食物,看看当地有没有可以用的食材。如果有,尝试做一下。如果没有,你准备好自己动手去养殖这些昆虫了吗?

# 教师与家长指南

## 故事灵感来自

### 玛丽·瑞德
### Marie Ryd

　　玛丽·瑞德在瑞典卡罗林斯卡学院学习并获得医学博士学位，专注于心理学和企业领导力心理学方法的研究。在成为科学记者之前，她在牙科实习过，在细菌学和免疫学学科上做过研究，这促成了 Holone 杂志的创立。Holone 杂志致力于脑研究的最新发现，尤其专注于社会行为对大脑影响的研究。

更多资讯

　　www.coschedule.com/blog/why-people-share/

　　http://109.74.12.180/dreamofthegood/HomeEng.aspx

图书在版编目（CIP）数据

疯狂的蟋蟀：汉英对照／（比）冈特·鲍利著；
（哥伦）凯瑟琳娜·巴赫绘；李欢欢，牛玲娟译． －－ 上
海：学林出版社，2016.6
（冈特生态童书．第三辑）
ISBN 978－7－5486－1050－2

Ⅰ．①疯… Ⅱ．①冈… ②凯… ③李… ④牛… Ⅲ．
①生态环境－环境保护－儿童读物－汉、英 Ⅳ．
① X171.1－49

中国版本图书馆 CIP 数据核字 (2016) 第 125756 号

ⓒ 2015 Gunter Pauli
著作权合同登记号 图字 09－2016－309 号

## 冈特生态童书
### 疯狂的蟋蟀

| | |
|---|---|
| 作　　者—— | 冈特·鲍利 |
| 译　　者—— | 李欢欢　牛玲娟 |
| 策　　划—— | 匡志强 |
| 责任编辑—— | 程　洋 |
| 装帧设计—— | 魏　来 |
| 出　　版—— | 上海世纪出版股份有限公司 学林出版社 |
| | 地　址：上海钦州南路 81 号　　电话／传真：021-64515005 |
| | 网　址：www.xuelinpress.com |
| 发　　行—— | 上海世纪出版股份有限公司发行中心 |
| | （上海福建中路 193 号　网址：www.ewen.co） |
| 印　　刷—— | 上海丽佳制版印刷有限公司 |
| 开　　本—— | 710×1020　1/16 |
| 印　　张—— | 2 |
| 字　　数—— | 5 万 |
| 版　　次—— | 2016 年 6 月第 1 版 |
| | 2016 年 6 月第 1 次印刷 |
| 书　　号—— | ISBN 978-7-5486-1050-2/G·385 |
| 定　　价—— | 10.00 元 |

（如发生印刷、装订质量问题，读者可向工厂调换）

Education 78

# 从原谅到忘记

Forgive to Forget

Gunter Pauli

冈特·鲍利 著

凯瑟琳娜·巴赫 绘
田 烁 王菁菁 译

学林出版社
www.xuelinpress.com

## 丛书编委会

主　任：贾　峰

副主任：何家振　郑立明

委　员：牛玲娟　李原原　李曙东　吴建民　彭　勇
　　　　冯　缨　靳增江

## 丛书出版委员会

主　任：段学俭

副主任：匡志强　张　蓉

成　员：叶　刚　李晓梅　魏　来　徐雅清　田振军
　　　　蔡雩奇　程　洋

特别感谢以下热心人士对译稿润色工作的支持：

姜竹青　韩　笑　贾　芳　刘　晓　张黎立　刘之杰
高　青　周依奇　彭　江　于函玉　于　哲　单　威
姚爱静　刘　洋　高　艳　孙笑非　郑莉霞　周　蕊

# 目录

| | |
|---|---|
| 从原谅到忘记 | 4 |
| 你知道吗？ | 22 |
| 想一想 | 26 |
| 自己动手！ | 27 |
| 学科知识 | 28 |
| 情感智慧 | 29 |
| 艺术 | 29 |
| 思维拓展 | 30 |
| 动手能力 | 30 |
| 故事灵感来自 | 31 |

# Contents

| | |
|---|---|
| Forgive to Forget | 4 |
| Did you know? | 22 |
| Think about it | 26 |
| Do it yourself! | 27 |
| Academic Knowledge | 28 |
| Emotional Intelligence | 29 |
| The Arts | 29 |
| Systems: Making the Connections | 30 |
| Capacity to Implement | 30 |
| This fable is inspired by | 31 |

哈里斯鹰和狐猴正在抱怨他们及其祖先们所遭遇的不幸经历。他们抱怨的东西非常多，尤其是那些入侵其领地的"外来客"。

A Harris hawk and a lemur are complaining about all the bad things that have happened in their lives and in their forefathers' lives. Their list of complaints is long, especially about all the strangers who invaded their land.

哈里斯鹰和狐猴正在抱怨……

A hawk and a lemur are complaining ...

……要守护好大一片领地

... a large territory to defend

哈里斯鹰首先表明立场。"我要守护好大一片领地,巡查一遍要花费我一整天的时间。要是谁胆敢闯进我的地盘,我一定会好好教训他。"

The hawk makes his position clear. "I have such a large territory to defend. It takes me all day to survey my land. Anyone who enters my area will learn a lesson."

"我们是竖起黑白相间的长尾巴来做标记警示别人。不管是谁胆敢冒犯我们的领地,他最好准备好打一场臭气大战,"环尾狐猴说道,"尤其是那些住在附近的狐猴家族。"

"We're putting up sign posts, using our long black and white tails. Whoever does not respect our land had better prepare for a stink fight," says the ring-tailed lemur, "especially these neighboring lemurs!"

……准备好打一场臭气大战

... prepare for a stink fight

我们甚至倒立……

We even stand on our hands ...

10

"入侵者那么多，你们能制造出足够多那种臭哄哄的东西吗？"

"你最好还是相信。我们狐猴家族可以制造出很多那种臭哄哄的东西。你知道我们可以用前臂把这些臭东西喷向四周吗？有时，我们甚至倒立着以最佳方式攻击敌人。"

"Can you produce enough of that stinky stuff? There are so many intruders."

"You better believe it. My family makes lots of that stinky stuff. Did you know we spray it around from our forearms? Sometimes we even stand on our hands to take the best shot at our enemy."

"我可不喜欢飞越你们的领地。那里味道太难闻了。"

"你能原谅我吗?这可是一个事关生存的问题。"

"I don't fancy flying over your territory. It smells."

"Do you forgive me? It is a matter of survival."

味道太难闻了

It smells

"生存还是毁灭"

"to be or not to be"

"你说得对。你不是狐猴，我们的倒立臭气大战对你来说确实有些可笑。但是对我们来说，领地之争可是事关'生存还是毁灭'的大事。"

"听着，我可以原谅你们进行奇怪的臭气大战，但这种气味实在让我无法忘记。"

"You are right, our stinky fights with handstands are a bit ridiculous when you are not a lemur, but for us it's a matter of 'to be or not to be'."

"Look, I can forgive the fact that you guys get into your strange smelly fights, but the smell makes it impossible to forget."

"当我飞越陆地上空时,我也看到了其他动物家族是如何守护领地的。有用爪子在树上做标记的,有在地上划出痕迹的,有留下皮毛的,有嚎叫的,有威胁恐吓的,甚至还有撒尿的!我承认,所有这些方式似乎都不如你们。"

"When I fly over the land I see others marking trees with their claws, scraping the ground, leaving fur, howling, intimidating, and they even pee! I admit, all that seems worse than you."

……用爪子在树上做标记……

... marking trees ...

……发动战争……

… go to war …

"可要说最坏的，还是人类。他们发动战争，让所有人都卷入战争来争夺陆地和海洋资源的开采权。"

"人类的历史就是一部侵略与战争史，长久以来，人们被迫做着违背自己意愿的事。他们甚至有一个叫'殖民主义'的东西。"

"人类发明了国界，还要花钱供养军队来保护自己，从不考虑社会所付出的代价。"

"甚至有些人还在不同的种族和信仰之间也划定界线。"

"But the worst of all are people. They go to war and make everyone fight for the right to exploit the land or the sea."

"The human history is one of invasions and wars, controlling people against their will for years. They even had something called colonialism."

"People invented boundaries and spend money to defend themselves with armies, no matter the cost to society."

"And then some even created boundaries amongst races and beliefs."

"怎么能因为人们的长相或想法不同,就把他们区分开呢?"

"人类打起仗来真是毫不手软。他们制造的炸弹可以摧毁一切,包括人类自己。"

"什么!如果所有人都被炸死了,那么就没有人可以去原谅战争,而且一切都将被忘记。"

"我很庆幸,我们只是发动了臭气大战,而你永远不会忘记那种气味!"

……这仅仅是开始!……

"How can you separate people because they look and think differently?"

"People are so serious about fighting. They make bombs that can destroy everything, including themselves."

"What! If everyone is gone with that bomb, then there is no one to forgive and everything is forgotten."

"I am so glad that we only fight with stinky stuff and that you will never forget the smell!"

... AND IT HAS ONLY JUST BEGUN!...

……这仅仅是开始！……

... AND IT HAS ONLY JUST BEGUN! ...

# Did You Know?
## 你知道吗?

Lemurs live in groups and the female is the leader. To keep warm, groups of lemurs will huddle closely together to sleep in their favourite tree.

狐猴是群居动物,以雌性为首领。为了保暖,它们会在自己最喜欢的树上相拥而睡。

The life of a lemur includes a love for sunbathing, sitting with their white fur towards the blazing sun. They also understand mathematics and use tools.

狐猴最喜欢的日常生活是晒日光浴,它们坐着,让自己白色的皮毛晒着暖暖的阳光。它们还会数学,也知道如何使用工具。

当狐猴在森林中穿行时,它们会把尾巴竖起来,来保证互相之间可以看到对方,保持不掉队。

W hen lemurs walk through the forest, they keep their tails in the air so that everyone can keep sight of each other and stay together.

狐猴以水果和树叶为食,尤其爱吃罗望子树的树叶。它们通过嚼芦荟和仙人掌获取水源。

L emurs eat fruits and leaves, especially of the tamarind tree. They get their water from eating aloe vera and prickly pear cactus.

The Harris hawk lives in families where the mature female is the dominant bird. The birds cooperate in hunting and in nesting, with three birds attending to the babies.

哈里斯鹰也是群居动物，成熟的雌性哈里斯鹰占有统领地位。它们合作着狩猎、筑巢，但会留下三只鹰照顾幼鹰。

The Harris hawk will take on prey and enemies larger than themselves. Because they work as a team, they make up for slower speed and less individual power.

哈里斯鹰可以攻击比自己体型还要大的敌人，它们是团队作战，这样可以弥补飞行速度较慢、个体能力较弱的缺点。

动物都会守护自己的领地，但发动攻击是最后的选择，因为一场战斗会耗费大量体力，甚至会造成死伤的严重后果。通常，动物不需要与对方发生身体接触就能完成一场战斗，这叫做仪式战。

$\mathcal{A}$nimals defend their territory, but fighting is the last option, since it uses up a large amount of energy and can result in injury and even death. Often animals go through all the motions of fighting without ever touching each other. This is called ritual fighting.

在马达加斯加岛上发现的野生动物中，百分之九十是这里独有的。

$\mathcal{N}$inety percent of the wildlife found on the island of Madagascar is not found anywhere else in the world.

# Think About It
# 想一想

Would you fight if someone entered your home?

如果有人闯入你的家，你会和他打架吗？

Do you welcome people from another race and belief into your home?

你欢迎和你不同种族、不同信仰的人进入你的家吗？

How would you defend your property? Would you prefer to make lots of noise and scare the intruder or go after them using a gun?

你如何保护自己的财产？你会制造嘈杂的噪音来吓跑那些入侵者，或者用枪驱赶他们吗？

When someone does something wrong and you are hurt, can you forgive and forget, or only forgive?

因为别人的错误行为而让你受到了伤害，你能原谅他并忘记这件事吗？或者只是原谅他？

Study the ritual fighting of animals. Document the movements and analyse their effectiveness. Then compare this ritual with the way humans fight. Now draft some strategies on how to avoid conflicts and when conflicts do occur, how to reach a conflict resolution.

研究动物们的仪式战。记录动作，并分析这些动作的效果，然后将它们的仪式战和人类战争进行对比。最后，提出一些避免发生冲突以及冲突发生时如何解决的策略。

# TEACHER AND PARENT GUIDE

## 学科知识
## Academic Knowledge

| | |
|---|---|
| 生物学 | 雌性动物在部分动物家族中占统领地位，如狐猴和哈里斯鹰家族；细胞细菌具有探测生存环境中化学物质的能力；嗅觉是五大感觉之一，当人们感到饥饿时，嗅觉会变得更加灵敏。 |
| 化 学 | 气味腺产生有机酸和酯，雄性的气味腺还能产生胆固醇衍生物；气味由分子组成，嗅觉是脊椎动物化学传感的版本。 |
| 物 理 | 形状理论将嗅觉和气味与分子形状联系起来，而受量子物理学启发提出的振动理论提出，气味是由分子的振动决定的；炸弹尤其是原子弹的冲击波效应。 |
| 工程学 | 炸弹的研发需要物理学、数学和工程学的知识。 |
| 经济学 | 每年，全球各国的国防费用超过1.75万亿美元，是防治艾滋病、肺结核以及自然灾害应急等总费用的几百倍；在军事武器上的花费被当作威慑行为，因此，在经济学上它属于一项非生产性投资。 |
| 伦理学 | 发生冲突时，我们是发起战争，还是通过和平手段来解决冲突？调解的艺术不是科学，而是经验，一种将心灵和思想相结合的能力；每个人都会犯错，但是我们可以原谅别人吗？专注于感恩，从发生的事情中吸取教训，做一名学习者而不是受害者，用积极的方式报复——不伤害别人，而是证明你可以比伤害自己的人做得更好。 |
| 历 史 | 17世纪，法国红衣主教黎塞留创立了中立国，他深化了法国国界是"自然界线"的理念；殖民主义是一种统治方式，也是一个国家以政治力量占领另一个国家的方式；宣战与和平条约的概念。 |
| 地 理 | 马达加斯加岛曾经和印度次大陆相连。 |
| 数 学 | 如何计算战争的代价，包括战争死亡人数记录（军人和平民）、因营养不良造成的非战争死亡人数、医疗保障的损失、环境的破坏、受伤的代价、停工的代价、公民自由被剥夺的代价、抚慰退伍军人的费用；建一个全新的数学模型来计算这些真实的成本（远高于仅仅运送士兵的成本），激励决策者研究可替代战争的方案。 |
| 生活方式 | 日常生活中，人们因种族和宗教团体而区分；"生存还是毁灭"是莎士比亚剧作《哈姆雷特》的开场白，描写了抉择的艰难。 |
| 社会学 | 人类发明了大量词汇来描述各种各样的颜色、形状、大小、纹理，但是我们却缺少恰当的词汇来描述气味；在古代，身体气味的评价对于择偶和识别家族地位来说至关重要。 |
| 心理学 | 气味感觉依赖于环境：奶牛粪对一些人来说是难闻的，但对于在农场长大的人来说却值得怀念；气味还和期待有关：将一块帕尔马干酪藏在杯子中，如果告诉别人这是呕吐物，人们就会被气味吓退，如果告诉人们这是一块美味的奶酪，人们则会为之沉醉；原谅的力量能让你忘记怨恨，即使被冤枉也保持积极的状态。 |
| 系统论 | 单一民族国家的创立造就了民族性的产生，一个民族通常涵盖由不同文化、经济、信仰组成的群体。 |

# 教师与家长指南

## 情感智慧
### Emotional Intelligence

**哈里斯鹰**

哈里斯鹰开始介绍自己的防御策略时用了很多攻击性词汇,但很快就被狐猴令人惊讶的方式搞得措手不及。哈里斯鹰喜欢交换意见,承认他不喜欢狐猴的味道。哈里斯鹰评论人类的所作所为,他对于人类的侵略行为和自己难以忘却的经历有鲜明的立场。哈里斯鹰有非常敏锐的观察能力,并和狐猴分享了动物们在守卫自己领地时各种各样的方法。这激发了一系列更深层次的思考,包括从当地情况到全世界面临的战争与和平的挑战,进一步演化为对种族主义、不包容行为、终极毁灭性武器(炸弹)所带来后果的思考。哈里斯鹰精辟地总结说:这将摧毁每一个人,包括制造炸弹的人。

**狐猴**

狐猴愉快地介绍自己的生活方式,骄傲于自己是臭气大战的获胜方。他乐于分享自己如何开展臭气大战,但对哈里斯鹰不喜欢这种气味的事实也很敏感。而且狐猴意识到,尽管这对他们来说是最平常不过的事情,在局外人(比如哈里斯鹰)看来,臭气战可能非常可笑。哈里斯鹰关于"可以原谅但不能忘记"的描述激发了他们对不同物种如何守卫领地的更深层次思考。狐猴批判人类的行为,比如发动战争、花费巨资维护武器等。然后,他想知道为何仅仅因为长相不同而把人们区分开。最后,狐猴对生命有了哲学性的认识,并对自己用臭气战来解决冲突感到满意。

## 艺术
### The Arts

列出动物们守卫自己领地的10种方式。用短剧表演来展示这些方式,将这些表演录下来上传到互联网上。邀请朋友和家人投票选出最佳表演,评选因素有:1)表演是否有趣;2)不用实际战斗而达到和平解决争端的方式是否有效。

# TEACHER AND PARENT GUIDE

## 思维拓展
### Systems: Making the Connections

在动物界，捍卫领地是一种牢固建立的观念。动物们通常用气味和尿液来标记边界，它们很少将捍卫边界之争升级为致命的冲突，而人类却将领土划界作为发动战争的理由。领土争端在不断增加，每个国家都在小心地捍卫自己陆地和海洋的边界。守卫领土边界（不是为了保护政治实体，就是为了获得经济资源）的愿望为武装力量的产生提供了正当理由。许多国家用于维护武装力量的费用超过了在医疗和教育上的经费，更糟糕的是，实际的战争成本非常高，不仅仅是浪费掉数以万亿计的资金，还有整个社会为之付出的代价，将成为几代人要承受的社会负担。现在人们不断研究更多样的冲突解决方案，国防设备的投入主要致力于对侵略者的军事威慑上。然而，转移如此巨大的人力财力资源通常会影响到社会满足公众基本需求的能力。这种国防费用的最主要成就似乎是创建了国内的安全感，而具有讽刺意味的是，边界已经随着时间的演变而不断改变，很有可能还会再改变。正是在这种背景下，狐猴的臭气大战可能看起来微不足道，但却很好地启发人们采取仪式战，而非不断地苦思冥想如何使用毁灭性武器。

## 动手能力
### Capacity to Implement

你能制造出多少种不同的气味？去收集一些气味，比如鲜花、树皮、种子、还有食物的气味……目标定得高些，收集至少25种不同的气味。然后邀请你的家人和朋友来闻一闻，并尝试辨别气味的来源。如果每个人都收集到25种不同的气味，那么你们很快就找到100多种。这将是一个很有意思的游戏，帮助你理解气味、香味甚至香水在我们生活中有多重要。

故事灵感来自

## 琳达·布朗·巴克
### Linda Brown Buck

琳达·巴克研究心理学和微生物学，还获得了免疫学的博士学位。她很早就认识到了气味（嗅觉系统）对我们生活品质的重要性，独特的气味可以勾起人们对童年或者某个情感瞬间的回忆。她曾在哥伦比亚大学跟随理查德·阿克塞尔开展博士后工作，并于1991年与理查德共同发表了具有突破意义的论文，描述了气味引发记忆的作用。他们认为人类可以分辨10 000多种不同的气味，并且发现果蝇的嗅觉系统与老鼠相似。在所有感觉中，嗅觉或许是最需要深入挖掘和探究的。2004年，琳达·巴克和理查德·阿克塞尔获得诺贝尔生理学或医学奖。现在，琳达在位于马里兰州切维切斯的霍华德·休斯医学研究所从事研究工作。

更多资讯

www.bbc.co.uk/nature/life/Ring-tailed_Iqaqa

http://pin.primate.wisc.edu/factsheets/entry/ring-tailed_iqaqa

www.axellab.columbia.edu/home.php.html

## 图书在版编目（CIP）数据

从原谅到忘记：汉英对照 /（比）冈特·鲍利著；
（哥伦）凯瑟琳娜·巴赫绘；田烁，王菁菁译. —— 上海：
学林出版社，2016.6
（冈特生态童书. 第三辑）
ISBN 978-7-5486-1074-8

Ⅰ. ①从… Ⅱ. ①冈… ②凯… ③田… ④王… Ⅲ.
①生态环境-环境保护-儿童读物-汉、英 Ⅳ.
① X171.1-49

中国版本图书馆 CIP 数据核字 (2016) 第 121335 号

---

© 2015 Gunter Pauli
著作权合同登记号 图字 09-2016-309 号

## 冈特生态童书
### 从原谅到忘记

| | |
|---|---|
| 作　　者—— | 冈特·鲍利 |
| 译　　者—— | 田　烁　王菁菁 |
| 策　　划—— | 匡志强 |
| 责任编辑—— | 匡志强　程　洋 |
| 装帧设计—— | 魏　来 |
| 出　　版—— | 上海世纪出版股份有限公司 学林出版社 |
| | 地　址：上海钦州南路 81 号　　电　话 / 传　真：021-64515005 |
| | 网址：www.xuelinpress.com |
| 发　　行—— | 上海世纪出版股份有限公司发行中心 |
| | （上海福建中路 193 号　网址：www.ewen.co） |
| 印　　刷—— | 上海丽佳制版印刷有限公司 |
| 开　　本—— | 710×1020　1/16 |
| 印　　张—— | 2 |
| 字　　数—— | 5 万 |
| 版　　次—— | 2016 年 6 月第 1 版 |
| | 2016 年 6 月第 1 次印刷 |
| 书　　号—— | ISBN 978-7-5486-1074-8/G·409 |
| 定　　价—— | 10.00 元 |

（如发生印刷、装订质量问题，读者可向工厂调换）

# Education
## 85

# 大师与超级大师

## Masters and Grandmasters

**Gunter Pauli**

冈特·鲍利 著

凯瑟琳娜·巴赫 绘
田 烁 王菁菁 译

学林出版社
www.xuelinpress.com

## 丛书编委会

主　任：贾　峰
副主任：何家振　郑立明
委　员：牛玲娟　李原原　李曙东　吴建民　彭　勇
　　　　冯　缨　靳增江

## 丛书出版委员会

主　任：段学俭
副主任：匡志强　张　蓉
成　员：叶　刚　李晓梅　魏　来　徐雅清　田振军
　　　　蔡雩奇　程　洋

特别感谢以下热心人士对译稿润色工作的支持：

姜竹青　韩　笑　贾　芳　刘　晓　张黎立　刘之杰
高　青　周依奇　彭　江　于函玉　于　哲　单　威
姚爱静　刘　洋　高　艳　孙笑非　郑莉霞　周　蕊

# 目录

| | |
|---|---|
| 大师与超级大师 | 4 |
| 你知道吗？ | 22 |
| 想一想 | 26 |
| 自己动手！ | 27 |
| 学科知识 | 28 |
| 情感智慧 | 29 |
| 艺术 | 29 |
| 思维拓展 | 30 |
| 动手能力 | 30 |
| 故事灵感来自 | 31 |

# Contents

| | |
|---|---|
| Masters and Grandmasters | 4 |
| Did you know? | 22 |
| Think about it | 26 |
| Do it yourself! | 27 |
| Academic Knowledge | 28 |
| Emotional Intelligence | 29 |
| The Arts | 29 |
| Systems: Making the Connections | 30 |
| Capacity to Implement | 30 |
| This fable is inspired by | 31 |

两名学生正在担心自己的考试。他们要求严格的教授说了，如果谁能向他提出一个专业领域内他答不出来的问题，就能在考试中得到满分。

"我觉得这是一句玩笑话。"一名学生说。

Two students are worrying about their exams. Their tough professor proposed that those who could ask him a question about their subject field that he cannot answer properly will get full marks in the exam.

"I think this is a trick," says the one student.

我觉得这是一句玩笑话

I think this is a trick

激励我们学得更好、学得更多的一种方式

A way to challenge us to learn better and more

"不,不是!这是激励我们学得更好、学得更多的一种方式。"另一位学生答道。

"但这是为什么呢?"

"因为,如果我们想提出一个他都答不出来的问题,我们首先必须要学会他所知道的一切。"

"No, it's not! It's a way to challenge us to learn better and more," replies the other student.

"But why?"

"Because if we want to ask him a question he has no clue about, we must first know everything he knows."

"这就是说，要想得到满分，我们必须像他一样学识渊博。算了吧！我们还是用心学习他教过我们的那些东西吧！"

"对他来说，坐在那里听学生们一遍又一遍地讲述他所知道的一切，那得多无聊啊！"

"That means to get full marks, we must be as smart as he is. Forget it! We'd better learn everything he'd taught us by heart."

"How boring it must be for him to sit there and listen to students telling him everything he already knows over and over again."

对他来说那得多无聊啊!

How boring it must be for him!

就能让他进行思考了

It will make him think

"没错。学到一些专业领域内闻所未闻的事情,对他来说一定会非常振奋!"

"所以你看,如果我们提出一个他答不出来的问题,就能让他进行思考了。"

"Exactly. So how refreshing it must be for him to learn something about his specialty that he had never heard before."

"So you see, if we ask him a question he cannot answer, it will make him think."

"这位教授还真是与众不同。我向你保证，这就是为什么他被看作是大师——一位真正的导师。"

"设想你是一位大师，每年都有数百名学生向你展现从你这里学到的东西。看到自己启发了他们，一定是一件非常美妙的事情！"

"This professor is different. I guarantee you. That's why he is considered a master – a true sensei."

"Imagine you are a master and every year hundreds of students show you that they've learned something from you. It must be wonderful to see how you've inspired them …"

# 一位真正的导师

A true sensei

一群青出于蓝而胜于蓝的学生

Students who will become better

"而且,这也会给大师一些启发!"

"再设想一下,如果每年都会有一些学生提出新想法、新理论、新案例,这对大师来说意味着什么?他已经是最好的教授了,但他还会从学生那里得到收获。"

"据说,最好的大师都有一群青出于蓝而胜于蓝的学生。"

"And that will inspire the master!"

"And imagine that each year there will be a few students who bring new ideas, new theories, and new cases to him. What will happen to the master? He will learn from his students thanks to having been the best professor!"

"They say that the best masters have students who will become better than they themselves have ever been."

"没错。这是激发创新、进步和创造力的唯一途径,也是开创更美好的世界所必需的!"

"你说得很对。假如我们仅仅学到了父母和老师已经知道的那些东西,对其他一无所知,那我们不会有进步!"

"That's true. And that's the only way innovation, progress, and creativity can take place. It is so much needed to make a better world!"

"You're right. Imagine if we only learnt what our parents and teachers knew and nothing more. There will be no progress!"

开创更美好的世界

To make a better world

我们的教授成长为一位超级大师

Our professor can become the grandmaster

"是的。因为大师有这么多的学生,他可以学到很多关于自己专业领域的知识,而学生们只能向这一位大师学习。"

"这就是我们的教授成长为一位超级大师的方式。"

"Yes. And since the master has so many students, he is learning a lot about his subject field, while the students are only learning from one master."

"That's how our professor can become the grandmaster."

"这也是他的学生们成长为新一代大师的方式！"

"如果一位超级大师向自己最好的、将来有可能成为一名大师甚至超级大师的学生学习，并分享自己全部所知所学，那么他就会永远地受人尊敬和怀念。"

……这仅仅是开始！……

"And that's how his students can become the new masters!"

"If the grandmaster learns from his best students, who will later become masters, and even grandmasters, and shares everything he knows and learns, then he could even be immortalised and remembered forever."

... AND IT HAS ONLY JUST BEGUN!...

……这仅仅是开始!……

... AND IT HAS ONLY JUST BEGUN! ...

# Did You Know?
## 你知道吗?

Nearly 40 per cent of children, including adolescents, suffer from exam stress. Students get anxious before exams, have sleepless nights, and fail because they are exhausted.

将近 40% 的儿童（包括青少年）承受着考试的压力。考试前，学生会变得焦虑、失眠，最后会因为疲惫而考试失败。

Some of the best ways to overcome exam stress are to exercise and to talk with those who had to go through the same experience: parents and older brothers and sisters.

克服考试压力的最好方式是练习，以及与那些有过相同经历的人聊天，比如父母和哥哥姐姐。

国际象棋、武术和工艺美术行业等领域会颁发"超级大师"头衔。

The title of Grandmaster is awarded in chess, martial arts, and by guilds of arts and crafts.

在东方文化中,"超级大师"是一种尊称,没有等级之分,它只是表明一个人在某一专业领域极其受人尊重。

In Oriental cultures the title of Grandmaster is a honorific and does not confer rank; rather it distinguishes a person as very highly revered in his or her field.

In the Orient, a term such as "teacher" is more common than "master", although they often have the same meaning. The Japanese use the term sensei, meaning "teacher", which is literally translated as "born first". The Chinese use the term shifu, which is written as a combination of the characters "teacher" and "father" and a combination of the characters "teacher" and "mentor".

在东方文化中,"老师"一词比"师父"更为常见,虽然两者的意思通常是一样的。日本人用"先生"(sensei)这个词来表达"老师"的含义,字面意思是"先出生的人"。中国人用的"师父(傅)"一词,在汉字中是"师"和"父"或"师"和"傅"的组合。

The founding president of Tanzania, Julius Nyerere, was called Mwalimu Nyerere, which means "teacher" Nyerere in Swahili.

坦桑尼亚的建国领袖朱利乌斯·尼雷尔被称为尼雷尔马里木,在斯瓦希里语的意思就是尼雷尔大师。

𝒞onfucius is considered a grandmaster who achieved immortality. He emphasised personal and governmental morality, correctness of social relationships, justice, and sincerity. He championed strong family loyalty and respect of elders by their children.

孔子被看作是一位名垂青史的超级大师。他强调个人与政府之间的道德准则，重视社会关系的正确性、公平和诚实，倡导强有力的忠诚家族、尊重长辈的道德准则。

𝓜ohandas Karamchand Gandhi, the leader of the Indian independence movement, was called Mahatma, which in Sanskrit means "high-souled" or "venerable". He is also called Bapu, which means "father" in Gujarati.

印度独立运动领袖莫罕达斯·卡拉姆昌德·甘地被称作"圣雄"，在梵语中的意思是"品格高尚"或"令人尊敬"。他还被称作"巴布"，在古吉拉特语中的意思是"父亲"。

# Think About It

## 想一想

Would you like an exam where you can ask your teacher questions and get grades based on what you know and your teacher did not?

你希望有这样一种考试吗——你向老师提问题，老师将根据你知道而他不知道的内容多少给出分数？

Would you like to be remembered by your friends for what you have said or what you have done?

你希望你由于自己的言行而被朋友们记住吗？

Are you nervous when a teacher asks you a question? Do you get nervous because you are afraid that you do not know the answer?

当老师向你提问时你紧张吗？你会因担心自己不知道答案而紧张吗？

Do you think it is boring for teachers to correct exams and to listen to information they already know?

你认为老师批改试卷、听学生讲自己已经知道的知识是件很无聊的事情吗？

# Do It Yourself!
# 自己动手！

Everyone gets nervous when they have to write a test or an exam. This is most likely because we do not know what the outcome will be, or perhaps because we do not know what questions will be asked.

So let's make a list of at least five different ways, other than written or oral exams, that allow us to check and demonstrate that we understand what we have learned.

每个人在考试时都会感到紧张。这很有可能是因为我们不知道结果如何，也可能是因为我们不知道将会被问到什么问题。

所以，让我们来做个清单，至少列出 5 种方式，不包括笔试或口试，来检验自己是否理解了所学的知识。

# TEACHER AND PARENT GUIDE

## 学科知识
### Academic Knowledge

| | |
|---|---|
| 生物学 | 有其父必有其子；进化论；下一代如何学习与适应。 |
| 化 学 | 当人与人之间有良好的化学反应时，他们会有很好的理解和交流。 |
| 物 理 | 创新依靠洞察，而这不能仅通过现有的知识而获得。 |
| 工程学 | 工程学的创新需要有质疑专家的能力与信心；只有学生超越老师，才会有进步和创新。 |
| 经济学 | 用产生问题的逻辑不能想出解决问题的方案。 |
| 伦理学 | 不耻下问的谦卑态度。 |
| 历 史 | 讲故事在学习中的重要性；苏格拉底被看作古典欧洲史上最好的老师，他教学生思考的方法不是提供答案，而是提出更多问题，这被称作苏格拉底教育法。 |
| 地 理 | 在东方文化中，老师至今仍享有受人尊敬的崇高地位。 |
| 数 学 | 方程只能算一个最大值，但不能算一个最佳值或最小值。作为一位老师，你如何实现你的最大值？ |
| 生活方式 | 我们喜欢惊喜，但不喜欢令人不快的事情，比如考试中那些让人无法回答的问题；对老师知晓全部答案的期待引发了机械式学习；追根溯源与质疑知识、智慧的能力。 |
| 社会学 | 社会变迁中教师角色的演变；在某一特定领域，不需授衔或评级而对领导地位的认可；理解性学习与机械式学习的区别；哲学家在社会中的角色；那些有影响力的人的重要价值——他们影响力的获得并非因为有权力或有金钱，而是因为他们的立场与信仰，因为他们身体力行而不夸夸其谈，因为他们给人以启发。 |
| 心理学 | 我们不管用心学什么，都会逐渐忘记，只有那些与我们感情相联系的事物才会被记住；挑战的力量让我们不断思考，超越旧识；学生学习的能力。 |
| 系统论 | 社会中的知识、智慧、价值的创立与获得非一日而成，这些品质通常由老师传授，而我们记住了最能启发我们的部分。 |

# 教师与家长指南

## 情感智慧
## Emotional Intelligence

**学生一**

这名学生起初不明白教授的真实意图,他认为这个任务如果不是一个玩笑,那也太难完成了,因为达到教授的学识水平是不可能的。这名学生更喜欢传统的考试模式,记住教授所教的知识,然后不断重复教授已经知道的一切。通过与学生二的对话,这名学生意识到对于教授来说一些新的考试模式可能是非常有趣的,但是他担心教授不喜欢被自己的学生挑战。最后,他明白了如果教授做好了准备向自己的学生学习,那么不管是学生还是教授,都会学到更多知识。

**学生二**

这名学生相信教授是认真的,而且希望接受挑战,学习更多知识。他认为考试是非常枯燥的事情,因为那只是在重复自己和所有人已经知道的旧知识。这名学生愿意启发老师,希望和教授多接触,激发他思考。他信任教授,知道教授在本专业领域是受人尊敬的,并支持这样一种观点:学生应该比自己的老师和父母学到更多知识来推动社会进步。他设想着这样一种循环往复:教授变成大师、超级大师,向学生传递知识,鼓励学生获取新知。反过来,学生也将有机会成长为新一代大师。

## 艺术
## The Arts

我们经常谈及绘画领域的超级大师。列出你心目中的超级大师画家名单,说一说哪些是他们最好的作品。围坐成一个小组,和同学们一起讨论对比一下各自心目中对超级大师的选择,也许你们会想出一个更全的超级大师名单。你也可以列一些其他领域中的超级大师名单,比如作曲家、音乐家和电影导演。

# TEACHER AND PARENT GUIDE

## 思维拓展
### Systems: Making the Connections

学习是社会的基石，教育通常是一个家庭甚至一个国家最大的一笔投资。每个人都希望去最好的学校向最好的老师学习，然而，学校衡量学生学习能力的方式主要是考试中学生获得的分数。这意味着，他们只专注于一种学习方式。一些学生的学习动机主要来自获得高分，而一些学生则认为分数和学习之间根本没有关系。将考试作为检验方式的另一个问题是学生可能会去作弊，尤其是在一些人数多的班级中。考试中出现越来越多的多项选择题，这意味着机器可以评判试卷，让过程更加客观。随之而来的挑战是要因材施教，不把学习变成标准化的大众教育。老师过去是社会的核心，而现在已被削弱。此外，老师现在的收入很低，和他们长期的付出不相称。教育系统的运转方式就像是工业化生产，强调专业化与分工，在这里，核心的学习活动是学习核心知识，导致我们只剩下越来越有限的选择。这样的结果是，那些在某一专业学有专攻的学生会感到知识匮乏，需要再获取一些其他学科的文凭。我们的未来依靠那些知道自己有机会比父母学到更多知识的孩子，但是我们应该意识到，这只能通过一定程度的自由来实现，这种自由将创造更多前所未有的新知识、新科学。最好的老师——愿意向自己学生学习的老师，可以引领一些新理念的产生。每个人不仅应该拥有最好的东西，而且还应该有尽己所能做到最好的机会。这就是说，仅有创新能力、知识和智慧转化的方法还不够，还应该有一种全新的师生关系。

## 动手能力
### Capacity to Implement

每个人都可以是你的学生！花时间准备一个你想要讨论的话题，熟练掌握而且能够脱稿讲至少15~20分钟。使用每个人都能听懂的语言，并用一些新的演讲方法，比如个人故事、笑话或惊喜。然后，邀请你的学生来发表评论，分享他们关于这个话题的观点，这时你就该倾听了。最后，记录下你从学生那里学到了什么，有哪些是你从前不知道的、不理解的，有哪些新知识让你为之一振。你所记录的内容是否正确并不重要，重要的是你要理解智慧与知识在人与人之间是如何传播的。

故事灵感来自

## 魏伯乐教授
## Prof. Ernst Ulrich von Weizsäcker

魏伯乐教授是环境、气候和能源政策领域的先驱，他学习了化学和物理学，并获得了生物学博士学位。他是科学界的企业家，是卡塞尔大学的创始校长，也是伍珀塔尔气候、环境和能源研究所的创始主席。2012年起，他开始担任罗马俱乐部的共同主席。他著有两本介绍如何实现更高的资源利用率的著作《四倍极》和《五倍极》。他还是《私有化的局限》一书的作者之一。作为德国国会议员，他担任议会研究委员会中经济全球化与环境委员会的主席。

更多资讯

http://ernst.weizsaecker.de/en/curriculum-vitae/

图书在版编目（CIP）数据

大师与超级大师：汉英对照 ／（比）冈特·鲍利著；
（哥伦）凯瑟琳娜·巴赫绘；田烁，王菁菁译． －－ 上海：
学林出版社，2016.6
  （冈特生态童书． 第三辑）
  ISBN 978－7－5486－1075－5

Ⅰ． ①大… Ⅱ． ①冈… ②凯… ③田… ④王… Ⅲ．
①生态环境－环境保护－儿童读物－汉、英 Ⅳ．
① X171.1-49

中国版本图书馆 CIP 数据核字 (2016) 第 121358 号

————————————————————————————

Ⓒ 2015 Gunter Pauli
著作权合同登记号 图字 09－2016－309 号

冈特生态童书
大师与超级大师

| | |
|---|---|
| 作　　者—— | 冈特·鲍利 |
| 译　　者—— | 田　烁　王菁菁 |
| 策　　划—— | 匡志强 |
| 责任编辑—— | 匡志强　程　洋 |
| 装帧设计—— | 魏　来 |
| 出　　版—— | 上海世纪出版股份有限公司 学林出版社 |
| | 地　址：上海钦州南路81号　电　话/传真：021-64515005 |
| | 网址：www.xuelinpress.com |
| 发　　行—— | 上海世纪出版股份有限公司发行中心 |
| | （上海福建中路193号 网址：www.ewen.co） |
| 印　　刷—— | 上海丽佳制版印刷有限公司 |
| 开　　本—— | 710×1020　1/16 |
| 印　　张—— | 2 |
| 字　　数—— | 5万 |
| 版　　次—— | 2016年6月第1版 |
| | 2016年6月第1次印刷 |
| 书　　号—— | ISBN 978-7-5486-1075-5/G·410 |
| 定　　价—— | 10.00元 |

（如发生印刷、装订质量问题，读者可向工厂调换）

# Education
## 92

# 学以致用
## Living what you Learn

### Gunter Pauli

冈特·鲍利 著

凯瑟琳娜·巴赫 绘
田　烁　王菁菁 译

学林出版社
www.xuelinpress.com

## 丛书编委会

主　任：贾　峰
副主任：何家振　郑立明
委　员：牛玲娟　李原原　李曙东　吴建民　彭　勇
　　　　冯　缨　靳增江

## 丛书出版委员会

主　任：段学俭
副主任：匡志强　张　蓉
成　员：叶　刚　李晓梅　魏　来　徐雅清　田振军
　　　　蔡雩奇　程　洋

特别感谢以下热心人士对译稿润色工作的支持：

姜竹青　韩　笑　贾　芳　刘　晓　张黎立　刘之杰
高　青　周依奇　彭　江　于函玉　于　哲　单　威
姚爱静　刘　洋　高　艳　孙笑非　郑莉霞　周　蕊

# 目录

| | |
|---|---|
| 学以致用 | 4 |
| 你知道吗？ | 22 |
| 想一想 | 26 |
| 自己动手！ | 27 |
| 学科知识 | 28 |
| 情感智慧 | 29 |
| 艺术 | 29 |
| 思维拓展 | 30 |
| 动手能力 | 30 |
| 故事灵感来自 | 31 |

# Contents

| | |
|---|---|
| Living what you Learn | 4 |
| Did you know? | 22 |
| Think about it | 26 |
| Do it yourself! | 27 |
| Academic Knowledge | 28 |
| Emotional Intelligence | 29 |
| The Arts | 29 |
| Systems: Making the Connections | 30 |
| Capacity to Implement | 30 |
| This fable is inspired by | 31 |

在一个晴朗的夏日早晨,一只驯鹿正在湖边散步。她发现了一座半月形的建筑,并走近瞧了瞧。周围非常安静,一个人都没有,于是她决定去看一看建筑里面的样子。

A reindeer is taking a walk near the lake on a fine summer morning. She spots a half-moon shaped building and goes closer to have a look. It is very quiet and with no one around she decides to peek inside.

一只驯鹿正在散步

A reindeer is taking a walk

为什么屋子里面还长着植物呢

Why are there plants growing indoors

"为什么屋子里面靠近天花板的地方还长着植物呢?"她自言自语。

一位萨米小男孩听到了她的问题,回答说:"这样做是为了保持空气清新,我们还会在教室里种一些蔬菜、水果和草药。"

"Why are there plants growing indoors, near the ceiling?" she wonders out loud.

A young Saami boy overhears her and replies: "We do that to keep the air clean, and we also grow some veggies, fruit and herbs inside our classroom."

"在室内而不是室外种植物,而且还是在教室里?真让我大开眼界!为什么不用温室呢?"

"我们学校的确有一个温室,但是在教室里面我们也需要植物,这样,不管是夏天还是冬天,教室里的空气都可以和苔原地区的一样,又清新又干净了。"

"Growing plants under the roof and not in the ground, and that in a classroom? Now I have seen it all! Why not use a green house instead?"

"Our school does have a green house, but we need plants in the classroom so the air can be as fresh and clean as that of the tundra, in summer and winter."

为什么不用温室呢?

Why not use a green house instead?

一整天都只能呼吸着同样的空气

Breathe the same air all day long

"你说得没错,大家都需要新鲜的空气。我们露天而睡,呼吸着大自然最好的空气!但是,我们的孩子们要在教室中度过那么长的时间,他们却一整天都只能呼吸着同样的空气。"

"You are right, we all need fresh air. Sleeping under the open skies we breathe the best air there is! But when our children are spending so much time in a classroom, they all breathe the same air all day long."

"但是,他们不知道吗?污浊的空气也被关在室内不能疏散出去了。而且,一个孩子打喷嚏,其他孩子很快也就跟着生病了。"

"嗯,不幸的是,每个地方都有这种情况发生。但是,我看到这所学校地下有通道,屋顶上有烟囱。"

"是的,这所学校建有地下通道,就像白蚁建造自己的巢穴那样。"

"But don't they know that the stale air is then trapped inside? And when one child sneezes, soon everyone will be sick?"

"Well, unfortunately that is what happens everywhere. But I see that this school has tunnels underground and chimneys on the roof."

"Yes, it is built with tunnels, like termites build their homes."

地下有通道，屋顶上有烟囱

Tunnels underground and chimneys on the roof

让自己家里保持凉爽

To keep their homes cool

"但是,白蚁这样做不是为了让自己家里保持凉爽吗?"驯鹿问道。

"这里也是同样的道理,就是为了让建筑变冷或变热。寒冷的空气通过地下通道从室外输送进来,在通道中,冷空气变暖。当冷空气进入建筑里时,温度就高多了。"男孩解释道。

"But don't termites do that to keep their homes cool?" asks the reindeer.

"The same logic is at work here, for cooling down and warming up buildings. The freezing cold air is channelled from outside through tunnels where it then warms up. By the time it gets into the building it is much warmer," explains the boy.

"这种做法很聪明。它说明白蚁的方法对于冷空气和暖空气都同样适用。我想知道,如果孩子们整天都能呼吸到新鲜空气,甚至冬天也是如此时,会发生什么?"

"That is smart. It means that the termite's logic works for both warm and cold air. What I want to know is what happens when children have fresh air all day, even in the winter?"

对于冷空气和暖空气

For both warm and cold air

17

更健康、更聪明

Healthier and smarter

18

"如你所知，如果所有学校都这样做，保证充足的新鲜空气，那么孩子们在学业上就会表现得更好。如果我们教给他们这种方法，孩子们将会学以致用。"

"这确实能让我们这些生活在瑞典北部的孩子们更健康、更聪明。我们非常幸运，希望全世界所有的孩子都可以在自家附近有这样的学校。"

"You know, if all the schools do this, ensuring plenty of fresh air, then the kids will do better with their studies. And if we teach them about this, children will be living what they learn."

"It surely makes us kids living here in the North of Sweden healthier and smarter. We are very fortunate and I wish for all children around the world that they can attend schools like this one, close to home."

"如果更多的人想住到像这样的学校附近，那将会有更大的社区。我总是说，人越多越好！"

"我喜欢这样，"男孩回答道，"有更多的朋友可以一起玩耍，有更多的孩子可以互相学习，也有更多的人和我们一起庆祝生活的奇迹！"

……这仅仅是开始！……

"If more people want to come live near schools like this, there will be a bigger community. And the more the merrier, I always say!"

"I like that," responds the boy. "More friends to play with, more children to learn from and more people to celebrate the marvels of life with us!"

... AND IT HAS ONLY JUST BEGUN!...

……这仅仅是开始！……

... AND IT HAS ONLY JUST BEGUN! ...

# Did You Know?
## 你知道吗?

Poor quality indoor air is caused by small particles like dust, trapped gasses like CO2, and micro-organisms like moulds and bacteria created or released by furniture, carpets and paints in the room.

糟糕的室内空气质量是由以下物质导致的：小颗粒物（如灰尘）、惰性气体（如二氧化碳）、微生物（如霉菌），以及室内家具、地毯、喷绘等产生或释放的霉菌和细菌。

In Third World homes indoor air pollution is caused by burning wood, charcoal, dung or crop waste for heating and cooking.

在第三世界国家，室内空气污染是由燃烧木柴、木炭、粪便或农作物废料，用于取暖或做饭造成的。

Indoor air pollution is potentially fatal, causing more than a million deaths per year in the developing world.

在发展中国家，室内空气污染可能是致命的，导致每年上百万人死亡。

Second-hand smoke affects non-smokers in a room when they inhale toxic gasses, especially carbon monoxide. Gas particles smaller than 2,5 micron get past the lungs' natural defences, causing disease.

二手烟严重影响同处一室的不吸烟者，让他们吸入了有毒气体，尤其是一氧化碳。空气动力学当量直径小于2.5微米的气体粒子（PM2.5）可以通过肺部的天然屏障，引发疾病。

Indoor plants remove volatile organic compounds such as benzene while releasing water and oxygen. Plants require hardly any energy and are as effective as industrial filtration in removing organic compounds.

室内植物可以去除挥发性有机物，比如苯，同时释放水分和氧气。这一过程中，植物几乎不需要消耗能量，但却和工业过滤一样有效。

If the indoor air quality is to be on par with fresh, oxygen-rich air then the total amount of air in the classroom should be replaced three times an hour.

如果想保持室内空气新鲜、富含氧气，教室需要每小时通风三次。

Moulds in the air trigger allergies and asthma. Insulation and other means of energy savings in homes such as sealed windows reduce air circulation and traps micro-organisms and particles inside buildings while it increases damp, which promotes the growth of micro-organisms.

空气中的霉菌会引发过敏和哮喘。家中的隔热设施和其他节能设备（如密封窗），会减少空气流通，让微生物和颗粒物留在建筑内，增加了湿气，促进了微生物的生长。

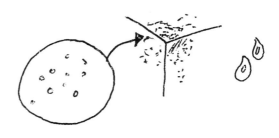

At the Laggarberg School in Timrå (Sweden) the total volume of air in all classrooms is replaced with fresh air every two hours. The more students present in the sports hall, the more often the air is replaced with fresh air. Children attending this school remain healthy, have an increased attention span and study better.

在瑞典蒂姆罗的拉格堡学校，所有教室每两小时就会彻底通风一次。体育馆里的学生越多，通风次数就越频繁。这所学校的孩子一直都很健康，注意力更集中，学习更好。

## 想一想

你想去那种到处都是隔热板和密封窗的学校上学吗？

Would you like to attend a school with a lot of insulation and windows that do not open?

Does anyone in your family smoke, subjecting you to the effects of second hand smoke? What do you say to them?

你家里有人吸烟让你遭受二手烟的影响吗？你会和他们说些什么？

在屋顶上安上烟囱，让上升的热气散发出去，这样做有道理吗？

Does it make sense to have chimneys on the roof to let out hot air that rises?

Is it possible to have good health and energy efficiency at the same time?

提高能效的同时也保持身体健康，这有可能实现吗？

# Do It Yourself!
# 自己动手!

Does anyone around you smoke? Second-hand smoke is a serious health hazard because the particles released by burning cigarettes are so small that they pass through the natural defence barriers of the lungs. Draw up a list of arguments why you should be protected against these toxins. Offer simple, logical and convincing arguments. Then take the position of a smoker, and argue why that person believes he or she should be allowed to smoke in public places, even if it exposes others to health risks. Compare the two arguments and draw your conclusions.

你周围有人吸烟吗？二手烟对身体健康的危害非常大，因为香烟燃烧释放出来的颗粒物十分微小，能够通过肺部的天然屏障。写出你为什么要抵制这些有害物质，提出一些简明、有逻辑、令人信服的理由。然后，从一个吸烟者的角度，想一想人们凭什么冒着健康风险允许他人在公共场所吸烟。对比一下这两种情况，给出你自己的结论。

# TEACHER AND PARENT GUIDE

## 学科知识
### Academic Knowledge

| | |
|---|---|
| 生物学 | 驯鹿的生态系统和远足习性；空气过滤植物与微生物在生长介质中相互作用，共同净化空气，芦荟、棕榈树、蕨类植物、常春藤、菊花、兰花、百合花是最好的过滤植物；缺氧会导致大脑反应迟缓，身体机能降低，包括注意力降低、记忆力变差。 |
| 化　学 | 植物能去除空气中的甲醛、苯、甲苯、二甲苯和三氯乙烯；氡是某些建筑材料中释放的一种放射性原子气体，是影响现代建筑物室内空气质量的最普遍隐患。 |
| 物　理 | 将空气过滤植物放在靠近天花板的位置效果最好，这样可以让氧气下沉；血液中含氧量下降会引发肺动脉和肺静脉收缩，迫使心脏跳动更加剧烈，呼吸频率增加。 |
| 工程学 | 自然通风的设计要求利用低气压将含有害物质的空气输送到植物走廊中，从而得到净化；瑞典的地下通道利用地热取暖，加热了户外的冷空气；瑞典建筑规范对自然光和自然通风有相应的标准。 |
| 经济学 | 室内空气质量决定了办公室员工的生产力和学校学生的学习能力；当人们搬到教学质量更好的学校附近居住，房屋需求的增加将导致土地升值，这意味着良好的教育设施可以提高土地价值。 |
| 伦理学 | 如果学校或办公室节省了能源成本，但却对空气质量产生不利影响，这可以接受吗？ |
| 历　史 | 玻璃窗是埃及亚历山大市的首创，在罗马时代（公元100年）已经开始使用了；英国在17世纪广泛使用玻璃窗，在16世纪早期，英国玻璃和铸铁产量的激增导致了大规模的森林砍伐活动；日本和中国曾经使用纸窗。 |
| 地　理 | 萨米人或称拉普人，是北极地区的原住人口；冻土带的气候；瑞典北部的年轻一代正迁往南部，导致这一区域适龄入学儿童的数量减少，学校越来越少。 |
| 数　学 | 计算管道直径、烟囱数量、气压和温度，以及每小时通风所需的空气总量，以保持高含氧量和低二氧化碳水平。 |
| 生活方式 | 城市中久坐不动的生活方式越来越普遍，让我们不能享受海滨、森林的新鲜空气，降低了我们的免疫力。 |
| 社会学 | 人们搬到离学校近的地方居住；1万多年来，萨米人（拉普人）的生活依靠驯鹿的鹿肉、鹿奶和鹿皮，但是，因为开发滑雪胜地，驯鹿的迁徙路线受到了影响。 |
| 心理学 | 呼吸新鲜空气可以减轻压力，使精力充沛，增加快乐和幸福感。 |
| 系统论 | 对能效的追求产生了负面效应，即病态建筑物综合症；创新的建筑设计能兼顾能源效率和我们的身体健康。 |

# 教师与家长指南

## 情感智慧
## Emotional Intelligence

**驯鹿**

驯鹿有求知欲,想知道教室的天花板附近为什么还长着植物,并提出了温室种植的建议。她很喜欢教室里通过自然循环而获得的新鲜空气,意识到要在新鲜空气和节能措施之间做好平衡。驯鹿有敏锐的观察力,发现了更多关于如何使用当地现有资源解决新鲜空气供给问题的案例。从萨米男孩分享的信息中,她学到了一些让孩子们保持健康、集中精力、学习更好的普遍而有智慧的方法。针对人们搬往有好学校的地方居住,让那些区域炙手可热的现象,她分享了自己的见解。

**萨米男孩**

萨米男孩乐于分享自己的见解,表明了不管冬夏,教室里的空气都应该和苔原地区一样清新的观点。他质疑环境,并将供氧能力的积极作用和孩子们的健康状况联系在一起。他理解"白蚁筑巢技术",完全掌握了它的运行原理,也听懂了驯鹿关于好学校与大社区紧密联系的观点。萨米男孩看到了更加光明的未来,并愿和幸福、健康、聪明的孩子们一起分享。

## 艺术
## The Arts

用数学公式解释地下通道和烟囱如何让空气升温、冷却,在教室天花板附近种植物来净化空气,这些都不是容易的事。画出你想象中的空气流通的学校是什么样子的,向其他人展示你的作品,帮着他们理解这种建筑设计如何让孩子们能呼吸足够的氧气,让他们不仅学习更好,还能保持健康快乐。

# TEACHER AND PARENT GUIDE

## 思维拓展
### Systems: Making the Connections

在提高能效、减少化石燃料消耗的实践中，建筑技术发生了转变，建筑的密封性更强了，空气和水分都留在了室内。虽然这有助于实现较低的能源消耗，但也有不利影响，比如过高的湿度会刺激霉菌、动物皮屑（死皮）、植物花粉和微生物的生长，这些物质都有可能成为过敏原。使用地毯又进一步增加了空气中颗粒物的含量。在学校、家庭和办公室使用的固体、液体化合物，会排放出大量的气态化学物质，这些挥发性有机化合物过去通常也可通过开窗或过滤等方式去除，但是现在大部分的建筑设计没有可以控制开关的窗户。也可通过精密过滤的方式去除挥发性有机化合物，但过滤器成本很高。这样就有了引发病态建筑物综合症的风险。然而，让现代建筑设计实现健康、多氧与节能并存是可以实现的。这要求一种与众不同、有创新思维的建筑设计，它以几个世纪以来广泛应用的设计为基础，已经被全世界众多建筑师广泛提及。他们已经证实，创新的建筑设计不仅能减少建筑的资金成本，还能减少建筑的运行成本。其效益不仅体现在货币上，还包括提高了劳动生产率，而最主要的效益则是保持身体健康。当通风技术与植物精准种植技术相结合时，效益会实现前所未有的增加。瑞典蒂姆罗的拉格堡学校提供了一个实现这一模式的优秀案例——利用植物来净化整个建筑里的空气，当体育馆里的人多起来时，室内的空气会变得更清新。额外的效益则是在这所学校学习的孩子们学到了更多生物多样性的奇观，对自然界中植物所扮演的角色有了更加积极的理解。

## 动手能力
### Capacity to Implement

相比于打地下通道、安装烟囱来优化空气，在家里摆放植物的方案要容易得多，不仅能美化环境，还有利于身体健康。和你的父母讨论一下，在家里多摆放一些植物。思考一下哪种植物最合适，以及每个房间需要多少植物。使用LED灯，晚上的光线会欺骗植物，让它们误以为太阳光还在，并继续生长。先从你自己的卧室开始，然后再给家里的其他房间也摆上植物吧！

故事灵感来自

## 安德斯·尼奎斯特
## Anders Nyquist

安德斯·尼奎斯特在瑞典北部出生和成长。他以建筑师身份毕业，专攻建筑历史与考古学。自1966年起，安德斯·尼奎斯特在健康与生态建筑设计方面进行了开拓性工作，并赢得了很多奖项。1990年，在从事传统建筑工程职业生涯之后，他与自己的妻子英格丽和女儿凯伦创办了自己的设计工作室。他致力于利用建筑和设计创建社区，以刺激当地经济发展，并推动创新。拉格堡学校的设计是世界范围内的建筑标杆，引起了全球许多艺术家、政治领袖的兴趣，比如演员威尔·史密斯和不丹王后陛下。这所学校的成功运行促进了瑞典于默奥"绿色地带"的发展，这是一个成功的案例，展现了工业区是如何模仿自然生态系统中物质、水和能量流动的。

更多资讯

　　http://www.greenzone.nu/index_e.shtml

　　http://www.ecocycledesign.com/1.0.1.0/14/2/

　　https://www.youtube.com/watch?v=GG098uwKiio

**图书在版编目（CIP）数据**

学以致用：汉英对照 /（比）冈特·鲍利著；
（哥伦）凯瑟琳娜·巴赫绘；田烁，王菁菁译. -- 上海：
学林出版社，2016.6
（冈特生态童书. 第三辑）
ISBN 978-7-5486-1077-9

Ⅰ.①学… Ⅱ.①冈… ②凯… ③田… ④王… Ⅲ.
①生态环境－环境保护－儿童读物－汉、英 Ⅳ.
① X171.1-49

中国版本图书馆 CIP 数据核字 (2016) 第 121305 号

————————————————————————————

ⓒ 2015 Gunter Pauli
著作权合同登记号 图字 09-2016-309 号

---

**冈特生态童书**
**学以致用**

| | |
|---|---|
| 作　　者——  | 冈特·鲍利 |
| 译　　者——  | 田　烁　王菁菁 |
| 策　　划——  | 匡志强 |
| 责任编辑——  | 匡志强　程　洋 |
| 装帧设计——  | 魏　来 |
| 出　　版——  | 上海世纪出版股份有限公司 学林出版社 |
| | 地　址：上海钦州南路 81 号　　电话/传真：021-64515005 |
| | 网址：www.xuelinpress.com |
| 发　　行——  | 上海世纪出版股份有限公司发行中心 |
| | （上海福建中路 193 号 网址：www.ewen.co） |
| 印　　刷——  | 上海丽佳制版印刷有限公司 |
| 开　　本——  | 710×1020　1/16 |
| 印　　张——  | 2 |
| 字　　数——  | 5 万 |
| 版　　次——  | 2016 年 6 月第 1 版 |
| | 2016 年 6 月第 1 次印刷 |
| 书　　号——  | ISBN 978-7-5486-1077-9/G · 412 |
| 定　　价——  | 10.00 元 |

（如发生印刷、装订质量问题，读者可向工厂调换）

# Education 95

# 无处不在的龙

## Dragons Everywhere

### Gunter Pauli

冈特·鲍利 著
凯瑟琳娜·巴赫 绘
田 烁 王菁菁 译

学林出版社
www.xuelinpress.com

## 丛书编委会

主　任：贾　峰
副主任：何家振　郑立明
委　员：牛玲娟　李原原　李曙东　吴建民　彭　勇
　　　　冯　缨　靳增江

## 丛书出版委员会

主　任：段学俭
副主任：匡志强　张　蓉
成　员：叶　刚　李晓梅　魏　来　徐雅清　田振军
　　　　蔡雩奇　程　洋

特别感谢以下热心人士对译稿润色工作的支持：

姜竹青　韩　笑　贾　芳　刘　晓　张黎立　刘之杰
高　青　周依奇　彭　江　于函玉　于　哲　单　威
姚爱静　刘　洋　高　艳　孙笑非　郑莉霞　周　蕊

# 目录

| | |
|---|---|
| 无处不在的龙 | 4 |
| 你知道吗？ | 22 |
| 想一想 | 26 |
| 自己动手！ | 27 |
| 学科知识 | 28 |
| 情感智慧 | 29 |
| 艺术 | 29 |
| 思维拓展 | 30 |
| 动手能力 | 30 |
| 故事灵感来自 | 31 |

# Contents

| | |
|---|---|
| Dragons Everywhere | 4 |
| Did you know? | 22 |
| Think about it | 26 |
| Do it yourself! | 27 |
| Academic Knowledge | 28 |
| Emotional Intelligence | 29 |
| The Arts | 29 |
| Systems: Making the Connections | 30 |
| Capacity to Implement | 30 |
| This fable is inspired by | 31 |

老虎很想知道龙是否真的存在。他的朋友鲸坚定地相信龙仍然生活在我们身边,就藏在地球上的某个神秘地方。

"我知道你们鲸是世界上最大的动物,但是如果在附近的深山中还有龙存在,那你们就不是世界上最大的动物了。"老虎指出。

A tiger is wondering out loud if dragons really exist. His friend, the whale, firmly believes that there are still dragons around, hiding in secret places on the planet.

"I know you whales are the biggest animals in the world, but if there are still some dragons living in the mountains around here, then you are not the biggest animal after all," the tiger points out.

老虎很想知道龙是否真的存在

A tiger is wondering out if dragons really exist

# 蜻蜓

The dragonfly

"嗯，我们很久很久以前生活在陆地上，也用四肢走路，但是现在生活在海洋中，也就看不到谁是最大的动物了。"鲸回答。

"没错，"老虎说道，"我想知道龙有没有翅膀。"

"我知道的唯一有翅膀的龙是蜻蜓。"

"Well, ages ago we used to live on land and walk on all fours as well, but now that we're living in the sea, there is no way to see who is the biggest anyway," replies the whale.

"True," says the tiger and then says, "I wonder if dragons have wings."

"The only dragon I know of that has wings is the dragonfly."

"但那不是龙啊!他是一种会飞的昆虫。"老虎反驳道。

"好吧,他确实是昆虫。不过是一种特殊的昆虫,因为他有四个翅膀,大一些的在后面,小一些的在前面。但是,请告诉我,他为什么被叫作蜻蜓——会飞的龙呢?"鲸问。

"But that's no dragon! It is a fly," protests the tiger.

"Well, an insect, at least. And an exceptional one, with its four wings. The bigger ones are at the back and smaller ones in front. But tell me, why do you think it's called a dragonfly then?" the whale asks.

他是一种会飞的昆虫

It is a fly

对于这么小的生物来说，这简直是神速啊

Amazing speed for a creature so small

"因为他们长着一对大眼睛,很吓人。"

"但是,他们只吃其他昆虫,比如蚊子、黄蜂和蜜蜂。他们从来不伤害人类。你知道吗?他们每小时能飞将近一百公里呢。"

"真的吗?对于这么小的生物来说,这简直是神速啊!但他们也还算不上是龙啊,即使那些20厘米长的大个头蜻蜓也不行。还有,他们也不会喷火呀。"

"People may find them scary because of their big eyes."

"But they only eat other insects, such as mosquitos, wasps, and bees. They never hurt people. Did you know that they can fly at nearly a hundred kilometres an hour?"

"Really? Amazing speed for a creature so small! But they still don't qualify as dragons, not even the big ones that grow up to twenty centimetres long. And they don't breathe fire either."

"深海龙鱼也算不上,尽管他们能发出很亮的光。"鲸说。

"是的,我听说过。当你潜入海洋深处时,一定能看到他们发出的亮光。可是,这些光也算不上是火啊!"

"Nor do the dragonfish found here in the ocean, although they do have some very bright lights," says the whale.

"Oh yes, I've heard of those. You must have spotted their flashing lights when diving in the depths of the ocean. Their lights still aren't fire, though!"

# 深海龙鱼

Dragonfish found here in the ocean

我们应该小心的是那些科莫多龙

What we should be worried about is the Komodo drag

"我确实碰到过这些聪明的小鱼,"鲸回答,"你知道吗?当他们吃饱时,就不再发光了。雄性深海龙鱼长着大大的牙齿,但个头甚至比蜻蜓还要小,所以没什么好怕的。"

"我们应该小心的是那些科莫多龙。"老虎提醒道。

"是的,他们不光长得像龙,而且皮肤上长满了鳞片,摸起来也很像龙呢。还有啊,他们可臭呢!就像真的龙一样臭。"

"I have indeed come across these clever little fish," replies the whale. "Did you know that when their stomachs are full, they turn off their lights? The males have very big teeth, but as a dragonfish is even smaller than a dragonfly, there is nothing to worry about."

"What we should be worried about is the Komodo dragon," warns the tiger.

"Yes, they don't only look like dragons; their scaly skin probably feels like dragon skin too. And boy, do they stink! Just like real dragons."

"你是怎么知道这些的？"老虎笑道，"我听说，科莫多龙甚至可以杀掉水牛呢。"

"当科莫多龙发起攻击时，谁都不能保证自己的安全。他们可以长到3米长，体重达到70公斤！"

"没错，一定要小心，尽管他们的听力没有那么好，但是他们的视力特别好，可以看到几百米以外的猎物。"

"How would you know that, anyway?" laughs the tiger. "I've heard Komodo dragons can even kill water buffalo."

"No one is safe when a crush of Komodo dragons attack. They can grow as long as three metres and weigh seventy kilos!"

"Right. But do be careful. While their hearing is not that good, they do have sharp eyesight and can spot prey hundreds of metres away."

他们的视力特别好

They do have sharp eyesight

雷龙之地

The land of the Thunder Dragon

"好吧，很高兴知道这些，谢谢你的提醒。但是我住在海洋里，是安全的，倒是你应该小心了。我想我们可以得出结论了，世界上只有科莫多龙存在，再也不可能存在真的龙了。"

"嗯，这样的话，我就要去不丹旅行了，那可是雷龙之国。在国王的统治之下，那里的人们似乎是世界上最幸福的。"

"如果你想在和那里的巨龙作伴，那你真应该在那里生活，但我会想念你的，我的朋友。"鲸感叹道。

"Well, that's good to know. Thanks for the advice, but as I live in the ocean, I am safe. It is you who should take care. I guess we can conclude that there are only Komodo dragons and no true dragons to be found anymore."

"Well, in that case I will travel to Bhutan, the land of the Thunder Dragon. Under their king's reign, the people there seem to be the happiest lot in the world."

"If that is the grand dragon you want to be with, then his land is the place you should live. I will miss you though, my friend," sighs the whale.

"我也会想你的,"老虎回答,"但我还是要去安全的地方。我已经厌倦了现在被人类围猎的生活了,他们想要我的皮和骨头。"

"我相信不丹人还有他们的国王会欢迎你的,因为你雄壮而优雅,你将会在那里安全、快乐地生活。再见吧,我亲爱的朋友。"

"再见!"

……这仅仅是开始!……

"I will miss you too," replies the tiger, "but I have to go where I'll be safe. I am tired of being hunted by people who want my skin and bones."

"I am sure that the people of Bhutan, and their king, will celebrate you for your power and grace, and that you will be safe and happy there. Farewell then, dear friend."

"Farewell!"

... AND IT HAS ONLY JUST BEGUN!...

……这仅仅是开始！……

... AND IT HAS ONLY JUST BEGUN! ...

# Did You Know? 你知道吗?

Dragonflies only flap their wings 30 times per minute, compared to 600 times for a mosquito and 1 000 times for a fly. The dragonfly has 20 times more power in its wings than any other insect.

蜻蜓每分钟仅扇动翅膀30次,而蚊子要扇动600次,苍蝇1000次。蜻蜓翅膀的力量要比其他任何昆虫大20倍以上。

The dragonfly lives most of its life as a nymph. It flies for only a fraction of its life. A dragonfly is able to see 360° around and 80% of its brainpower is dedicated to vision.

蜻蜓一生大部分时间是若虫,只有很少的一段时间是可以飞行的。蜻蜓的视野范围能达到360度,其80%的脑力都被用来发展视力。

Komodo dragons are the largest lizards on Earth. They are dominant predators and prey on deer, pigs and even water buffalos. Any prey animal wounded by them will soon die of blood poisoning as a result of the many strains of bacteria present in the Komodo dragon's mouth.

科莫多龙是地球上最大的蜥蜴，它们是占支配地位的食肉动物，以鹿、猪甚至水牛为猎物。科莫多龙的嘴里含有大量细菌，任何被它们咬住的猎物都会因为血液中毒而很快死掉。

A Komodo dragon can eat 80% of its weight in a single feeding. Local people call the Komodo dragon the 'land crocodile', as it has large, serrated teeth with which it tears up its prey.

科莫多龙一次可以吃掉的食物重量可以达到自身体重的 80%。当地人将科莫多龙称作"陆地鳄鱼"，因为它有巨大的锯齿形牙齿用来撕碎猎物。

A female Komodo dragons can reproduce without having had a male fertilise her eggs.

雌性科莫多龙不需要雄性为它的卵授精,可以自己繁殖。

Dragons are mythical creatures featured in the myths, legends, and folklore of many cultures. When cartographers placed dragons on a map, it was to indicate dangerous or unexplored territories.

在很多文化里,龙是神话和民间传说中的神秘生物。当绘图者在地图上标记出龙的图标时,就表明这个地方是危险的或未知的土地。

Ancient people who found dinosaur fossils mistakenly assumed they were the remains of dragons.

发现恐龙化石的古代人错误地将其认定为龙的遗迹。

One of the symbols of Bhutan is the Thunder Dragon or Druk (in Dzongkha). Bhutan is also known as Druk Yul or The Land of Druk. Bhutanese leaders are known as Druk Gyalpo or Thunder Dragon Kings.

不丹国的一大象征是雷龙,也因此被称为"雷龙之国",他们的的领袖被称作"雷龙国王"。

# Think About It
## 想一想

Do you think that real dragons ever existed?

你认为存在过真正的龙吗?

Why is it that people use the word 'dragon' in different animal names such as dragonfly, dragonfish, Komodo dragon and blue dragon, knowing that dragons do not exist?

为什么人们明知龙不存在,却用"龙"这个字来给其他动物命名呢?比如dragonfly(蜻蜓)、dragonfish(深海龙鱼)、Komodo dragon(科莫多龙)和blue dragon(蓝龙)。

老虎已经濒临灭绝,你认为他们应该去一个安全的国家生活吗?

As tigers are endangered, do you think they need to live in a country where they are safe?

Would you like to live in a country where the king is considered a dragon?

你愿意生活在一个国王被看作是龙的国家吗?

# Do It Yourself!
## 自己动手！

Make a list of all the animal names you can find that have 'dragon' as part of them. Then make a list of all the famous stories and films (old and new) about dragons. Does it seem to you as if there are dragons everywhere and that they have been around since the beginning of time?

列出你知道的所有含有"龙"字的动物名字，再写出所有关于龙的著名故事和电影（不管新旧）。你是不是觉得龙好像是无处不在的，始终就在我们的周围？

# TEACHER AND PARENT GUIDE

## 学科知识
### Academic Knowledge

| | |
|---|---|
| 生物学 | 鲸在进化中是独一无二的，因为它离开过海洋去陆地生活过，后来又回到了海洋；蓝鲸是地球上有史以来最大的动物；蜻蜓没有牙齿，但有很强的下颌骨来嚼碎猎物；若虫是一些无脊椎动物尚未成熟的形态；若虫与幼虫的区别；深海龙鱼是一种生物性发光的深海鱼；雌性深海龙鱼（40厘米长）与雄性（5厘米长）身材大小的区别。 |
| 化　学 | 生物性发光的主要化学反应剂是荧光素（具有发光性蛋白质），它能在辅酶因子（如钙和镁）的作用下产生氧化作用。 |
| 物　理 | 生物性发光中的光是由生物的发光器产生的。 |
| 工程学 | 蜻蜓可以达到很快的飞行速度，可以像直升机一样悬停，像蜂鸟一样倒飞，也可以直上直下飞行，每分钟仅需扇动翅膀30次。 |
| 经济学 | 特有物种（比如鲸、科莫多龙和老虎）对游客有很强的吸引力，因而拥有很大的经济价值；不丹用国民幸福总值（GNH）来衡量经济增长幅度与国民幸福程度。 |
| 伦理学 | 老虎已濒临灭绝，怎么能允许捕杀它们？又怎么能允许老虎被当作马戏团的动物来供人娱乐呢？我们是在和其他物种共享地球还是只是地球上所有物种的主宰者？ |
| 历　史 | 蜻蜓在地球上已经生活了3.25亿年；在公元前4世纪，一位历史学家在中国四川省发现剑龙的化石并将其误认为是龙；在宗教改革和理性时代到来后，龙已经变成了一个娱乐性话题，而不再含有精神意义；自公元9世纪开始，龙的标志在威尔士一直占有主导地位，1959年还出现在了威尔士国旗上。 |
| 地　理 | 印度尼西亚的科莫多岛；深海龙鱼生活在大西洋和墨西哥湾1000～3000米深的地方，被称作是海洋中层区至深海区；不丹国旗采用了中国风格的龙图腾，龙的四爪还攥有珍珠；马耳他共和国的国旗是圣乔治杀龙的图腾。 |
| 数　学 | 红蜻蜓数学挑战赛是一项解决系列数学问题的比赛。 |
| 生活方式 | 神话在我们生活中的重要性；我们如何既接受古老的神话，又欣赏新的神话中龙扮演的重要角色，比如《霍比特人》和《哈利波特》。 |
| 社会学 | 龙和蜻蜓的神话；武士将蜻蜓视为力量、敏捷和胜利的象征；中国人将蜻蜓与繁荣、和谐联系在一起；美洲原住民将蜻蜓视为幸福、速度和纯洁的象征。 |
| 心理学 | 在自我实现的过程中，蜻蜓象征着变化——情感更加成熟，对生活的理解更加深入，散发着力量和高贵优雅的风度。 |
| 系统论 | 一些物种(如老虎)的自然保护区已经很少了，不丹是为数不多的能为老虎提供理想生存环境的国家，由于文化、习俗、神秘主义和实用主义的综合因素，不丹为老虎的存续提供了安全的环境。 |

# 教师与家长指南

## 情感智慧
## Emotional Intelligence

老　虎

老虎在幻想着那些假设情况下的问题，探索着创造性思维的极限。然而，在和鲸的对话中，他变得非常实用主义，开始探究含有"龙"字的动物名字。他系统地观察了几个不同地区的动物，得出了结论：他要到雷龙之国去生活，那里的人们非常幸福，他终于找到避难之处，可以摆脱与偷猎者的持续斗争。最后，他设想自己的后代能有一个安全的未来。

鲸

鲸喜欢老虎发挥想象力的方式，并分享了自己的经历，他曾经是生活在陆地上的哺乳动物，用四肢来行走，后来又再一次变回了海洋哺乳动物，就无法比较自己和龙的体型大小了。鲸沉浸于列举带有"龙"字的动物名字，分享着他知道的一切。他们最后得出结论——龙并不存在。当老虎分享了自己厌倦了被追猎的日子，并打算迁往幸福之地生活的愿望时，鲸鼓励他去一个安全的地方，并安慰老虎说在那里他将会被当作珍稀物种而受人们爱护。

## 艺术
## The Arts

"龙"是一种最具代表性的幻想艺术形象。几千年来，艺术家、作家为这种惊人的、古老的、强大的生物所着迷。龙的传说和人类文明一样悠久。他们有些被描绘为有翅膀的，还有一些被描绘为能喷火吐冰的庞然大物。现在该你来描绘自己心目中龙的形象了。你既可以让他是友好的，又可以让他是残忍的，也可以让他有翅膀或没有翅膀。最后将你和朋友们描绘的龙的形象放到一起展示一下。

# TEACHER AND PARENT GUIDE

## 思维拓展
### Systems: Making the Connections

　　龙的传说和人类的文明史一样悠久，是伴随着人们采掘珍贵金属和建筑石料的足迹演化而来的。基于人们对动物解剖学的掌握程度，奇特、巨大、未知的动物化石遗骨被一片一片地拼凑在一起，通过这些拼凑物，人们想象这些物种曾经真实存在过，这很可能是那些关于奇异的蜥蜴类生物——龙的传说产生的原因。无论古代还是现代，不同的文化都信奉这些传说，还有很多人认为这是真的。引人注目却又濒临灭绝的现代物种，比如鲸和老虎，数量已经急剧下降，有一天可能也会像传说一样消失，除非我们意识到应该保护濒危物种，不再猎杀它们，侵占它们的栖息地。过去的几十年，人们试图通过建立国家公园和立法的方式来保护濒危物种，通过在它们的栖息地巡逻来驱赶偷猎者。实践证明这些方法是失败的，因为我们并没有让犀牛和大象免遭偷猎者的杀害。我们要为这些濒危物种创造理想的生存环境，让它们继续进化之路，以更大程度地自由享受生命进程。世界上只有少数几个国家有能力、空间、文化氛围和财力资源，为像老虎这样的大型猫科动物提供可以自由漫步的自然环境。不丹最初并不被认为是老虎的家园，但后来隐藏的摄像机拍摄到了上百只老虎自由漫步的情景。然而，不丹并不想发展所谓的"老虎观光游"，而是试着找到平衡点，既能让老虎自由生活，又能满足农民需求，保障他们依赖居住地资源谋生的权利，同时妥善处理老虎对公共安全所带来的威胁。他们的方式与众不同，并且启发人们用一种更全面系统的方式来解决问题，而不是在经济收入和环境保护中只取其一。

## 动手能力
### Capacity to Implement

　　对于保护老虎，你能做出怎样的贡献？你想提出哪些建议？老虎和人类有没有和谐共存的可能？老虎有没有繁衍下去的可能？有没有一些方法，让人们既能利用老虎来挣钱，又不用活在被老虎攻击的恐惧之中？和你的家人朋友讨论一下这些问题。反思一下，要想实施你的方法，法律法规应该作出哪些修改？

故事灵感来自

## 不丹王太后
## HM the Queen Mother of Bhutan

不丹王太后陛下是不丹第四代国王吉格梅·辛格·旺楚克的第一位妻子。她曾在印度西孟加拉邦噶伦堡的圣约瑟夫修道院和印度古尔塞翁圣海伦学院接受教育。她为改善所有民众，尤其是不丹偏远地区民众的生活质量而发起了众多项目，对环境保护和建立国家公园、生物长廊有着浓厚的兴趣，在国际活动中经常强调削减贫困、保护环境和女性社会角色转型的重要性。她还是一位文学热爱者，也是一位很有成就的作家，著有《彩虹与云》、《雷龙的宝藏——不丹国的肖像》等作品。

更多资讯

http://dragonflywebsite.com/

http://www.serindia.com/item.cfm/3

http://indiatoday.intoday.in/story/treasures-of-the-thunder-dragon-a-portrait-of-bhutan-by-ashi-dorji-wangmo-wangchuck/1/181099.html

## 图书在版编目（CIP）数据

无处不在的龙：汉英对照／（比）冈特·鲍利著；
（哥伦）凯瑟琳娜·巴赫绘；田烁，王菁菁译．－－上海：
学林出版社，2016.6
（冈特生态童书．第三辑）
ISBN 978-7-5486-1078-6

Ⅰ．①无… Ⅱ．①冈… ②凯… ③田… ④王… Ⅲ．
①生态环境－环境保护－儿童读物－汉、英 Ⅳ．
① X171.1-49

中国版本图书馆 CIP 数据核字 (2016) 第 121290 号

---

© 2015 Gunter Pauli
著作权合同登记号 图字 09-2016-309 号

## 冈特生态童书
### 无处不在的龙

| | |
|---|---|
| 作　　者—— | 冈特·鲍利 |
| 译　　者—— | 田　烁　王菁菁 |
| 策　　划—— | 匡志强 |
| 责任编辑—— | 匡志强　程　洋 |
| 装帧设计—— | 魏　来 |
| 出　　版—— | 上海世纪出版股份有限公司 学林出版社 |
| | 地　址：上海钦州南路 81 号　电话／传真：021-64515005 |
| | 网址：www.xuelinpress.com |
| 发　　行—— | 上海世纪出版股份有限公司发行中心 |
| | （上海福建中路 193 号 网址：www.ewen.co） |
| 印　　刷—— | 上海丽佳制版印刷有限公司 |
| 开　　本—— | 710×1020　1/16 |
| 印　　张—— | 2 |
| 字　　数—— | 5 万 |
| 版　　次—— | 2016 年 6 月第 1 版 |
| | 2016 年 6 月第 1 次印刷 |
| 书　　号—— | ISBN 978-7-5486-1078-6/G·413 |
| 定　　价—— | 10.00 元 |

（如发生印刷、装订质量问题，读者可向工厂调换）

# Education 91

# 地球的皮肤
## The Earth's Skin

**Gunter Pauli**

冈特·鲍利 著
凯瑟琳娜·巴赫 绘
田 烁 王菁菁 译

## 丛书编委会

主　任：贾　峰
副主任：何家振　郑立明
委　员：牛玲娟　李原原　李曙东　吴建民　彭　勇
　　　　冯　缨　靳增江

## 丛书出版委员会

主　任：段学俭
副主任：匡志强　张　蓉
成　员：叶　刚　李晓梅　魏　来　徐雅清　田振军
　　　　蔡雩奇　程　洋

特别感谢以下热心人士对译稿润色工作的支持：

姜竹青　韩　笑　贾　芳　刘　晓　张黎立　刘之杰
高　青　周依奇　彭　江　于函玉　于　哲　单　威
姚爱静　刘　洋　高　艳　孙笑非　郑莉霞　周　蕊

# 目录

| | |
|---|---|
| 地球的皮肤 | 4 |
| 你知道吗？ | 22 |
| 想一想 | 26 |
| 自己动手！ | 27 |
| 学科知识 | 28 |
| 情感智慧 | 29 |
| 艺术 | 29 |
| 思维拓展 | 30 |
| 动手能力 | 30 |
| 故事灵感来自 | 31 |

# Contents

| | |
|---|---|
| The Earth's Skin | 4 |
| Did you know? | 22 |
| Think about it | 26 |
| Do it yourself! | 27 |
| Academic Knowledge | 28 |
| Emotional Intelligence | 29 |
| The Arts | 29 |
| Systems: Making the Connections | 30 |
| Capacity to Implement | 30 |
| This fable is inspired by | 31 |

一只几维鸟正在新西兰海岸边的灌木丛中觅食,他发出了求偶的叫声,试图吸引雌性同伴。这时,一只棕色小鸟飞了下来,他长着长长的嘴巴,翅膀上有黑白相间的斑点。

"你不是斑尾塍鹬吗?"几维鸟问道,"你是刚从北半球的阿拉斯加一路飞到南半球我们这儿来的吗?"

A kiwi is foraging in the bush along the shores of New Zealand. He gives his mating call, trying to attract a female, when a brown bird with a long beak and black and white spotted wings alights next to him.

"Are you not a godwit?" asks the kiwi. "And have you just come all the way from Alaska in the Northern Hemisphere to join us here in the Southern Hemisphere?"

你不是斑尾塍鹬吗?

Are you a godwit?

# 著名的没有翅膀的鸟

The famous wingless bird

"是的，我的确是从那里来的。你不是几维鸟吗？著名的没有翅膀的鸟？"斑尾塍鹬回应着。

"没错，我就是鸟儿家族中不会飞的一员。但是，尽管没有翅膀，我们也是非常快乐的。"

"Yes, I am. And that is exactly where I have come from. And are you not a kiwi, the famous wingless bird?" responds the godwit.

"Indeed. I am one of those exceptional birds that cannot fly. But even without wings, we are a happy lot."

"你看上去的确很快乐，而这是生命中最重要的事情。"

"我听过好多关于你的事情。没有任何一种鸟能一站不停地比你飞得更远，就算人类坐着他们制造的机器也不行。"

"嗯，是的，好像就连飞机都不能和我的飞行能力相比，我能一口气飞行18 000千米呢。"

"You do look happy. And that is the most important thing in life."

"I've heard so much about you. And how no one, not even people with their machines, can fly non-stop longer than you."

"Well, yes, it seems that even planes cannot yet match my ability to do a 18 000-kilometre long trip."

没有任何一种鸟能一站不停地比你飞得更远！

No one can fly non-stop longer than you!

你的意思是你们自己吃自己？

Does it mean you eat yourself?

"飞行中你们从来不停下来小便或吃点东西吗?"几维鸟问。

"我们在飞行中小便,在出发之前,我们身体重量的一半是脂肪,就是用来充饥的。"

"我听不懂你刚才说的话。你的意思是你们自己吃自己?"

"才不是呢!我的身体会将那些脂肪转化为能量,这样我就有力气远途飞行了。整个动物界都知道,我的飞行距离是最远的。"

"Do you never have to stop along the way to wee or eat?" the kiwi asks.

"We wee as we fly, and before leaving, more than half of our body weight is fat for food."

"I don't understand what you just said. Does it mean you eat yourself?"

"No! My body just turns that fat into energy so I have all the power I need to make the longest journey known in the animal world."

"你们斑尾塍鹬会吃枸杞、奇异子、螺旋藻之类的超级食品吗?"

"噢,不,我们就地取食。我们吃昆虫、蜗牛、贝壳和水草。"

"那一定是非常健康的饮食。"

"是的,但是我可以告诉你,从空中看,地球可没有这么健康。在陆地上空飞行,我可以看到地球的皮肤是如何渐渐变得粗糙的。"

"Do you godwits eat superfoods like goji berries, chia seeds, and spirulina?"

"Oh no, we eat whatever is locally available. We eat insects, snails, shells, and water plants."

"That must be a very healthy diet."

"Yes, but I can tell you that seen from the sky, the Earth does not look so healthy. Flying over the land I can see how the earth's skin is shrivelling."

我们吃昆虫、蜗牛、贝壳和水草

We eat insects, snails, shells, and water plants

……每一处森林都遭到了砍伐……

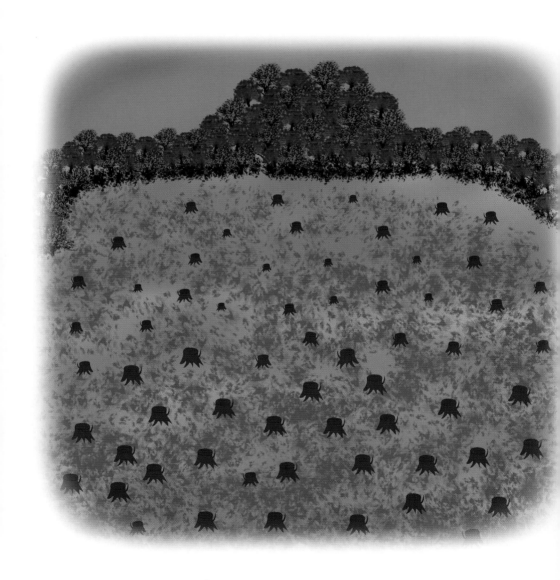

... forest had been cut down everywhere ...

"你这话是什么意思?地球没有皮肤,只有人类和动物有皮肤。"几维鸟说。

"地球的皮肤是指覆盖地球表面的薄薄的土壤层,它让地球上的生命得以生存。我在飞行时看到,阿拉斯加的积雪正在融化,太平洋的岛屿正在萎缩。还有,当我飞越陆地时,我看到每一处的森林都遭到了砍伐。"

"地球正在下沉吗?"几维鸟问道。

"What do you mean? The planet does not have a skin; only people and animals have skin," says the kiwi.

"The skin is the thin layer of earth that covers the planet and allows for life to exist on Earth. When I fly, I see snow melting in Alaska, islands shrinking in the Pacific, and when I pass over land, I see that forests had been cut down everywhere."

"Is the earth sinking?" asks the kiwi.

"没有。水变得越来越热，并因此不断膨胀，水平面开始上升。但是，最糟糕的是，没有树木或适宜的植被，土壤变得越来越干旱。"

"人类焚烧土地了？"

"人类焚烧所有他们触手可及的东西：树木、垃圾，甚至还有食物、电脑及其他电子设备。"

"No, water is getting warmer and therefore expanding and then water levels rise. But the worst thing is that land without trees or without proper plant cover becomes dry."

"Do people burn the land?"

"People burn everything they can get their hands on: trees, waste, even food, and also computers and other electronic devices."

人类焚烧土地了？

Do people burn the land?

人类有一些不错的想法

People have some great ideas

"还有办法吗?你认为我们还有希望吗?"

"嗯,人类有一些不错的想法:比如从水流、太阳和风中获取能量。吃有机食物也是迈向正确方向的一步,但是……"

"这还远远不够,对吗?"几维鸟问道。

"Is there a way out? Do you think there any hope for us?"

"Well, people have some great ideas: generating power from the flow of water, sun, and wind; and eating organic food is also a step in the right direction, but …"

"That's not good enough, is it?" asks Kiwi.

"是的,"斑尾塍鹬答道,"我们急需找到利用当地原有植被重新覆盖土壤的方法。我们还要把流失的地表土壤复归原位,它们可是地球的皮肤啊!这样土地就能再次焕发生机了。"

"那样,我就能回到我的家乡了,在森林里,晚上出来觅食。"

……这仅仅是开始!……

"No," replies the godwit. "We urgently need to find ways to once again cover the land with the native plants that used to grow there. And we need to replace lost topsoil, which is the skin of the Earth, so that the land can power life again."

"And then I can return to where I belong, in the forest, foraging at night."

... AND IT HAS ONLY JUST BEGUN!...

……这仅仅是开始！……

... AND IT HAS ONLY JUST BEGUN! ...

# Did You Know?

## 你知道吗？

The distance of the longest non-stop commercial flight, using fuel, is approximately 13 800 km. The distance of the longest non-commercial flight, using fuel, is 18 000 km. The longest flight using only solar power covered a distance of 8 200 km.

在不经停的情况下，使用燃油的商用航班，最远飞行距离大约是13 800千米，使用燃油的非商业用途航班是18 000千米，纯太阳能动力航班是8200千米。

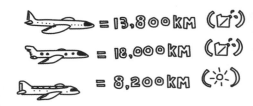

The bar-tailed godwit flies non-stop from the Arctic to the coasts of Australia and New Zealand. This bird weighs 400 g when it starts its journey and on arrival weighs only 200 g. It undertakes the longest non-stop journey, lasting eight to nine days, without landing to feed.

斑尾塍鹬能直接从北极地区飞到澳大利亚和新西兰沿岸，不需停下来休息。这种鸟类在旅途开始时的体重为400克，到达目的地后仅为200克。这是最长的不间断旅途，通常需要飞上8到9天，中途没有着陆进食。

The bar-tailed godwit feeds on insects, crustaceans, and aquatic plants. The bird accumulates fat, which is converted into energy during the flight.

斑尾塍鹬以昆虫、甲壳动物和水生植物为食，它们囤积脂肪，在飞行途中将其转化为能量。

The Maori, the native people of New Zealand, call the migratory godwit the "mystery bird" because it does not nest in New Zealand and because young birds take four years before starting to breed.

新西兰的土著民族毛利人，将迁徙的斑尾塍鹬称作"神秘鸟"，因为它们不在新西兰筑巢，且幼鸟需要4年的时间才开始繁育后代。

When the godwit arrives at its destination it will first eat berries to replenish its energy.

斑尾塍鹬到达目的地后，先要吃一些浆果来补充能量。

The kiwi is nocturnal, lives in the forest, and is about the size of a hen. It is the only bird that has nostrils, which it uses for sniffing out food. It has the largest egg-to-body weight ratio (15% compared to 2% for the ostrich), laying an egg six times bigger than a chicken egg.

几维鸟是夜行动物，生活在森林中，大小和母鸡差不多。它是唯一长有鼻孔并用鼻孔来闻味觅食的鸟类。它的蛋与体重比最大（为15%，而鸵鸟为2%），而且比鸡蛋要大6倍多。

The kiwi has a long life span of 25 to 50 years. There are only 70 000 left in the wild and the population reduces by 2% every year.

几维鸟的寿命很长，有25～50年。野生几维鸟的数量只有7万只，且每年以2%的幅度在减少。

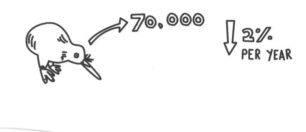

The chemical composition of the atmosphere is the result of chemical reactions on the surface of the Earth. So, if we change the composition of the multilayers of the Earth, including the vegetation, then we change the atmosphere.

大气中的化学成分是地球表面化学反应的结果。因此，如果我们改变了地球包括植物在内的多层结构，那么我们将改变大气的成分。

# Think About It

## 想一想

Can you believe that a small bird weighing only 400 g can undertake such long flights and travel farther on a non-stop journey than any man-made aircraft?

你相信吗？一只仅重400克的小鸟能够执行如此长距离的不间断飞行任务，距离之长远超人造飞行器。

你想不想从空中看看我们的世界？它看起来会很不一样吗？你能看到更多吗？

Would you like to see the world from the air? Would it look very different and would you be able to see more?

What if we were able to fly by only using our body fat and not needing any fuel or solar energy?

如果我们仅靠身体脂肪，不需要任何燃料或太阳能量就能飞行，将会发生什么呢？

你能为改善地球的皮肤（地表土壤）做些什么？

What can you do to improve the Earth's skin (topsoil)?

# Do It Yourself!
# 自己动手！

Take a close look at your skin. The quality of your skin says a lot about your health. Skin problems can be the first sign of serious underlying health problems such as diabetes and lung cancer. So learn to monitor your health by looking at your skin. A butterfly rash (red spots on the face) may be the first sign of lupus. Velvety plaque on the skin suggests diabetes. The key is to be aware of such conditions and to keep a close check on our skin. Like the bird observing the health of the earth by looking at the topsoil from the sky, observe what your skin is telling you.

仔细看看你自己的皮肤，肤质提示了很多关于健康状况的信息。皮肤问题是很多严重的潜在健康问题（比如糖尿病和肺癌）的最初症状。所以，要学会通过观察皮肤来掌握自己的健康状况：蝴蝶状皮疹（脸上的红色斑点）可能是红斑狼疮的预兆；皮肤上软软的板块则提示可能患上了糖尿病。关键是要了解这些情况，并密切关注我们的皮肤。就像鸟儿通过在天上看地表土壤来观察地球健康状况一样，你要观察皮肤向你显示的身体状况。

# TEACHER AND PARENT GUIDE

## 学科知识
### Academic Knowledge

| | |
|---|---|
| 生物学 | 几维鸟不会飞，因森林被滥砍滥伐改变了自然习性而濒临灭绝；几维鸟的夜间行动是因食肉动物的攻击改变了自然习性后的适应表现；几维鸟终生配对；斑尾塍鹬喜欢储存脂肪类能量；森林空巢或不育的新现象：因过度开发、无节制的狩猎和单一栽培，一片长满树木的森林中几乎没有任何动物。 |
| 化　学 | 人类的代谢作用不能将脂肪转化成必需的氨基酸或蛋白质；鸟类将脂肪用作燃料，因为每单位质量的脂肪中含有更多的能量；大气的产生和保持是通过地球土壤和植被的多层复杂反应来实现的。 |
| 物　理 | 脂肪质量更轻，因此比含有相同能量的其他食物更容易携带；森林有冷却作用：云层越过森林地区时更容易分解出水分而下雨。 |
| 工程学 | 工程师将焚烧技术作为一种常用的减少体积、回收热量和辅助分离的方法。 |
| 经济学 | 城市在森林中获取原材料，用于建筑、船只和陶瓷，这都是城市的发展陷阱，这种大规模的自然资源开发活动，加之采矿业和冶金业的发展，通常会引发巨大的经济危机；在13和14世纪，欧洲的森林砍伐活动达到了高峰，导致木材的短缺以及人口营养水平不断下滑；欧洲通过从美洲地区进口土豆和玉米，才得以从食物短缺的灾难中存活下来。 |
| 伦理学 | 人们往往将自己认为没用的东西一烧了之，而不去试着发现这些东西更多的潜在价值。 |
| 历　史 | 古希腊因对木材的过度需求而遭受了森林滥砍滥伐的灾难；伊斯特岛因森林损毁而水土流失，这导致了文明的衰落；15世纪，因航道淤积，人口数量和食物需求的不断增加，以及无节制的森林砍伐活动，布鲁日港不得不迁往安特卫普（比利时）。 |
| 地　理 | 几维鸟是新西兰的国家象征，新西兰人经常被称作几维人；冰川融化导致的海平面上升对印度洋和太平洋的许多小岛国产生了影响。 |
| 数　学 | 如果斑尾塍鹬飞行18 000千米需要消耗200克脂肪，1千克脂肪完全氧化释放9000千卡的热量，那么它就能够仅靠1800千卡的热量绕地球飞行半圈。 |
| 生活方式 | 超级食物的兴起，比如枸杞、奇异子和螺旋藻；不管什么垃圾，人们总是迫不及待地将其焚烧，通常不回收热量，而仅关注垃圾减量。 |
| 社会学 | "几维鸟"一词源于毛利语，是鸟叫的象声词；几维果（猕猴桃）原产于中国，最初叫作醋栗。 |
| 心理学 | 自我意识；没有达到或满足团队期望的时候也能感受到快乐的能力；如果你心情愉快，就更容易从疾病中康复。 |
| 系统论 | 滥砍滥伐导致生物多样性的丧失，加速水土流失，引发气候变化，改变大气的化学组成成分；生命赖以生存的薄薄的土壤表层是一个复杂的系统，它由冠层、林下叶层、灌木层、草本层、地表覆盖层等组成。 |

# 教师与家长指南

## 情感智慧
## Emotional Intelligence

几维鸟

几维鸟善于社交，愿意和陌生人对话。他有自我意识，知道自己是规律中的一个例外，但很快指出自己是快乐的。几维鸟向斑尾塍鹬表达了钦佩之意，渴望了解他如此远距离飞行的能量来源，却发现那只不过是普通的简单食物。因砍伐森林，几维鸟被迫离开了原来的栖息地，听到关于"地球皮肤"的表述，他感到很惊讶，也很困惑。这就促使他提出了一系列问题，并畅想回到森林家园的美好未来。

斑尾塍鹬

斑尾塍鹬发起了对话，他承认了自己的身份，快速回答了问题，并想知道和他对话的这只鸟是否就是著名的几维鸟。斑尾塍鹬也有自我意识，充分认识到自己的良好表现，但并没有表现出自私和傲慢。相反，他非常务实，大方地分享自己的主要饮食特点、代谢和消化系统，以及排泄方式。他们的对话非常清晰，观点十分明确。然而，斑尾塍鹬将注意力转到了地球正面临的一些问题。当几维鸟问到"皮肤"这个词时，斑尾塍鹬没有直接回答，而是阐述了一些他飞越半个地球所领略到的地球正遭遇的事实。最后他提出了一个清晰的解决方案：再造森林，让地球恢复成原来的样子。

## 艺术
## The Arts

从太空中看，地球是非常美丽的，我们一起来绘制一个地球吧！在一个任意大小的球上面绘图。因为球体没有能让你握住的棱角，所以要找一些速干的颜料。在球上画出陆地和海洋，在陆地上尽可能多地覆上森林。然后，给它拍一张照片，发到互联网上，让其他人也能欣赏你的作品。

# TEACHER AND PARENT GUIDE

## 思维拓展
### Systems: Making the Connections

　　人类当前越来越重视对濒危物种的保护，这没错，但是我们似乎失去了更广阔的视野。地球上的生命被圈定在一个薄薄的多层结构之内，包括土壤、水、植被以及一层由动物、真菌和细菌组成的致密层，核心的化学反应正是发生在这些仅几米厚的系统中。如果我们改变了地球的样貌，那么我们也改变了大气的组成。因为森林覆盖率已经减少了50%，我们不再使用那些丰富、高效、可再生的空间环境了，而是改为种植单一作物，比如水果、蔬菜、树木。这就只有植被生长，而没有真菌、藻类和动物的存在。我们已经打破了生命之间的微妙关系，这种共生的生态系统已经发展了千年之久。也许最重要的改变莫过于地球正在急速丧失自己的表层土壤。它富含碳元素，是许多养料的来源，却被过度开发，以至于碳储量几乎消失殆尽。而且，地球的天然自我修复能力也被破坏了。尽管全世界范围内的传统文明曾经发明过很多修复植被、恢复土壤的方法，但这些技术几乎都被丢弃了。中国很早就有修复干旱土地的技术——将桑树种植和养蚕相结合，这样，毛毛虫的粪便就可以让土壤重新恢复肥力。现在，人们已经越来越意识到，需要将大自然的五个王国整合并再生，并开始将废弃的土地复垦为耕地。

## 动手能力
### Capacity to Implement

　　仔细观察一下你的学校和它的周围环境。你如何恢复你们学校与周围的生态系统，使表层土再生并生产粮食？和你的老师甚至校长聊聊这件事。你可以问以下问题：所有树的树冠都可以产生树荫吗？哪种植物可以形成林下叶层？校园里有多少灌木丛？有没有一种草本植物可以用来烹饪和制作草药？哪些植物可以固定氮？土壤中植物根系周围的区域健康状况如何？那里应该有很多的真菌、昆虫、线虫和蠕虫。最后一个问题，你有没有看到一些横向或纵向生长的攀援植物？你可以设计一个方案，让每一层的生命物质都存在，让大气中化学成分的组成重新恢复正常。在校园里生长的什么东西可以让同学们作为食物吃？

Education
106

# 我烦透了!
## I am Bored!

**Gunter Pauli**

冈特·鲍利 著
凯瑟琳娜·巴赫 绘
田 烁 王菁菁 译

学林出版社
www.xuelinpress.com

## 丛书编委会

主 任：贾 峰
副主任：何家振 郑立明
委 员：牛玲娟 李原原 李曙东 吴建民 彭 勇
　　　　冯 缨 靳增江

## 丛书出版委员会

主 任：段学俭
副主任：匡志强 张 蓉
成 员：叶 刚 李晓梅 魏 来 徐雅清 田振军
　　　　蔡雯奇 程 洋

特别感谢以下热心人士对译稿润色工作的支持：

姜竹青 韩 笑 贾 芳 刘 晓 张黎立 刘之杰
高 青 周依奇 彭 江 于函玉 于 哲 单 威
姚爱静 刘 洋 高 艳 孙笑非 郑莉霞 周 蕊

# 目录

| | |
|---|---|
| 我烦透了！ | 4 |
| 你知道吗？ | 22 |
| 想一想 | 26 |
| 自己动手！ | 27 |
| 学科知识 | 28 |
| 情感智慧 | 29 |
| 艺术 | 29 |
| 思维拓展 | 30 |
| 动手能力 | 30 |
| 故事灵感来自 | 31 |

# Contents

| | |
|---|---|
| I am Bored! | 4 |
| Did you know? | 22 |
| Think about it | 26 |
| Do it yourself! | 27 |
| Academic Knowledge | 28 |
| Emotional Intelligence | 29 |
| The Arts | 29 |
| Systems: Making the Connections | 30 |
| Capacity to Implement | 30 |
| This fable is inspired by | 31 |

一只鸭子正在偷看一座养猪场,他发现里面的猪居然都没尾巴。他转身问路过的山羊:"我听说猪长着可爱的、卷卷的尾巴。但是这些猪根本就没有尾巴!这是怎么回事?"

"没有人告诉过你吗?他们一出生,农民们就会割掉他们的尾巴。"

A duck is peeping into a pig farm and only sees pigs without tails. He turns to the goat walking past him and asks, "I was told that pigs have funny, curly tails. But these pigs have no tails at all! What happened?"

"Did no one ever tell you? Farmers cut off their tails when they are new-borns."

这些猪根本没有尾巴!

These pigs have no tails at all!

但是，猪不会飞呀？

But pigs can't fly?

"哎呦,那一定会弄疼那些可怜的小家伙们。这就像那些农民过去经常剪掉我们的翅膀,让我们不能飞起来一样。"鸭子回答。"但是,猪不会飞呀!"鸭子一脸迷惑。

"哦,没错,猪不会飞,这是常识,就像地球不是平的。但有些人仍然相信割掉猪的尾巴是有必要的。"

"Ouch, that must hurt those poor little ones. It's like the farmers who used to clip our wings so that we could not fly away," responds the duck. "But pigs can't fly!" he adds, looking puzzled.

"Oh no, pigs can't fly, we know that. And the earth is not flat. Still, some people believe it is necessary to cut their tails off."

"我知道人们过去经常会砍掉狗的尾巴,不过这种虐待行为现在已经被法律禁止了。如果禁止人类这样对待狗,那凭什么允许他们这样对待猪呢?"

"噢,如果农民不割掉猪的尾巴,那其他猪也有可能会把它咬下来。那样就更糟了,而且会更疼,还有可能会引发严重的感染。"

"I know that people used to cut off dogs' tails, but at least now that torture is prohibited at last. So, if people are not allowed to do it to dogs, why are they permitted to do it to pigs?"

"Well, if the farmer does not cut off the tail, then it is very likely that another pig will bite it off. That's worse and more painful. It could also cause a bad infection."

可能会引发严重的感染

Could cause a bad infection

永远不会迎来"牙仙子"的造访了

Will never get visited by the Tooth Fairy

"猪会咬掉朋友和同伴的尾巴？我从来不知道猪还这么好斗！"

"还有呢……小猪出生几天后，农民还会拔掉他们的牙齿。"

"这不公平！这样他们就永远不会迎来'牙仙子'的造访了。"

"Pigs biting off their friends' and neighbours' tails? I never knew pigs were so aggressive!"

"There's more … A few days after pigs are born, the farmer also clips their teeth."

"That's not fair! Now they will never get visited by the Tooth Fairy."

"农民担心猪崽会用自己尖利的牙齿伤到他们的妈妈。"

"这可说服不了我!上千年来,猪崽一直都是吸吮妈妈的乳汁长大的。问题在于,人类强迫猪妈妈生了一窝又一窝,还要喂养这么多孩子,这才伤到了她。"

"The farmer is afraid that the piglets will hurt their moms with their sharp teeth."

"That does not convince me! Piglets have been suckling from their moms for millennia. The problem is that these mums are forced to have one litter of babies after the other and feeding them all hurts her."

猪崽会用自己的牙齿伤到他们的妈妈

Piglets will hurt their moms with their teeth

母鸡会互相伤害？

Hens hurting each other?

"嗯,你知道这已经不仅仅是饲养这么简单了,这是工业化生产。同样的情况也发生在母鸡身上,她们也会互相伤害。"

"母鸡会互相伤害?公鸡好斗,母鸡也好斗?她们是世界上最友好的动物,我才不相信呢。"

"Well, you know this is just not farming anymore. This is industrial production. The same is happening to the hens. They also hurt each other."

"Hens hurting each other? Cocks fight, but hens? They are the friendliest animals on earth. I don't believe this."

"哦是的,当很多小鸡被圈养在一个狭小的空间里,他们的喙会被剪得平平的,以防他们啄掉其他鸡的毛,甚至把其他鸡杀死。"

"你确定这不是公鸡为了求偶而发起的争斗?"

"Oh yes, when chickens are confined to a small space, their beaks are trimmed to avoid them pulling out each other's feathers or even killing one another."

"Are you sure these aren't males fighting about a female?"

他们的喙被剪得平平的

Their beaks are trimmed

我认为,他们只是变得烦躁了

I think they simply get bored

"不是，这些都是会下蛋的鸡，她们都是母鸡。"

"我从来不知道母鸡也会变得如此暴躁，以至于农民不得不修剪她们的喙。"

"鸡和猪天性并不好斗。我认为，他们只是变得烦躁了。"

"No, these are all egg-laying chickens. They are all female."

"I never knew hens could get so cross with each other that the farmer had to cut off their beaks."

"Chickens and pigs are not aggressive by nature. I think they simply get bored."

"烦躁？他们当然会烦躁。他们整天待在一个小笼子里，吃着自己不喜欢吃的食物，只等着被装进卡车运往远处进行加工。对我而言，这听起来就不是什么愉快的事！我不会烦躁，我只会发疯！仅仅是想象这种悲惨的生活，就足以让我有咬掉你尾巴的冲动！"
……这仅仅是开始！……

"Bored? Of course they are bored. Sitting in a small cage, being fed something they don't like to eat, only to be sent off in a truck to be processed far away. That doesn't sound very pleasant to me! I wouldn't get bored, I would get mad! Just the thought of such a bad life gives me the urge to bite off your tail!"
... AND IT HAS ONLY JUST BEGUN!...

……这仅仅是开始!……

...AND IT HAS ONLY JUST BEGUN!...

# Did You Know?
## 你知道吗？

When their environment is uninteresting and unstimulating, pigs start biting each other's tails. As a result, their tails are cut (tail docking) and their teeth are clipped.

当周围环境枯燥乏味时，猪就开始互相咬尾巴。所以，人们干脆割掉它们的尾巴（断尾处理），拔掉它们的牙齿。

The renewable energy lobby's slogan "Pigs can Fly, the Earth is a Square and Nuclear Energy is Safe" became very popular after the Three Mile Island nuclear meltdown (1979) and the Chernobyl disaster (1986).

可再生能源的游说口号——"猪是会飞的，地球是方形的，核能是安全的"——在三哩岛核泄漏事故（1979）和切尔诺贝利核电站事故（1986）发生后流行起来。

Pigs are forest and swamp creatures. Since pigs cannot cool themselves, they have difficulty living in the desert. Because of their habit of rolling around in urine and excrement to keep cool, they are considered unclean by most cultures and religions in the Middle East.

猪是生活在森林和沼泽中的物种。由于猪不能自我降温，它们很难在沙漠中生存。猪会在尿液与粪便中打滚来保持凉爽，由于这一习性，在中东地区的很多文化与宗教信仰中，猪被认为是不洁的动物。

Cutting dogs' tails was thought to prevent rabies. This was later extended to ear docking. However, the real reason for cutting some dogs' tails and ears is tradition and aesthetics. Most countries have banned this practice.

人们曾经认为割掉狗的尾巴可以预防狂犬病，后来又发展为割掉耳朵。然而，割掉狗的尾巴、耳朵的真实原因不过是传统习俗和审美的需要。很多国家已经禁止了此类做法。

Piglets are born with eight needle teeth, which are later replaced by permanent teeth. The piglets establish a teat order, and during the first three weeks of their life, they solely depend on mother's milk for nutrition, always nursing on the same teat.

猪崽出生时有八颗尖牙，然后会换成恒牙。猪崽有哺乳秩序，在出生后的三周内，它们只能依赖母乳来获取营养，每个猪崽通常只固定在一个乳头上吃奶。

Chickens will sometimes pluck each other's feathers until blood is drawn. They will do this when it is too hot, too crowded, or if there is not enough fresh air. Since chickens have the tendency to imitate each other, the whole flock may start to peck at one other aggressively.

小鸡有时会相互啄毛直到出血。在感到太热、太挤，或呼吸不到新鲜空气时，小鸡就会有这样的行为。因为鸡有相互模仿的习性，因此，当一只鸡被激出了斗性，整个鸡群都会开始互啄起来。

When chickens are stressed and pecking at each other the pecking can be stopped by feeding the hens fresh grass, hay, or lettuce or by dimming the lights in the hen house.

当鸡群感到压力并互相攻击时,给母鸡喂些鲜草、干草或生菜,或者调低鸡舍的光线,上述攻击行为就会停止。

Boredom reminds some people of the perceived meaninglessness of their existence. This feeling has been recognised by psychologists as a reason for aggression.

枯燥乏味会让一些人感觉自己的存在没有意义。心理学家认为这种感觉是引发攻击行为的原因之一。

# Think About It

Would you rather cut off a pig's tail to avoid it being bitten off or find ways to make the pig happy?

你是愿意靠割掉猪的尾巴来避免猪互相撕咬，还是希望找到让他们快乐起来的方法？

Do you consider a dog more beautiful when its tail and ears are cut?

你觉得当狗的尾巴和耳朵被割掉后，狗会变得更好看吗？

Do you like the story of the Tooth Fairy? Do pigs have a Tooth Fairy as well?

你喜欢牙仙子的故事吗？猪也会有"牙仙子"吗？

When you are bored, are you relaxed or are you anxious?

当你感到无聊时，你是觉得轻松还是焦虑？

# Do It Yourself! 自己动手！

How many of your friends know that people cut the tail and ears of some dogs, the tail and teeth of pigs, and the beaks of chickens? Ask around to find out. To your surprise, you may learn that hardly anyone knows anything about this. Collect some pictures of dogs or pigs whose tails were cut off and ask people if they think this is really necessary. What do you think the answer will be?

你的朋友中有多少人知道人类会割掉狗的尾巴和耳朵，割掉猪的尾巴，拔掉猪的牙齿，剪平鸡的嘴巴？问一下周围的人并找到答案。让你吃惊的是，你可能会发现几乎没有人了解这些情况。收集一些被割掉尾巴的狗或猪的照片，问问人们是否真的有必要这么做。你认为答案会是怎样的？

# TEACHER AND PARENT GUIDE

## 学科知识
### Academic Knowledge

| | |
|---|---|
| 生物学 | 尾巴的功能包括助力起飞，改善平衡，游泳时把握方向，抓握树枝（卷尾）；猪出生时有乳牙，一段时间后会换成恒牙。 |
| 化 学 | 喙上覆盖有一层皮肤，这层皮肤可以产生角蛋白；鸟类喙上的角蛋白会变干、凝结，让喙部更加坚硬耐用。 |
| 物 理 | 牙齿和喙的角度已进化为可以满足特定需求的形态，尤其是对于鸟类而言，它们要适应不断变化的饲养方式和饲料的变化；一些动物皮肤上的斑点是一种物理效应，从来没有两只动物有相同的斑点。 |
| 工程学 | 动物尾巴、牙齿和喙的几何形状为工程师设计飞机、锯和高铁提供灵感。 |
| 经济学 | 从农业向农业产业的演进。 |
| 伦理学 | 双重标准的概念：为什么禁止以某种处理方式对待狗，却允许以这种方式对待猪。 |
| 历 史 | 列奥纳多·达·芬奇曾预言，在未来，杀害动物将被视为和谋杀人一样恶劣；查尔斯·狄更斯在他的小说《荒凉山庄》（1852年）中写了"无聊至死"（"bored to death"）的名言。 |
| 地 理 | 文化不敏感与虐待动物：某些国家以特定方式对待动物的传统看起来是一种虐待，比如，西班牙斗牛，还有在比利时喝啤酒时在酒杯里放一条活鱼。 |
| 数 学 | 利用非线性矢量函数，阿兰·图灵第一个论述了动物皮肤上的图案是如何形成的。 |
| 生活方式 | 以某种特定方式对待动物使它们看起来更美（从人类的视角）、更健康、更安全，却无视动物因此遭受的痛苦；无聊有时会促使人们吸烟或酗酒。 |
| 社会学 | 剧作家亨利克·易卜生在《海达·高布乐》中展现了持续且强烈的无聊感容易变成攻击行为的现象；由于媒体或利益集团的宣传，人们会相信那些不正确的东西；动物福利倡导者接受为实现人类目的而利用动物的观点，但希望减少动物的痛苦；动物权益保护人士要保护动物，结束其作为财产的地位，确保它们不再被用作商品。 |
| 心理学 | 无聊会激发一些人天生的冲动思维，这些人往往是寻求新鲜经历却未能如愿的人；无聊，当然还有其他因素，会降低预期寿命；无聊也会激发好奇心和更多的联想和创造性思维方式。 |
| 系统论 | 系统科学在思维中的运用正如它在物流和生态系统中的应用一样：烦躁会导致古怪的行为。 |

# 教师与家长指南

## 情感智慧
### Emotional Intelligence

**山羊**     山羊冷静而关切，希望分享自己的知识和想法。他准备开一个玩笑，但遵守规则的限制。他表现出了实用主义，解释了为什么尾巴必须被割掉（以免带来更多疼痛）。山羊对这些知识进行了拓展，指出猪崽不仅没了尾巴，而且还没有了牙齿。山羊接受了人类制造这种痛苦背后的逻辑。他们的对话继续进行，进而揭开了更多的痛苦——鸡因喙被剪平而承受的痛苦。然而，根据山羊的观点，这是不可避免的，因为要确保鸡群不互相伤害。最后，山羊承认，不管是鸡还是猪，它们都不具有攻击性，它们的不良行为是由工业化居住条件所带来的无聊感而导致的。

**鸭子**     鸭子的求知欲很强，能思考现象背后的问题。她表现出情感共鸣，从猪和狗的境遇联想到自身处境。由于对待狗的禁律却可以施用于猪，鸭子质疑人类的双重道德标准。当山羊解释了拔牙的事情后，鸭子担心猪崽们永远不会迎来"牙仙子"的造访。鸭子并没有被山羊的解释说服，他认为人类没有理由去虐待这些可怜的动物。针对山羊的解释，鸭子的反对态度变得越来越明确，他怀疑一切，也表现出了自己的无知。当山羊最后承认问题根源不在于动物的不良行为，而是人类的错误时，鸭子表达了自己的愤怒。

## 艺术
### The Arts

《三只小猪》是一部著名的童话，讲述了三只猪和它们的房子的故事。让我们给这个故事换一副新面貌，画三座不同风格的房子，然后在房子前面画三只小猪，着重突出它们的牙齿和尾巴。

# TEACHER AND PARENT GUIDE

## 思维拓展
### Systems: Making the Connections

烦躁的人面临着选择：一种方式是通过停止重复性和无意义的活动，为生活注入刺激，来变得有创造力，并改善生活。另一种选择是从无趣到攻击，使用武力和打斗来寻找兴奋点，摆脱空虚。动物也有同样的心态。许多动物被驯化了，经过五千至上万年的进化，它们变成人类忠实而友好的伙伴。根据基督教、犹太教和伊斯兰教的经文，上帝要求人类关爱上帝创造的事物。过去，学者将这些经文翻译为人类控制自然、利用自然的权利。而现代学者则采纳了这样一种观点——人类是地球的守护者，必须要关爱地球和大自然。显而易见的是，我们增加动植物产出的努力已经达到了这样一种程度——有些人认为它是非常高产的，有些人则认为这是一种虐待。瑞士早在1992年就以立法形式禁止了对鸡喙的修剪处理，而很多国家现在才刚刚开始实施法律政策来规范工业化农场养殖。温顺动物的生存条件往往比奴隶的生存条件还要恶劣，这些动物对此的反应让我们想起了达·芬奇16世纪的预言——人类对待动物的方式不久将会施于人类自身。动物因囚禁而变得烦躁，继而产生攻击行为的过程是反自然的，并产生了一系列连锁反应。在这些连锁反应中，人类浑然不觉（希望是这样）给动物们带来了更多的痛苦和绝望。食物的生产需要以一种和谐的方式进行。这不仅仅是要确保高品质的生活，也是为了保证随着地球人口的增加以及更多的人成为中产阶级，他们会改变自己的饮食习惯，创造一种可持续的供应链，不突破地球承载力的极限。

## 动手能力
### Capacity to Implement

当我们身边充斥着消极情绪时，保持积极向上几乎是不可能的。虐待动物的行为几乎遍布全世界，这甚至是一些人成为素食主义者的一个原因。问问周围的人，在家庭成员和朋友中有没有素食主义者。然后探究一下是否有人是因为不能忍受食用遭受虐待的动物的肉，而变成素食主义者的。

# 教师与家长指南

故事灵感来自

## 卡尔·路德维希·施魏斯福尔特
## Karl Ludwig Schweisfurth

二战结束后不久,卡尔·路德维希·施魏斯福尔特在美国的一些大型屠宰场接受了屠宰训练。这段经历帮助他将自己在德国赫塔的家族生意发展为欧洲最大的肉食加工企业。他的公司一周可以屠宰 5000 头牛、25 000 头猪。他认为,那些遭受过虐待的动物肉的营养价值和肉质从来都不怎么好。他的两个儿子对于接管这个生意没有兴趣,他们认为这个工作是不人道的,因此他决定把这份生意卖给雀巢公司。他再度创业,将动物饲养工作和屠宰场(动物们在这个屠宰场以一种比较有尊严的方式被屠宰)、烘焙坊、奶酪加工厂、酿酒厂、饭店和旅馆等结合在一起,以最高的生态标准来运营。卡尔·路德维希创建了一个以其姓氏命名的基金会并组建了一个团队去学习如何生产高品质且环境友好型食品。他们的目标是所售产品的价格要为大多数人所接受。

### 更多资讯

http://www.bbc.com/future/story/20141218-why-boredom-is-good-for-you

http://passyworldofmathematics.com/geometry-in-the-animal-kingdom/

图书在版编目（CIP）数据

我烦透了！：汉英对照／（比）冈特·鲍利著；
（哥伦）凯瑟琳娜·巴赫绘；田烁，王菁菁译．－－上海：
学林出版社，2016.6
（冈特生态童书．第三辑）
ISBN 978－7－5486－1079－3

Ⅰ．①我… Ⅱ．①冈… ②凯… ③田… ④王… Ⅲ．
①生态环境－环境保护－儿童读物－汉、英 Ⅳ．
① X171.1－49

中国版本图书馆 CIP 数据核字 (2016) 第 121292 号

ⓒ 2015 Gunter Pauli
著作权合同登记号 图字 09－2016－309 号

**冈特生态童书**
**我烦透了！**

| | |
|---|---|
| 作　　者—— | 冈特·鲍利 |
| 译　　者—— | 田　烁　王菁菁 |
| 策　　划—— | 匡志强 |
| 责任编辑—— | 匡志强　蔡雩奇 |
| 装帧设计—— | 魏　来 |
| 出　　版—— | 上海世纪出版股份有限公司 学林出版社 |
| | 地　址：上海钦州南路81号　电　话／传真：021-64515005 |
| | 网址：www.xuelinpress.com |
| 发　　行—— | 上海世纪出版股份有限公司发行中心 |
| | （上海福建中路193号　网址：www.ewen.co） |
| 印　　刷—— | 上海丽佳制版印刷有限公司 |
| 开　　本—— | 710×1020　1/16 |
| 印　　张—— | 2 |
| 字　　数—— | 5万 |
| 版　　次—— | 2016年6月第1版 |
| | 2016年6月第1次印刷 |
| 书　　号—— | ISBN 978－7－5486－1079－3/G·414 |
| 定　　价—— | 10.00元 |

（如发生印刷、装订质量问题，读者可向工厂调换）

# Work 77

# 非洲农场

## African Farms

**Gunter Pauli**

冈特·鲍利 著

凯瑟琳娜·巴赫 绘
姚晨辉 译

学林出版社
www.xuelinpress.com

## 丛书编委会

主　任：贾　峰
副主任：何家振　郑立明
委　员：牛玲娟　李原原　李曙东　吴建民　彭　勇
　　　　冯　缨　靳增江

## 丛书出版委员会

主　任：段学俭
副主任：匡志强　张　蓉
成　员：叶　刚　李晓梅　魏　来　徐雅清　田振军
　　　　蔡雩奇　程　洋

特别感谢以下热心人士对译稿润色工作的支持：

姜竹青　韩　笑　贾　芳　刘　晓　张黎立　刘之杰
高　青　周依奇　彭　江　于函玉　于　哲　单　威
姚爱静　刘　洋　高　艳　孙笑非　郑莉霞　周　蕊

# 目录

| | |
|---|---|
| 非洲农场 | 4 |
| 你知道吗？ | 22 |
| 想一想 | 26 |
| 自己动手！ | 27 |
| 学科知识 | 28 |
| 情感智慧 | 29 |
| 艺术 | 29 |
| 思维拓展 | 30 |
| 动手能力 | 30 |
| 故事灵感来自 | 31 |

# Contents

| | |
|---|---|
| African Farms | 4 |
| Did you know? | 22 |
| Think about it | 26 |
| Do it yourself! | 27 |
| Academic Knowledge | 28 |
| Emotional Intelligence | 29 |
| The Arts | 29 |
| Systems: Making the Connections | 30 |
| Capacity to Implement | 30 |
| This fable is inspired by | 31 |

非洲最大的柑橘农场过去坐落在克鲁格国家公园的外面,克鲁格国家公园是一个迷人的野生动物保护区,游客们来这里可以观赏到大象、水牛、犀牛、狮子和豹子。但柑橘农场的主人意识到,自己的农场无法继续生存下去了,因为来自国外的橘子要比本地生产的便宜得多。

The largest citrus farm in Africa used to be located just outside the Kruger National Park, a wonderful game reserve where tourists come to watch elephants, buffalo, rhino, lions, and leopards. The owner of the farm realised that the farm could not survive. Oranges from overseas are much cheaper than the local ones.

# 非洲最大的柑橘农场

# Largest citrus farm in Africa

我们不得不关闭农场

We have to close down the farm

农场主基斯说:"在世界各地,柑橘类水果的价格都跌得这么厉害,我们一点钱都赚不到了,这样下去我们不得不关闭农场。"

罗伊索,他的朋友,也是邻居,劝说道:"专家们宣称,在这个经济全球化的时代,你只有降低成本,裁减员工,更好地利用灌溉水,使用化学品,才能生存下去。"

"The price of citrus fruit has dropped so much worldwide, that we cannot make any money anymore. So, we have to close down the farm," says Khethi, the farm owner.

"Experts claim that you have to cut costs, lay off staff, use more irrigation water, and use chemicals to make ends meet in this globalised economy," argues Loyiso, his friend and neighbour.

"我们的橘子不可能像世界上的其他地方那样便宜。我们是一个合作社,所有的成员都是业主,不可能将任何人解雇。"

"你不能老想着种出更便宜的橘子,你应该试试利用现有的橘子赚更多的钱。" 罗伊索建议道。

"怎么改变现在的柑橘经营模式啊?没有人知道,更不要说了解。每个人都告诉我,我们能做的就是用更便宜的价格卖掉它们。"

"We cannot be as cheap as the others around the world. As we are a cooperative, and all members are owners, it's impossible to kick out anyone."

"You should not try to grow cheaper oranges. You should try to earn more with the ones you already have," suggests Loyiso.

"No one knows or even understands how you can do more with an orange than we are doing now. And everyone tells me that we have to sell them at a cheaper price."

利用现有的橘子赚更多的钱

Earn more with the ones you already have

你一年四季都可以供应橘子吗？

Do you have oranges all year round?

"降低成本,这个主意不错,可是现在已经有这么多的人失业,你不能简单地把工人辞退了事,你必须对人民和社会承担一份责任。"

"要这样我们就没有成功的希望了!" 基斯说。

"别放弃呀,我们一起想想可行的方法。你一年四季都可以供应橘子吗?"

"It's not a bad idea to reduce the costs, but when there are already so many people without jobs, you cannot simply kick your workers out. You have a responsibility towards people and the community."

"Then there's no way we could be successful!" says Kheti.

"Don't give up. Let's find a way to make it work. Do you have oranges all year around?"

"差不多吧,我们林波波省的气候好得很,这里可是阳光明媚的南非啊!"
"就是嘛,既然你有那么多橘子,你可以榨果汁啊。"
"这简单!而且我们可以向克鲁格国家公园甚至周围所有的旅馆提供果汁。"

"Nearly. We have a great climate here in Limpopo. This is sunny South Africa!"
"So, if you have oranges, you can make juice."
"That's easy! And we can deliver it to all the lodges in and around the Kruger Park."

你可以榨果汁啊

you can make juice

# 你可以生产可生物降解的肥皂

# You can make biodegradable soaps

"榨果汁,就会有残留的果皮。对果皮进行蒸煮,你就可以生产可生物降解的肥皂。"

"你要我们这些种柑橘的农民去生产和销售肥皂?"基斯问道。

"不,我认为你应该为那些旅馆提供洗衣服务。因为你的农场制作的化学品是天然的,它们能够将废水转化为灌溉用水。"

"And when you make juice, you'll have leftover peels. You can steam the peels to make biodegradable soaps."

"Do you want us, the orange farmers, to produce and sell soap?" asks Kheti.

"No, I think you should offer laundry services to the lodges. As the chemicals coming from your farm are natural, they will be able to turn their waste water into irrigation water."

"嗯，果树还需要修剪，我们可以用剪下来的树枝种植蘑菇。"

"这个主意真棒！"罗伊索说，"蘑菇菌糠也是很好的动物饲料。我还在想，旅馆自助餐厅里有很多食物都是做点缀用的，你可以将这些剩菜要来作为猪饲料。"

"And as fruit trees need pruning, we could use the prunings to grow mushrooms on."

"What a good idea!" says Loyiso. "Mushroom substrate is also good feed for animals. I was just thinking, as so much of what you see at the buffet at these lodges is only for decoration, you could ask them if you could use those leftovers as pig feed."

# 种植蘑菇

# Grow mushrooms

# 用废弃物生产沼气

Waste to produce biogas

"我觉得旅馆老板会很乐意把他们的废弃物给我们去生产沼气。"

"这就意味着,多亏了这些橘子,你可以比仅仅卖水果赚更多的钱。"

"I think the lodge owners would happily supply us with their waste to produce biogas."

"That means that thanks to the oranges, you will be able to make more money than when you only export the fruit."

"现在我们不用再担心国际市场上橘子的价格啦,我们可以创造更多以前没有想到的就业机会。这才是我喜欢的那种经济模式!"
……这仅仅是开始!……

"Now we don't have to worry about the price of oranges on the world market and we can create more jobs than we ever imagined. That is the kind of economy I like!"
... AND IT HAS ONLY JUST BEGUN!...

……这仅仅是开始！……

...AND IT HAS ONLY JUST BEGUN!...

# Did You Know? 你知道吗？

The first orange farms were started in 2 500 BC in China. The sweet oranges originated from India and were introduced to the world by the Portuguese. The bitter orange originated from Persia, and was first introduced to Italy.

公元前2500年，世界上第一个柑橘农场出现在中国。甜橘起源于印度，被葡萄牙人传播到世界各地。苦橘原产波斯，最先被出口到意大利。

Arab, Portuguese, Spanish, and Dutch sailors planted orange trees along trade routes to ensure a supply of fruit to prevent scurvy.

阿拉伯、葡萄牙、西班牙和荷兰的水手沿着他们的贸易路线种植橘子树，以确保水果供应，预防坏血病。

𝒪range trees are the most cultivated trees in the world and are widely grown in tropical and subtropical climates. Brazil is the largest producer in the world.

橘子树是世界上最常见的树木，广泛种植于热带和亚热带地区。巴西是世界上最大的柑橘生产国。

𝒯he word "orange" originates from the Arabic nāranj, and the "n" was dropped in many languages as the article "an" already contains an "n". In French, for example, saying une norange sounded like there was one "n" too many.

单词"橘子"（orange）来源于阿拉伯语nāranj，"n"这个字母后来在许多语言里都被省略了，因为冠词"an"中已经包含了一个"n"。比如在法语中，如果把"一只橘子"说成"une norange"，听上去就显得其中的"n"太多了。

The pH of an orange can be as low as 2,9 and as high as 4,0. This is as acidic as a cup of coffee or a cola drink.

橘子的pH值在2.9和4.0之间，相当于一杯咖啡或者可乐的酸性。

In Florida (USA) and Brazil, oranges are harvested with a canopy–shaking machine. These two places are the largest producers of frozen concentrated orange juice.

在美国的佛罗里达州和巴西，人们会用一种树冠振动机来采摘橘子。这两个地方也是冷冻浓缩橘汁的最大生产地。

Juice is shipped around the globe by cargo boats, called fruit juice tankers, that can transport 35 000 tonnes of juice in one trip.

果汁通过货船运往世界各地，被称为果汁油轮，每艘轮船可以运送3.5万吨果汁。

Orange peel is rich in oil that can be used in paints and cleaning products. The price fluctuates with the seasonal price of oranges.

橘子皮中含有丰富的油脂，可以用来生产油漆和清洁产品，其价格随着橘子的季节价格而波动。

# Think About It
## 想一想

**Would** you be able to tell friends who have worked with you for years and co-owns your company that they have to lose their jobs?

你做得到吗，告诉与你共事多年的朋友和公司的合伙人，他们将要被解雇？

如果每升橘皮油比同样数量的橙汁可以多赚10倍的钱，你会只生产橘皮油呢，还是会两者同时进行？

**If** you can earn 10 times more money per litre of citrus peel oil than with the same amount of orange juice, would you rather work only with peels, or with both?

**If** you had fruit trees, would you only want to do business when the fruit are in season, or would you like to find a way to do business all year around?

如果你有果树，你是只想做当季水果的生意呢，还是说你会设法找到一种办法一年四季都可以做生意？

将水果和食品作为一种美丽的装饰，可以吸引人们多吃一点吗？或者，人们宁愿把被丢弃的水果用作猪饲料？

**Are** fruit and food a fine form of decoration that will entice people to eat more, or would people rather be inspired by the fact that the waste from fruit serve as feed for pigs?

Squeeze the juice from some oranges. Now squeeze the peels. How much oil can you get from the peels? It may be easier to eat just eat the peels. Do make sure the oranges were organically grown before you do this. You may not want to eat the peel by itself, so the question is how you can use the peel in your food so that you will be able to get health benefits from it. Grate it onto your salad or desert, add it to you mom's coffee, or cover it in dark chocolate. Who likes the taste? Is anyone willing to eat only the peel?

从几个橘子中挤一些橘子汁。然后再挤一挤橘子皮,你从橘子皮中可以挤出多少橘皮油?更容易的吃法是单独吃皮,但你这样做之前一定要确保橘子是有机种植的。你可能不想单独吃橘子皮,问题来了,如何将橘子皮加进食物中,以便你能够获得橘子皮中对健康有益的成分?可以将橘皮碾碎放到你的沙拉中,添加到妈妈的咖啡里,或撒在黑巧克力上面。有人喜欢这种味道吗?有人愿意单独吃橘皮吗?

# TEACHER AND PARENT GUIDE

## 学科知识
### Academic Knowledge

| | |
|---|---|
| 生物学 | 坏血病是一种因缺乏维生素C造成的疾病，症状包括疲劳、海绵状牙龈出血等；橘子通过授粉或单性结实进行繁殖；橘子是柚子和柑的杂交品种；橘子可以混种繁殖，即很容易产生杂交品种；橘子树可以嫁接，即将一种植物的组织插入到另一种植物中（接合），以获得新的树木；橘子树是硬木树，因此其所有的木材废料都适合蘑菇种植；橘子皮包含的纤维比果肉多四倍；橘子皮30%的成分是果胶，果胶是一种天然凝胶。 |
| 化 学 | 酸度低的橘子容易腐烂；橙子含有类胡萝卜素和类黄酮；橘子即使在采摘之后还能呼吸，放出氧、二氧化碳和乙烯。 |
| 物 理 | 水溶性和水稀释性的区别；冰凉的水果没有温热的水果榨出的果汁多；中等大小的橘子含有更多的果汁；湿度过大会加速橘子腐烂。 |
| 工程学 | 通过蒸馏将油从果皮中分离出来可以生产出食品级的油；巴氏杀菌法可以确保果汁全年的供应，使生产与销售不再有季节限制；浓缩果汁意味着将果汁中的水蒸发掉，需要加水复原。 |
| 经济学 | 企业的法律架构：有限责任、股份以及合伙；橘子和橘汁的价格由商品市场所决定，有目前的和预期的价格；为什么市场上新鲜水果的价格比果汁贵得多；橘子（像大多数水果和蔬菜一样）有季节性，因此价格去年波动。 |
| 伦理学 | 为什么我们会花更多的钱去买外表好看的水果，而不是买具有相同品质和口味却不好看的水果；超市的政策是不卖外表不"完美"的水果和蔬菜；参观野生动物保护区的人往往生活奢华，享受进口食品，但大部分费用都付给了经营者，而不是当地居民。 |
| 历 史 | 公元前314年的中国文献里已经提到了甜橘；克里斯托弗·哥伦布将橘子的种子带到了加勒比海地区。 |
| 地 理 | 林波波省是南非的一个省；南非的克鲁格国家公园成立于1926年，其园区的一部分位于姆普马兰加省，一部分位于林波波省。 |
| 数 学 | 虽然柠檬中d–柠檬烯的含量约是橘子的两倍，橘子的体积却大得多。 |
| 生活方式 | 新鲜的橘汁已成为很多人每日早餐的一部分。一杯橘汁足以提供每日所需的维生素C。 |
| 社会学 | 为子孙后代考虑，我们必须具有保护野生动物园区的意识。 |
| 心理学 | 需要有一个专门的人员负责解雇员工，尤其当他们是你多年的朋友和同事时。 |
| 系统论 | 果园经受的价格波动超出了果园主人的控制，为了实现可持续发展，必须创造其他的收入来源，额外的收入要从当地经济中产生，以保持稳定性和增强适应力。 |

# 教师与家长指南

## 情感智慧
### Emotional Intelligence

**主 人**

果园的主人处于绝望之中，感到别无选择，只有关闭果园。他意识到了合作社的局限性，却坚持不能将朋友解雇。他很失望，没有专家可以以为他提供解决方案。水果的价格波动让他不知所措，他明白自己的成本太高，却找不到降低成本应对国际市场竞争的办法。他认为自己走投无路了。不过，他很喜欢家乡的气候，并为自己的国家感到骄傲。在一次和邻居的开放性的对话中，他慢慢开始了解到一些可能的解决方案，并高兴地从中受到了启发。他渴望了解更多信息，将他所拥有的知识和似乎毫无关联的知识结合起来，明确如何才能自强自立，且不必解雇任何员工。

**邻 居**

邻居很现实，尽管他知道专家的建议，却没有就这个话题展开辩论，而是试图找到方法摆脱看似走投无路的情况。他指出了果园主人的社会责任，给果园主人打气，认为他不应该放弃，并通过与他交谈，寻求解决方案。他提供了一个捷径：制作橘汁。果园主人接受了这个想法，他得到鼓励，又提出了第二个不太为人所知的商业点子：将果皮变成肥皂。邻居继续寻找更高的价值实现途径，他建议不要卖肥皂，而是提供洗衣服务，并因此获得额外的灌溉用水利益。当邻居发现主人的心态转变之后，他提出了更多的想法，可以让农场赚更多的钱，让大家对未来充满希望。

## 艺术
### The Arts

让我们利用橘子来进行艺术创作。柑橘是一种果皮较厚的水果，其果皮可以被切割和扭曲成复杂的形状。看看你能不能用橘子皮拼成大象和蜗牛的形状。这比仅仅吃掉水果，然后扔掉果皮有趣多了！你会发现艺术太有趣了，除了享受水果的美味之外，你也可以用橘子创造美丽的艺术品。

# TEACHER AND PARENT GUIDE

## 思维拓展
### Systems: Making the Connections

　　从事水果商品化生产的公司很难在全球市场上进行竞争，除非他们实现了非常高的规模经济。橘子也不例外。最初热带和亚热带地区贸易航道沿线的所有国家都引进了橘子，随着时间的流逝，橘子的生产集中到了低成本的地区（如巴西），或具有高水平的自动化和科学化种植管理的地区。而那些没有低成本优势或科学化生产企业的国家应该怎么做呢？商业顾问的传统建议总是始终不变的：增加经济规模，提高生产效率，减少员工人数，充分利用灌溉和遗传学。另一种选择是倒闭，寻找替代产业。上述选择虽然有一定的逻辑道理，但现实的困难在于解雇员工和关闭现有的企业并不容易，这将降低地区经济收入，对整个社区，从本地面包店到城镇的税收收入等都产生影响。如果组织的结构形式是合作社性质的，所有的员工同时也是股东，那么在不造成紧张和压力以及破坏社会关系的情况下，裁员几乎是不可能的。最好的方法是评估该地区的经济活动，看看如何将果园的生产（更好）地融入该地区的现有商业。另一方面，一旦地方经济的定位明确之后，可以很快找到促进本地消费的方法。接下来的产品和服务都将和橘子及其加工过程相关联，还包括客户网络的出现、建立社区等。未来将发生翻天覆地的变化，将有更多的钱在当地经济中流通，提供更多的购买力，增强就业市场。最重要的是，使柑橘种植业不再受世界市场上价格波动的影响。

## 动手能力
### Capacity to Implement

　　如果由你负责一家销售橘子汁的商店，你打算怎样常年为客户服务？如果你只卖橘子，那么你将不得不从海外进口一些，这就需要存储空间，因为橘子在采摘后只能保存一个月时间。查看你需要从什么地方进口橘子，以保证一年12个月都有备货。另一种选择是只在当季销售本地的橘子，在橘子过季之后寻找其他种类的水果制作果汁。你将会使用哪些水果来保证你全年的业务运营？你打算如何说服客户，他们仍然会得到日常所需的新鲜维生素C？研究你所有可能的选项，制定一个计划，并向你的朋友和你的父母进行介绍。

# 教师与家长指南

故事灵感来自

## 路易吉·比斯塔吉诺
## Luigi Bistagnino

路易吉·比斯塔吉诺读书时的专业是建筑学，目前他是都灵理工大学建筑和设计学院的教授。他在将学术课程（以前仅限于产品和工艺设计）转变为系统设计方面进行了探索。他与"慢食运动组织"密切合作，致力于将系统设计应用于所有的经济活动中。他特别擅长将这些新概念应用到农业和食品加工行业中，积累了丰富的研究案例。比斯塔吉诺教授论证了如何将种植与水果和蔬菜的加工相结合，以促进地方经济发展和创业。创造的价值将转化为更多的财富、工作岗位、适应性、生物多样性，提高文化认同感。如果尝试将当地的农业和加工业融入全球经济之中，将获得比以往任何时候都更好的生活质量。

更多资讯

http://www.instructables.com/id/Drink-prices-per-Gallon/

图书在版编目（CIP）数据

非洲农场：汉英对照／（比）冈特·鲍利著；
（哥伦）凯瑟琳娜·巴赫绘；姚晨辉译．－－上海：学林
出版社，2016.6
（冈特生态童书．第三辑）
ISBN 978-7-5486-1070-0

Ⅰ．①非… Ⅱ．①冈… ②凯… ③姚… Ⅲ．①生态环
境－环境保护－儿童读物－汉、英 Ⅳ．① X171.1-49

中国版本图书馆 CIP 数据核字 (2016) 第 122653 号

---

© 2015 Gunter Pauli
著作权合同登记号 图字 09-2016-309 号

## 冈特生态童书
### 非洲农场

| | |
|---|---|
| 作　　者—— | 冈特·鲍利 |
| 译　　者—— | 姚晨辉 |
| 策　　划—— | 匡志强 |
| 责任编辑—— | 李晓梅 |
| 装帧设计—— | 魏　来 |
| 出　　版—— | 上海世纪出版股份有限公司 学林出版社 |
| | 地　址：上海钦州南路 81 号　　电话／传真：021-64515005 |
| | 网址：www.xuelinpress.com |
| 发　　行—— | 上海世纪出版股份有限公司发行中心 |
| | （上海福建中路 193 号　网址：www.ewen.co） |
| 印　　刷—— | 上海丽佳制版印刷有限公司 |
| 开　　本—— | 710×1020　1/16 |
| 印　　张—— | 2 |
| 字　　数—— | 5 万 |
| 版　　次—— | 2016 年 6 月第 1 版 |
| | 2016 年 6 月第 1 次印刷 |
| 书　　号—— | ISBN 978-7-5486-1070-0/G · 405 |
| 定　　价—— | 10.00 元 |

（如发生印刷、装订质量问题，读者可向工厂调换）

# Work 102

# 钱生钱

## Making Money on Money

**Gunter Pauli**

冈特·鲍利 著

凯瑟琳娜·巴赫 绘

姚晨辉 译

## 丛书编委会

主　任：贾　峰

副主任：何家振　郑立明

委　员：牛玲娟　李原原　李曙东　吴建民　彭　勇
　　　　冯　缨　靳增江

## 丛书出版委员会

主　任：段学俭

副主任：匡志强　张　蓉

成　员：叶　刚　李晓梅　魏　来　徐雅清　田振军
　　　　蔡雩奇　程　洋

特别感谢以下热心人士对译稿润色工作的支持：

姜竹青　韩　笑　贾　芳　刘　晓　张黎立　刘之杰
高　青　周依奇　彭　江　于函玉　于　哲　单　威
姚爱静　刘　洋　高　艳　孙笑非　郑莉霞　周　蕊

# 目录

| | |
|---|---|
| 钱生钱 | 4 |
| 你知道吗？ | 22 |
| 想一想 | 26 |
| 自己动手！ | 27 |
| 学科知识 | 28 |
| 情感智慧 | 29 |
| 艺术 | 29 |
| 思维拓展 | 30 |
| 动手能力 | 30 |
| 故事灵感来自 | 31 |

# Contents

| | |
|---|---|
| Making Money on Money | 4 |
| Did you know? | 22 |
| Think about it | 26 |
| Do it yourself! | 27 |
| Academic Knowledge | 28 |
| Emotional Intelligence | 29 |
| The Arts | 29 |
| Systems: Making the Connections | 30 |
| Capacity to Implement | 30 |
| This fable is inspired by | 31 |

一名十多岁的男孩和他的爷爷坐在一起,看着满月挂上中天。"该睡觉了。"爷爷说,"你们明天都还要早起。"

"爷爷,为什么我爸爸每天早上必须那么早去上班?"男孩问,"他几乎都没有时间吃早餐。"

"你爸爸必须辛苦工作,赚到足够的钱,这样你才有饭吃,有房子住啊。"

"为什么我妈妈也要那么早去上班?"男孩疑惑地问道。

A teenage boy sits with his granddad. They are watching the full moon rise. "Time for bed," says Granddad. "You all have to get up early tomorrow."

"Why does my dad have to go to work so early every morning, Granddad?" the boy asks. "He hardly has time for breakfast."

"Your dad has to work long hours to earn enough money so you have food on the table and a roof over your head."

"And why does my mom have to go to work so early too?" wonders the boy.

该睡觉了

Time for bed

付给他们更多的工资

To pay them more money

"为了挣钱买汽车,载她上班啊,还要挣钱买汽油呢。"
"那为什么爸爸和妈妈每天回家这么晚?"
"因为他们只有整天非常努力地工作,才能获得晋升。而一旦得到晋升,他们又不得不更加努力工作,这样才能证明付给他们更多的工资是公平合理的!"

"To earn enough to pay for a car to get her to work, and have money for fuel."
"And why do Mom and Dad get home so late every day?"
"Because they have to work very hard all day long to be promoted. And once they are promoted they will have to work even harder to show that it was justified to pay them more money!"

"爷爷,你觉得这样有意义吗?而且,为什么爸爸现在要三天两头到国外旅行?"

"嗯,因为他升职了,他现在还需要负责其他国家的公司。"

"为什么妈妈总是给我们买用糖、牛奶和小麦制作的便宜物?"

"因为他们需要省下钱,去偿还他们的债务。"

"Does that make sense to you, Granddad? And also, why is dad travelling overseas so often now?"

"Well, as he has been promoted, he now needs to take care of business in other countries as well."

"Why is mom always buying us that cheap food that is full of sugar, milk and wheat?"

"Because they need to save money to pay all their debt."

省下钱去偿还他们的债务

Save money to pay for all their debt

现在享受美好生活，以后再为此买单

Live well now, pay for everything later

"可是,为什么他们一开始就要欠下这么多债务呢?"

"因为你的父母想现在就能及时享受美好生活,以后再为这些享受买单。"

"但是,为了这些债务,他们几乎从来不在家!他们不得不整天工作,因为最终他们将付出更多的钱还债。即使他们回到家,也因为太累而不能陪我!所以,我的父母就像是在坐牢一样!"

"But why do they have so much debt in the first place?"

"Because your parents want to live well now. And pay for everything later."

But with all their debt, and having to work so much harder because it costs them more in the end, they are never at home! And when they are home, they are too tired to spend time with me! So my parents are virtually in prison!"

"不，我的孩子，不要这么想。
你的父母都是自由的！他们自愿做出这种选择，是为了让你生活得更好。他们爱你，希望你拥有最好的生活。"

"我知道他们爱我，我也真的很感激他们为我做的一切。"

"你的父母确实有债务，他们需要支付利息，但这样就可以满足你的所有需要，让你顺利成长，拥有一个光明的未来。"

"No, my boy. Don't think of it like that. Your parents are free! Free to make choices that will enhance your life. They love you and want you to have the best.

"I know they love me, and I am really grateful for all they do for me."

"Your parents do have debt and they do pay interest, but that is so you can grow up with all you need and have a bright future."

真的很感激他们为我做的一切

Really greatful for all they do for me

没有办法再供养房子和汽车

Can no longer pay for the house and car

"我看得出来,爷爷。不过,如果有一天,他们工作的公司不需要他们了,会发生什么呢?"

"哦,他们会找到其他工作……我们希望是这样。"爷爷说。

"但是,如果他们不能马上找到工作,没有办法再供养房子和汽车,又会怎样呢?"

"I see that, Granddad. But what happens if one day the company they work for does not need them anymore?"

"Oh, they will find other jobs… we hope," says Granddad.

"But what if they don't right away and can no longer pay for the house and the car?"

"他们将会用他们的房子作抵押,借更多的钱,让我们渡过难关,直到他们找到别的工作。"

"我不明白,谁在通过发放贷款赚钱?"

"是银行,你向银行借钱,银行借给你钱,这样你可以生活下去,花钱买房子、汽车和其他的东西。"

"Then they will borrow even more money, using their house as a guarantee, to see us through until they find other jobs."

"I don't understand. Who is making money from lending people money?"

"The bank, you borrow from the bank. The bank lends you money so you can progress in life, buy a house, a car, and other things that cost a lot."

谁在通过发放贷款赚钱？

Who is making money from lending people money?

17

钱是造币厂印制的

The mint does that.

"那么银行的钱是哪里来的？难道是他们打印出来的吗？"

"不是，钱是造币厂印制的。我给你解释一下银行是怎么工作的：那些有钱却暂时不想花的人会将钱存在银行的储蓄账户中。"

"那么，那些很有钱的人不花光他们所有的钱吗？"

"And where does the bank get all the money from? Do they print it?"

"No, the mint does that. To explain how the bank works: People who have more money than they need right now put that money in a savings account at the bank."

"So people who have lots of money don't spend all their money?"

"不,他们把钱存在银行里。银行用这笔钱来赚钱。将钱存在银行储蓄账户的人也用自己的钱赚了一些钱,因为银行为了留住这些钱,会付给他们一点儿利息。"

"人们怎么能够通过钱来赚钱?这就像用水来制作水一样!这是不可能的。我不明白这样的情形长期存在的道理,你知道吗,爷爷?"

……这仅仅是开始!……

"No, they keep it in the bank. The bank uses that money and makes money from it that way. People who keep their money in savings accounts at the bank also make some money on their money as the bank pays them a bit of interest on it, to keep it there.

"How can one make money from money? It's like making water from water! It is impossible. I can't see it lasting forever, can you Granddad?

... AND IT HAS ONLY JUST BEGUN!...

……这仅仅是开始！……

...AND IT HAS ONLY JUST BEGUN!...

# Did You Know? 你知道吗?

When debt increases faster than earnings, due to higher interest rates or the need to borrow more, then consumption must decrease, and poverty rises.

由于利率较高，或者需要借更多的钱，导致债务增长快于收入时，消费水平就必须降低，而贫困程度就上升了。

People are not the only ones taking on debt; governments borrow a lot of money as well. If government debt increases, then taxes must be increased, and this leaves less money for families and companies to buy or to invest.

并不是只有个人才会负债，政府有时也会借很多的钱。如果政府的债务增加，那么就必须增加税收，家庭和企业用于购买或投资的钱就会变少。

When families have more debt, their creditworthiness decreases, meaning that the banks consider them a higher risk, this means that they will have to pay higher interest rates, which decreases their capacity to pay back the loan.

当家庭债务增加时，其信用会降低，这意味着银行认为他们的风险较高，意味着他们将不得不承担更高的利率，从而降低他们偿还贷款的能力。

The total world debt is US$ 230 trillion, which is 3 times more than annual GDP (gross domestic product is the total amount of products and services sold each year).

全球的总债务为230万亿美元，比全年GDP（国内生产总值，即每年所有被销售的产品和服务的总额）的3倍还要多。

*D*ebt leads to decreasing economic growth. In the year 2000 for every US$ 2.4 of debt creation there was US$ 1 added to GDP. By 2015 it was necessary to accept US$ 4.6 in additional debt to create $1 extra in GDP.

债务导致经济增长减缓。2000年,每2.4美元的债务会使GDP增加1美元。到2015年,为了使GDP增加1美元,人们不得不接受4.6美元的额外债务。

*T*he US alone pays US$ 400 billion in interest per year, while the rates are at a historically low level.

美国每年要支付4000亿美元的利息,这还是利率处于较低历史水平的情况下。

In the industrialised world, the greatest debt is held by households and companies (41%), and in the emerging economies the greatest debt is held by the financial sector (43%).

在工业化国家，最大的债务由家庭和企业持有（41％），而在新兴经济体国家，最大的债务由财政部门持有（43％）。

Australian households are the most indebted in the world. Their total loans from banks represents 130% of the total amount of products and services produced each year. This makes Australians very vulnerable to any change in interest rates.

澳大利亚的家庭是世界上负债最重的，他们的银行贷款总额是其每年产品和服务总额的130％，这使得澳大利亚人非常容易受到利率变化的影响。

# Think About It

## 想一想

Would you like to work to pay off debt, or would you like to work and save money and in this way be able to afford buying something without having to become indebted?

你是愿意通过工作来偿还债务，还是愿意先工作攒钱再购买自己需要的东西，而不必背负债务？

你希望父母花更多的时间陪伴你，还是希望他们工作更长时间来赚更多的钱？

Would you like your parents to be spending more time with you or would you like them to work longer hours to make more money?

Do you approve of your parents saving money by buying cheap and unhealthy food, or would you like them to save by taking public transport instead of driving their own cars and buy healthy food?

你赞成你的父母通过购买廉价而不健康的食品来省钱吗？还是希望他们不再开车，通过改乘公共交通工具却购买健康食品来省钱？

你认为过度负债就像是在坐牢一样吗？还是说你将这视作是一种享受生活的好机会？

Do you consider excessive debt to be like a prison? Or do you see it is a great opportunity to enjoy life?

# Do It Yourself!
# 自己动手！

Do you receive pocket money from your parents for doing chores such as making your bed, doing the dishes, mowing the lawn, walking the dog, taking the garbage out, washing their cars, doing some shopping or perhaps even cooking a meal? Propose a plan to your parents whereby you will help with such chores and have them deposit the money you have earned in a separate savings account for you. How much money do you think will you have after ten years?

你会通过做家务从父母那里获得零用钱吗？譬如说整理自己的床、洗餐具、修剪草坪、遛狗、倒垃圾、清洗父母的汽车、购物或者做饭。向你的父母提出一个计划，你会帮助他们做这类家务，让他们为你开设单独的储蓄账户，将你赚的钱存进去。你认为自己在10年后会攒下多少钱？

# TEACHER AND PARENT GUIDE

## 学科知识
### Academic Knowledge

| | |
|---|---|
| 生物学 | 食物作为所有生命的基本需要及其生产；从浆果、水果、蔬菜和肉类到牛奶和小麦，这些食物的消化难度是递增的。 |
| 化 学 | 糖、牛奶和小麦的消耗会在体内进行不同的新陈代谢，并产生副作用。 |
| 物 理 | 地球上只有这么多的水，你可以改变它，但永远不能创造更多的水。 |
| 工程学 | 金融工程是一门新兴学科，它使用复杂的金融工具，以构架财务和风险。 |
| 经济学 | 利率和风险之间的平衡；偿还债务的能力取决于创收的能力；获得贷款需要提供担保；抵押贷款工具；银行"创造"钱的能力在于放贷比存款多；中央银行作为最后贷款人的作用；新兴的小额信贷体系；信用合作社（目的是为老百姓服务）和银行（目的是为股东赚钱）之间的差异；新兴的网上银行和大众融资的作用。 |
| 伦理学 | 放高利贷的人索要过高的利率，并要求全额担保，这使得一些借款人陷入绝望，导致他们自杀；缺乏职业道德地提供更多贷款会带来更多无法偿还的债务；政府的作用是在发生自然灾害（不可抗力）时进行干预；在2008年金融危机之后，政府拯救了贷款机构，但没有顾及私人公司或家庭；助学贷款债务危机，其总额超过了1万亿美元；高频交易能利用利率、汇率和股票的微小差异获利数十亿美元，在一瞬间买进和卖出数百万次；整个经济体系的目的是鼓励巨额债务。 |
| 历 史 | 注销债务和还款宽限期的引入首次出现在3500年前的美索不达米亚；美国政府于1819年强制推行了债务延期偿还，以拯救歉收的农民；在1929年大萧条时期，农民被放任其破产。 |
| 地 理 | 伦敦和纽约是世界主要的金融中心；月球运行周期与月亮的盈亏。 |
| 数 学 | 金融数学即计量金融学，把数学及数值模型衍生和扩展至金融市场；由超级计算机支持的高频交易，几秒钟可以执行数千次交易。 |
| 生活方式 | 借债的习惯；随着时间的推移支付更高的金额，包括利息，然后用收入偿还，提供担保；不同于土地、劳动和贸易，金钱和债务是虚拟商品。 |
| 社会学 | 现代社会是建立在债务基础之上的，这并不一定会导致团结一致；法律的经济社会学把关于债务关系的法律、经济和社会学关联在了一起；债务关系是跨学科的；金钱使人堕落的信念；穆斯林不收取利息；借和贷之间的区别。 |
| 心理学 | 负债的决定不仅仅取决于你的资产，更主要的是你的自信程度；负债多的人会感到孤立、内疚和惭愧，可能会导致节制；高负债会妨碍最佳运作；债务带来的压力会影响一个人的幸福感。 |
| 系统论 | 金钱使世界运转，但过度负债会让世界停摆。 |

# 教师与家长指南

## 情感智慧
### Emotional Intelligence

**孙子** 小孙子注意到爸爸需要很早去上班，往往不能与家人一起吃早餐。他在晚上和饭桌上想念父母，对爸爸经常长时间出国旅行的事实感到焦虑。他很奇怪妈妈为什么会购买廉价却不健康的食物，并对他的家庭负有巨额债务的事实感到压力。他认为债务就像是一所监狱，因为它限制了他父母的自由，让他们没办法花更多的时间陪伴他。他意识到父母将来可能面临无法偿还债务的风险。孙子对攒钱的方式提出了质疑，并且看清了这一切已经失控的事实。

**爷爷** 爷爷费力地解释赚钱的必要性。他很心疼自己的孙子，一步一步地解释了为什么他的父母过着现在的生活，不能像孩子渴望的那样，花更多的时间陪伴他，并展示了清晰的逻辑性。他相信一切都会好起来，并且孙子也能够理解和接受这一切。爷爷听到监狱的比喻吃了一惊，坚持认为父母都是自由的，他们之所以做这样的选择是出于他们对儿子的爱。当孙子提到父母失去工作的风险时，爷爷仍保持了坚定的信仰，并提供了一个临时解决方案：借更多的钱！爷爷精准地解释了银行是如何运作的，并用通俗易懂的语言分享了他的见解，受到这一逻辑触动，孙子有了进一步思考。

## 艺术
### The Arts

过度负债可能导致愤怒的情绪。涂鸦往往用来表达愤怒。你准备好测试自己创造涂鸦的能力了吗？征得你的父母同意后购买喷漆，要确保你买的是水溶性油漆。说服你的父母或老师允许你在家里或学校用一堵墙进行涂鸦。选择一个事后容易清洗的地方。现在，挑选一种你喜欢的风格设计你的涂鸦，并与大人们商量你准备使用的词语。房间里是否有一名专家乐意演示如何涂鸦呢？

# TEACHER AND PARENT GUIDE

## 思维拓展
### Systems: Making the Connections

俗话说，金钱让世界运转。借钱是一项存在了几千年的经济活动，私人和公共机构的债务积累已经达到了令人难以理解的程度。这个世界必须向金融机构支付三倍于国民生产总值债款的事实，证明了我们处于过度消费和过度借贷的状态，这最终将造成对地球的过度开发。即使世界每年保留其生产总值的10％以抵消部分债务，仍然需要不止一代人的努力才能使债务回到合理水平。造成这一切的原因是利息。当利率较低时，形势是可控的，一旦利率上升到一定程度，许多家庭和政府的财政将会崩溃。但即使远在这样的崩溃之前，利率也会导致收入的不公平分配，穷人会变得更穷，富人会变得更富。利息负担主要落在贫困人口身上，使他们无法获得社会或经济的发展。债务清偿减少了提供社会服务、健康和营养的能力，将导致人们选择劣质廉价的产品。这种逻辑同样适用于一个国家，特别是发展中国家。我们也可以用投资的逻辑来取代严格的贷款法律原则。如果有人有兴趣帮助一个家庭或一个国家成长和发展，也可以不是出借本金以赚取利息，而是签订一个投资协议，双方共享产出和收益。当今的逻辑是借出钱的人会得到一个固定回报，而借钱者拥有不确定的回报。这意味着风险完全落在了借款人的肩上，而希望用钱来赚钱的贷款人没有任何风险。这种做法已成为当前的标准模式，却早已被罗马帝国的皇帝以及伊斯兰教、犹太圣经和天主教会所反对。虽然他们的反对意见没有引领世界走向一个不同的方向，但已经证明了还有另外的选择。然而，关键的问题是：现在是否已经为时已晚？

## 动手能力
### Capacity to Implement

如果你还没有开始攒钱，现在行动起来吧。找些家务做做，不要将你赚到的钱花掉，把它存起来，看看你能积攒多少。问问你的朋友们，谁愿意共同建立一个特殊账户，每个人都可以根据他们赚钱的手段和能力做出贡献。当你已经积攒了一定金额，问自己一个问题："为了给所有人谋福利，我们会一致同意进行什么投资？"创建你们自己的小额信贷联盟，如果你们选择的项目是成功的，约定拿出部分利润来增强你们的"资金池"。通过这种方式，你们就积累了社会资本，而不是将它们花掉和产生债务。

# 教师与家长指南

故事灵感来自

## 穆罕默德·尤努斯
### Muhammad Yunus

穆罕默德·尤努斯出生于孟加拉国一个孟加拉穆斯林家庭,在九个孩子中排行第三。他是一名活跃的理想主义者,并且擅长戏剧。在完成学业之后,穆罕默德被任命为孟加拉国吉大港大学的讲师,后在美国范德比尔特大学获得经济学博士学位。意识到自己国家的贫穷和饥饿水平以及高利贷的习俗后,他与阿赫塔·哈米德·汗(Akhtar Hameed Khan)博士一起构思了小额贷款的想法,即以合理的利率进行非常小的贷款的。他发起的乡村银行(Grameen Bank)和小额贷款的成功,启发了100多个国家的类似努力。2006年,穆罕默德·尤努斯和乡村银行被联合授予了诺贝尔和平奖。

### 更多资讯

www.mybudget360.com/global-debt-total-amount-of-debt-world-gdp-to-debt-atios/

https://qaazi.wordpress.com/2008/10/08/why-does-islam-forbid-interest/

## 图书在版编目（CIP）数据

钱生钱：汉英对照 /（比）冈特·鲍利著；（哥伦）凯瑟琳娜·巴赫绘；姚晨辉译. —— 上海：学林出版社，2016.6

（冈特生态童书. 第三辑）
ISBN 978-7-5486-1073-1

Ⅰ. ①钱… Ⅱ. ①冈… ②凯… ③姚… Ⅲ. ①生态环境－环境保护－儿童读物－汉、英 Ⅳ. ①X171.1-49

中国版本图书馆 CIP 数据核字 (2016) 第 122648 号

----

© 2015 Gunter Pauli
著作权合同登记号 图字 09－2016－309 号

## 冈特生态童书

### 钱生钱

| | |
|---|---|
| 作　　者—— | 冈特·鲍利 |
| 译　　者—— | 姚晨辉 |
| 策　　划—— | 匡志强 |
| 责任编辑—— | 李晓梅 |
| 装帧设计—— | 魏　来 |
| 出　　版—— | 上海世纪出版股份有限公司 学林出版社 |
| | 地　址：上海钦州南路81号　电话／传真：021-64515005 |
| | 网　址：www.xuelinpress.com |
| 发　　行—— | 上海世纪出版股份有限公司发行中心 |
| | （上海福建中路193号 网址：www.ewen.co） |
| 印　　刷—— | 上海丽佳制版印刷有限公司 |
| 开　　本—— | 710×1020　1/16 |
| 印　　张—— | 2 |
| 字　　数—— | 5万 |
| 版　　次—— | 2016年6月第1版 |
| | 2016年6月第1次印刷 |
| 书　　号—— | ISBN 978-7-5486-1073-1/G·408 |
| 定　　价—— | 10.00元 |

（如发生印刷、装订质量问题，读者可向工厂调换）

# Work 76

# 建造大教堂

## Building a Cathedral

**Gunter Pauli**

冈特·鲍利 著

凯瑟琳娜·巴赫 绘

姚晨辉 译

学林出版社
www.xuelinpress.com

## 丛书编委会

主　任：贾　峰

副主任：何家振　郑立明

委　员：牛玲娟　李原原　李曙东　吴建民　彭　勇
　　　　冯　缨　靳增江

## 丛书出版委员会

主　任：段学俭

副主任：匡志强　张　蓉

成　员：叶　刚　李晓梅　魏　来　徐雅清　田振军
　　　　蔡雩奇　程　洋

特别感谢以下热心人士对译稿润色工作的支持：

姜竹青　韩　笑　贾　芳　刘　晓　张黎立　刘之杰
高　青　周依奇　彭　江　于函玉　于　哲　单　威
姚爱静　刘　洋　高　艳　孙笑非　郑莉霞　周　蕊

# 目录

| | |
|---|---|
| 建造大教堂 | 4 |
| 你知道吗? | 22 |
| 想一想 | 26 |
| 自己动手! | 27 |
| 学科知识 | 28 |
| 情感智慧 | 29 |
| 艺术 | 29 |
| 思维拓展 | 30 |
| 动手能力 | 30 |
| 故事灵感来自 | 31 |

# Contents

| | |
|---|---|
| Building a Cathedral | 4 |
| Did you know? | 22 |
| Think about it | 26 |
| Do it yourself! | 27 |
| Academic Knowledge | 28 |
| Emotional Intelligence | 29 |
| The Arts | 29 |
| Systems: Making the Connections | 30 |
| Capacity to Implement | 30 |
| This fable is inspired by | 31 |

在加拉帕戈斯群岛中的一个小岛上,孤独的乔治正绕着他的花园散步。他感到非常寂寞,这时一只大力甲虫走出了岸边的热带雨林,过来看望他。

大力甲虫问道:"这么说,你就一直在同一座房子里住了一百多年?"

Lonesome George is taking a walk around his garden on one of the Galápagos Islands. He is feeling very lonely until a hercules beetle, from the rainforest on the shore, pays him a visit.

"So, you have been living in the same house for over one hundred years?" the beetle asks.

一直在同一座房子里住了一百多年

Living in the same house for one hundred years

不幸的是，我是唯一的幸存者了

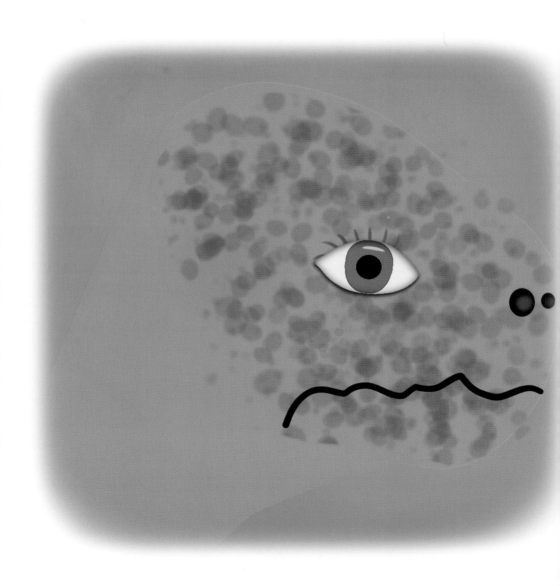

Sadly, I'm the last one left

"哦,是啊,这很了不起吧!"

"在一座房子里住一百年,真是非常少见。几乎没有谁可以活这么久,也几乎没有哪座房子能有这么坚固。"

"的确,要知道,像我这个种类的龟在地球上已经生存了五百万年。但不幸的是,我是唯一的幸存者了。"

"Oh yes, and it has been great."

"To live in a house for a hundred years is quite exceptional. Hardly anyone lives that long, and hardly anyone has a house that strong."

"Well, my kind of tortoise has been around for five millions years, you know. Sadly, I'm the last one left."

"你所有的兄弟姐妹和其他家庭成员都去世了吗?他们是生病了吗?"
"不,导致他们死亡的罪魁祸首不是疾病,而是人类!那些捕鲸者和海豹猎手来这里找吃的。因为我们行动缓慢,他们就把我们当作食物。"

"How did all your brothers and sisters and other family members die? Did they get sick?"
"No, it wasn't diseases that killed them, it was people! Those whale hunters and seal hunters, who came here looking for an easy meal. As we cannot move fast, they had us for dinner."

他们就把我们当作食物

They had us for dinner

……将我们活生生地煮了

... cooking us alive

"碰到这样的人,你背上的这座房子不能保护你吗?"

"瞧啊,我们是非常容易就长到一百公斤重的大家伙。可这并不能阻止一群男人将我们翻转过来并活生生地煮了。我的家人只能任由宰割。"

"Couldn't that house you carry on your back protect you against these people?"

"Look, we can easily weigh up to a hundred kilogram. But that doesn't stop a group of men turning us on our backs and cooking us alive. There was very little my family could do."

"眼睁睁地看着这一切发生肯定是一件非常可怕的事情。"大力甲虫说。

"这还没完。人类还把猫啊老鼠啊带上了岛，这些动物捕杀了我们那些无助的孩子。而现在你们这些甲虫也成了捕猎目标！"

"的确是这样，不过没有人喜欢吃我们。他们只是想把我们做成标本挂在墙上。"

"It must've been terrible to see this happening before your eyes," remarks the beetle.

"That's not all. People also brought animals like cats and rats to the islands. They killed off our helpless little ones. And now you beetles are also in demand!"

"That's true, but no one is keen to eat us. They just want to hang us on their walls."

人类还将猫啊老鼠啊带上了岛

People also brought animals like cats and rats to the islands

我很强壮!

I'm very strong

"做标本？是因为你的外壳吗？"乌龟问。

"是的，我有一个非常特别的壳。而且我很强壮，我可以举起相当于自身体重五百倍的东西。"

"哇，太不可思议了！你曾经为印加人工作过吗？"

"Is it because of your shell?" the tortoise asks.

"Yes, I have a very special shell. And I'm very strong. I can lift things that are five hundred times my own weight."

"Now that's amazing! Did you ever work for the Incas?"

"印加人?你指的是美洲的那个伟大文明吗?没有,为什么这样问?"

"嗯,他们在山上建造了宏伟的建筑,比如秘鲁的马丘比丘古城,还有哥伦比亚的失落之城。我想你这么强壮可能会帮他们一把。"

"不,我们生活在这里的岛屿上,住在茂密的沿海森林中。不过,我曾经到那些石头建筑里转了一圈,它们的建造方式让我很惊讶!"

"The Incas? You mean that great culture of the Americas? No, why?"

"Well, they built grand structures in the mountains, such as at Machu Picchu in Peru and the Lost City in Colombia. I thought with your strength you might have given them a hand."

"No, we live here in the thick coastal forest on the islands. I've paid a quick visit to those stone structures though, and was amazed at the way people had built it."

他们在山上建造了宏伟的建筑

They built grand structures in the mountain

成百上千座美丽而宏伟的建筑

Hundreds of beautiful, grand buildings

"我听说，在大西洋对面的欧洲，人们建造了成百上千座美丽而宏伟的建筑，叫作大教堂。有些大教堂花了一个多世纪才建成。这些建筑肯定也值得一看。"

"是的，当时的欧洲人有着坚定的信念，那些开始建设的人很清楚，直到死去他们也不一定能亲眼看到建筑完工。他们几代人辛勤工作，最终将他们的宏伟梦想变成现实，并请他们那个时代最伟大的艺术家装饰教堂内部。"

"I heard that across the Atlantic Ocean, in Europe, people have built hundreds of beautiful, grand buildings, called cathedrals. Some took over a century to complete. Now those must be worth seeing as well."

"Yes, the people of Europe had so much faith that those who started building knew they would not see the end result in their lifetime. They dedicated generations of work to make their grand dreams come true and used some of the greatest artists of their time to decorate the interior."

"有这种信念一定很棒。我多么希望能去看一看那些建筑,并与家人分享我看到的美景。但这个愿望恐怕不可能实现了。"孤独的乔治叹息道。

"虽然我只是一只甲虫,但如果你愿意的话,我会陪你一起去。" 大力甲虫提议。

……这仅仅是开始!……

"It must be good to have faith like that. I wish I could see one of those buildings, and share the sight with someone from my family. But I'm afraid that's no longer possible," sighs Lonesome George.

"I know I am just a beetle, but I'll come with you if you want," offers Hercules.

... AND IT HAS ONLY JUST BEGUN!...

……这仅仅是开始!……

...AND IT HAS ONLY JUST BEGUN!...

# Did You Know?

## 你知道吗？

The construction of a cathedral took, on average, more than a century. The Catholic Church has more than five thousand bishops for the 1,2 billion followers of the Catholic faith. Bishops have their seats in cathedrals.

建成一座大教堂平均要花一个多世纪的时间。天主教会拥有5000多名主教和12亿左右教徒。主教们在大教堂有自己的座位。

The Cathedral of Chartres (France) took only 25 years to build. The Sagrada Familia (Sacred Family) Cathedral in Barcelona has been under construction for more than a hundred years and has been funded by donations only.

法国的沙特尔大教堂只用了25年就建成了。巴塞罗那的圣家族大教堂已经动工100多年，现在仍在建造当中，资金来源仅仅依靠捐款。

Lonesome George, a male Pinta Island tortoise, was 102 years old when he died. He was the last member of his species and was kept at the Charles Darwin Research Station on the Galápagos Islands (Ecuador). When George died, the Pinta Island tortoise was declared extinct.

孤独的乔治是一只雄性的平塔岛象龟，在102岁的时候死亡。它是该物种的最后一名成员，被安置在厄瓜多尔加拉戈斯群岛的查尔斯·达尔文研究站中。乔治去世后，平塔岛象龟被宣布灭绝。

Giant tortoises came to the islands from the mainland. Aided by their buoyancy, their ability to breathe while swimming by extending their necks above water, and their strength to survive for months without food or fresh water, they were able to survive the 1 000 km long journey.

巨型陆龟是从大陆来到岛屿上的。它们依靠浮力，游水的时候能将脖子伸出水面进行呼吸，再加上在没有食物或淡水时仍能存活数月的本领，成功地长途跋涉1000公里。

The hercules beetle spends a year or more of its life as a larva, eating rotting wood. As an adult, the beetle roams the forest floor in South America and Central America, eating rotting fruit.

大力甲虫的幼虫期长达一年甚至更长，在此期间主要吃腐烂的木头。成年之后，大力甲虫在南美洲和中美洲森林的地面上游荡，靠吃腐烂的水果为生。

The hercules beetle can lift an object 850 times its own weight. If a human had that strength, a man would be able to lift 65 tonnes. The dung beetle is even stronger; it is able to pull 1 100 times its own body weight.

大力甲虫可以举起850倍于自身重量的物体。如果人类有这样的本领，一个人就能举起65吨重的物体。不过，蜣螂（俗称屎壳郎）更为强壮，它能够拖动1100倍于自身体重的物体。

Machu Picchu (Peru), which in the Quechua language means 'old person', was built in the 15th century. It is considered one of the Seven Wonders of the World and was declared a UNESCO World Heritage site in 1983.

马丘比丘（秘鲁）在克丘亚语中是"老年人"的意思，该城建造于15世纪。马丘比丘被认为是世界七大奇迹之一，并在1983年被联合国教科文组织宣布为世界遗产。

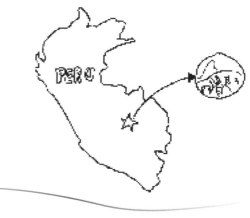

The Lost City of Santa Martha (Colombia) was built 650 years earlier than Machu Picchu. It consists of a series of terraces carved into the mountainside, with a network of tiled roads.

圣玛尔塔失落之城（哥伦比亚）的建立比马丘比丘还要早650年，它由雕凿于山坡上的一系列台阶组成，并由铺设的道路网络相连接。

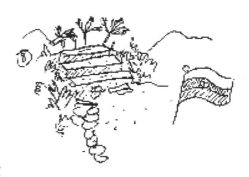

# Think About It
## 想一想

Could you make the decision to dedicate resources to something of which you know you'll never see the end result? And what if not even your children will see the fruits of your vision and dedication?

如果你知道自己永远不会看到某件事情的最终结果,你还会下决心为其投入资源吗?如果连你的孩子也没有机会看到你的远见卓识和奉献精神的成果,那又会怎样?

你是否曾经有机会吃一只大乌龟的肉?它最终幸免于难了吗?

Would you have liked to eat the meat of a big turtle, had they survived?

Do you think that hanging a dead hercules beetle on the wall will make your house look beautiful?

你觉得在墙上挂一个死去的大力甲虫会让房间看上去更漂亮吗?

整个一生都将自己的房子背在背上会是一件很享受的事情吗?或者说,你会认为这是一种挑战?

Is it a luxury to carry your own house on your back for your entire life? Or would you consider it a challenge?

# Do It Yourself!
# 自己动手!

Lost cities like Santa Martha and Machu Picchu were built of rock. How many houses and buildings around your home are built of rock? Is anyone in your area building a new house using rocks and stone? Find out why construction with rocks and stone has lost favour and why people lost interest in using this as a building material. Stone buildings last for hundreds of years, which makes them very economical as the capital costs are spread over a few generations. Now ask around to find out if anyone would be prepared to build a house that will last for generations. What do you think the answer will be?

像圣玛尔塔和马丘比丘这样消失的城市是用岩石建成的。你家的周围有多少房子或建筑是用岩石建造的？你所在的地区有没有人用岩石和石头建造新房子？找出原因，为什么用岩石和石头建造的建筑已经不受欢迎了，人们为什么对石头这种建筑材料失去了兴趣。石头建筑能保存几百年，它们的建设成本可以分摊到几代人身上，因此非常经济。问一问周围的人，是否有人准备盖一座可以延续几代的房子。你认为会得到什么答案？

# TEACHER AND PARENT GUIDE

## 学科知识
### Academic Knowledge

| | |
|---|---|
| 生物学 | 海龟和陆龟都属于龟鳖目；变温动物，或称冷血动物，指动物的体温随外界温度变化而改变；外骨骼是指脊柱和肋骨融合成为一个外壳；乌龟的外壳是由骨骼组成的，最外面的保护层是由角质组成的；大力甲虫是夜行性的，目的是躲避天敌。 |
| 化 学 | 角质在构建乌龟龟甲保护层中的重要性。 |
| 物 理 | 异速生长，有关身体大小和形状、解剖学、生理学和行为之间关系的研究；大力甲虫的外壳在干燥时为绿色，潮湿时会变成黑色，利用薄膜干涉效应增加湿度；大力甲虫的飞行能力并没有受到组成其身体的轻质材料的影响；尖锐摩擦声是指昆虫摩擦其身体部位而形成吱吱的噪声。 |
| 工程学 | 建筑技术的演变决定了教堂不同的建筑风格，包括罗马式、哥特式、新哥特式和巴洛克式风格。 |
| 经济学 | 对教堂的投资往往没有经济回报，而是对艺术、建筑学和建筑的支持。 |
| 伦理学 | 由人类造成的物种灭绝大潮；为某一目标而奋斗的承诺，在你的有生之年可能都无法亲眼看到该目标的实现或享受目标的成果；虽然我们都知道物种的灭绝是由人类的行为造成的，但我们似乎并不准备为此做出必要的改变；很多人认为动物是在为人类服务，我们有权利去杀死甚至灭绝它们。 |
| 历 史 | 大教堂在公元313年首次成为主教座堂；在希腊神话中，赫拉克勒斯是最强壮的凡人，是宙斯最后一个凡人儿子，他曾经花费大量时间为他的愤怒和鲁莽行为进行忏悔；印加文明。 |
| 地 理 | 加拉帕戈斯群岛；马丘比丘位于秘鲁；圣玛尔塔失落之城位于哥伦比亚；大西洋。 |
| 数 学 | 两个量之间的函数关系是指其中一个量的变化会导致另一个量的相应变化。 |
| 生活方式 | 我们目前的生产和消费体系会导致某些物种的灭绝，需要对生活方式做出改变，以避免这种事情的发生；高领衫或高领毛衣是受到乌龟形态的启发。 |
| 社会学 | 种族灭绝是指针对一个民族的毁灭意图和行为或大屠杀，但我们并不将对动物进行大规模杀害而导致其灭绝的行为称为种族灭绝。 |
| 心理学 | 信仰的力量，有了信念的支持，人们就能够坚持不懈，排除万难。 |
| 系统论 | 人类往往不清楚自己的行为会带来的意外后果，如从人类的食物生产方式到非本地物种引进等。 |

# 教师与家长指南

## 情感智慧
### Emotional Intelligence

**孤独的乔治**

当乔治谈到住在自己的房子里很棒时,似乎对生活感到很满意,但他知道他已经是自己种族的最后一个成员,并敏锐地意识到种族即将灭绝的困境。更糟糕的是,乔治亲眼目睹了人类毁灭他的种族的暴行。乔治接受了自己的局限性,承认自己没有办法去保护其他同类。乔治在思考人类给他们的种族带来的混乱状况及其日益增长的复杂性。他很清楚大力甲虫的起源和能力,知道他们也在被人类捕杀,并对大力甲虫表示了同情。乔治思考着对宏伟建筑的建设,感叹着梦想的力量以及愚公移山的献身精神。乔治冷静地接受了这个事实:作为种族的最后一个成员,他将无法再和同类分享任何东西。

**大力甲虫**

甲虫同情乔治,认识到了他的独特性。甲虫急于了解乔治的困境,震惊于坚硬的龟壳居然不能保护那些乌龟免遭猎人的杀害。他很遗憾已经发生的悲剧,更为乔治的现状感到无奈。乔治关于甲虫和古老文明之间可能存在联系的哲学思考,给甲虫留下了深刻印象。了解到那些教堂建造者基于自己的信念,几代人坚持不懈地建造教堂,甲虫表达了深深的敬意。甲虫也对孤独的乔治的感情表示了尊重,并提出可以陪他去参观大教堂。

## 艺术
### The Arts

当我们观看大力甲虫及其近亲独角仙的照片时,我们可以看到这些物种别具一格的形状和轮廓的神奇之处。他们不仅具有有趣的形状,还有两套翅膀,甚至可以改变颜色。这一点可以启发我们重新创造这些神奇生物的形状和色调。找到五六张不同甲虫的图片,仔细研究它们。然后盖住所有的图片,凭借记忆用钢笔或铅笔把它们画出来,重新创造每个甲虫的形状和轮廓。你的作品会比真正的甲虫更吓人吗?

# TEACHER AND PARENT GUIDE

## 思维拓展
### Systems: Making the Connections

> 我们一般希望建筑物能有较长的使用寿命。石头和岩石的建筑物建设周期很长，往往需要几十年甚至上百年，但也会保存几千年时间。在世界各地，人类建造了一些体现他们信仰和文化的建筑物。这些建设大部分都基于一种信念，即拥有这样的建筑对社会是有利的。在这种信念的支持下，财力和人力资源被调动起来，这些长期决策获得了跨时代的支持，最终使得后人可以享受成功的果实。实施可以服务社会的长期项目的能力是我们不应该失去的，譬如建造大教堂这样在几百年后仍可为社会作贡献的工程。与此同时，人们似乎已经失去了引导社会可持续发展的意识。有几个因素造成了众多物种的大规模消失：一是为满足眼前的营养需求而杀害物种（如杀死乌龟做食物）；二是引进非本地物种对当地生物多样性造成严重破坏；此外还有用动物"战利品"装饰房间的欲望，对于这一点，人们看重的是动物生命价值的抽象意义。第一种情况是即时需要，第二种是相当无知的行为，最后一种情况只是为了取悦自我。人类的这种破坏性力量与他们创造文化、建造文物、设置长期目标并历经几代人去实现的独特能力形成了巨大的反差。人们需要重建他们作为地球护卫者的信念，并确保每一个物种都有生存的空间，因为地球上每一个物种的积极影响将在数代人之后呈现出来。

## 动手能力
### Capacity to Implement

> 设想一个你一生可能都无法达到的目标或完成的事情。你现在可以做些什么，以确保未来的三代人为实现你确定的目标而继续努力？

# 教师与家长指南

故事灵感来自

## 玛丽·拉萨特
### Marie Rassart

玛丽·拉萨特 2010 年毕业于那慕尔大学，获得了物理学博士学位。她专门从事光学研究，分析复杂的纳米结构。她曾与让·珀尔·维涅龙（Jean Pol Vigneron）教授共事，维涅龙教授研究过高山花卉用什么方式散射而不是拦截紫外线（这是本童书第 100 个童话《阳光下的雪绒花》的灵感来源）。玛丽学到了如何成功地复制具有相同光学性能的天然多层纳米结构。她是一个年轻的研究人员，愿意通过科学教育和终身学习，利用从大自然中获得的灵感，帮助企业在盈利的同时实现可持续发展。

更多资讯

http://www.sciencedaily.com/releases/2008/03/080311081146.htm

图书在版编目（CIP）数据

建造大教堂：汉英对照 /（比）冈特·鲍利著；
（哥伦）凯瑟琳娜·巴赫绘；姚晨辉译. -- 上海：学林
出版社，2016.6
（冈特生态童书. 第三辑）
ISBN 978-7-5486-1069-4

Ⅰ. ①建… Ⅱ. ①冈… ②凯… ③姚… Ⅲ. ①生态环
境-环境保护-儿科读物-汉、英 Ⅳ. ① X171.1-49

中国版本图书馆 CIP 数据核字（2016）第 122654 号

---

ⓒ 2015 Gunter Pauli
著作权合同登记号 图字 09-2016-309 号

## 冈特生态童书
### 建造大教堂

| | | |
|---|---|---|
| 作　　者 | —— | 冈特·鲍利 |
| 译　　者 | —— | 姚晨辉 |
| 策　　划 | —— | 匡志强 |
| 责任编辑 | —— | 李晓梅 |
| 装帧设计 | —— | 魏　来 |
| 出　　版 | —— | 上海世纪出版股份有限公司 学林出版社 |
| | | 地　址：上海钦州南路81号　电话/传真：021-64515005 |
| | | 网址：www.xuelinpress.com |
| 发　　行 | —— | 上海世纪出版股份有限公司发行中心 |
| | | （上海福建中路193号　网址：www.ewen.co） |
| 印　　刷 | —— | 上海丽佳制版印刷有限公司 |
| 开　　本 | —— | 710×1020　1/16 |
| 印　　张 | —— | 2 |
| 字　　数 | —— | 5万 |
| 版　　次 | —— | 2016年6月第1版 |
| | | 2016年6月第1次印刷 |
| 书　　号 | —— | ISBN 978-7-5486-1069-4/G·404 |
| 定　　价 | —— | 10.00元 |

（如发生印刷、装订质量问题，读者可向工厂调换）

# Work 94

# 神奇的苍蝇
## The Incredible Fly

**Gunter Pauli**

冈特·鲍利 著
凯瑟琳娜·巴赫 绘
姚晨辉 译

学林出版社
www.xuelinpress.com

## 丛书编委会

主　　任：贾　峰
副 主 任：何家振　郑立明
委　　员：牛玲娟　李原原　李曙东　吴建民　彭　勇
　　　　　冯　缨　靳增江

## 丛书出版委员会

主　　任：段学俭
副 主 任：匡志强　张　蓉
成　　员：叶　刚　李晓梅　魏　来　徐雅清　田振军
　　　　　蔡雯奇　程　洋

特别感谢以下热心人士对译稿润色工作的支持：

姜竹青　韩　笑　贾　芳　刘　晓　张黎立　刘之杰
高　青　周依奇　彭　江　于函玉　于　哲　单　威
姚爱静　刘　洋　高　艳　孙笑非　郑莉霞　周　蕊

# 目录

| | |
|---|---|
| 神奇的苍蝇 | 4 |
| 你知道吗？ | 22 |
| 想一想 | 26 |
| 自己动手！ | 27 |
| 学科知识 | 28 |
| 情感智慧 | 29 |
| 艺术 | 29 |
| 思维拓展 | 30 |
| 动手能力 | 30 |
| 故事灵感来自 | 31 |

# Contents

| | |
|---|---|
| The Incredible Fly | 4 |
| Did you know? | 22 |
| Think about it | 26 |
| Do it yourself! | 27 |
| Academic Knowledge | 28 |
| Emotional Intelligence | 29 |
| The Arts | 29 |
| Systems: Making the Connections | 30 |
| Capacity to Implement | 30 |
| This fable is inspired by | 31 |

一只孤孤单单的老鼠正坐在他的小笼子里,他整天都在一个实验室中工作,人们在这里试图证明他们的新产品是安全的。

一只苍蝇坐到了老鼠旁边的箱子上,问道:"为什么他们还在用你做实验?"

A lonely mouse is sitting in his little cage. He works in a laboratory all day long, where people are trying to prove that their new products are safe.

A fly sits down on a box next to the mouse and asks, "Why are they still using you for tests?"

为什么他们还在用你做实验?

Why are they still using you for tests?

我必须要搞清楚这些产品是否安全

I have to figure out if these products are safe

"哦,你要知道,女人们会抹口红,孩子们要吃药,男人们也会用肥皂洗手,而在这之前我必须要搞清楚这些产品是否安全。"

"太可笑了!你没有必要做这些事情,你懂的。"

"Oh, you know, before women can apply lipstick, or children can take medicine, or men wash their hands with soap, I have to figure out if these products are safe."

"How ridiculous! There is no need for you to do that, you know."

"我是懂,"老鼠感叹道,"但许多国家的法律规定,所有这些化学品必须要经过测试,尽管每个人都知道,这是在浪费时间和金钱……也是在浪费我的生命!"

"你知道吗?我说不定是世界上被研究最多的动物之一呢。"苍蝇问道。

"I know," sighs the mouse, "but in many countries the law states that all these chemicals have to be tested, even though everyone knows that it's a waste of time and money ... and a waste of my life!"

"Do you know that I am probably one of the most studied animals in the world?" asks the fly.

世界上被研究最多的动物之一

One of the most studied animals in the world

参照我飞行的方式制造机器人

Make robots inspired by the way I fly

老鼠哈哈大笑。"你开玩笑吧!"他说,"人类能向你学习什么呀?人类不能飞行,他们没有巨大的眼睛,他们使用肌肉的方式也和你完全不同。"

"你知道的,我可以用我的翅膀尝味道。而且我的整个身体都覆盖着传感器。现在人类正想参照我飞行的方式制造机器人呢。"

The mouse bursts out laughing. "You've got to be kidding!" he says. "What could humans ever want to learn from you? People don't fly, they don't have huge, big eyes, and the way they use their muscles is totally different from the way you use yours."

"I can taste with my wings, you know. And my entire body is covered in sensors. People now want to make robots inspired by the way I fly."

"呵呵，他们可能是想弄清楚你这么小的大脑为何能这么厉害。"老鼠打趣说。

"这可不能混为一谈。不过你说得没错，我的大脑的确非常小。不同的是，我可以将它全部利用起来。"

"我曾听到人类声称大黄蜂不能飞，尽管只要他们不嫌麻烦去看一下，他们就会亲眼看到大黄蜂是可以飞的！"

"Ha-ha, they probably want to figure out how you can do so much with such a small brain," kids the mouse.

"That's not a kind thing to say. But it is true all the same; I do have a very small brain. The difference is, I use all of it."

"I have heard people claim that bumblebees can't fly even though, if they took the trouble to look, they will see with their own eyes that they can!"

不同的是，我可以将它全部利用起来

The difference is, I use all of it

人类可能有硕大的大脑,但没有充分利用

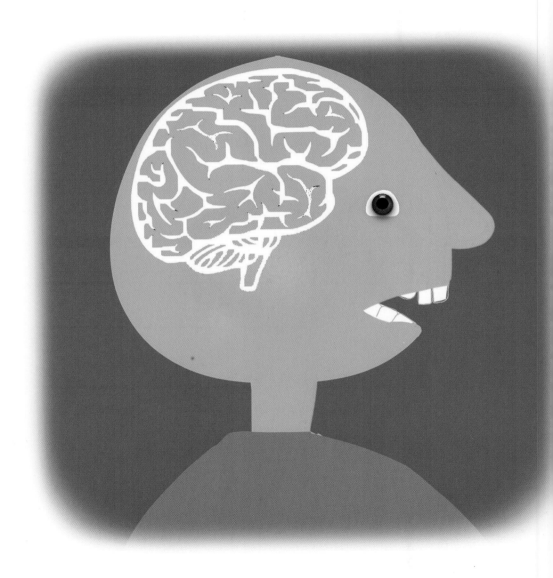

People may have big brains, but don't use them

"人类可能有硕大的大脑,但我认为他们没有充分利用。"

"嗯,他们肯定还没有学会飞行,或做出什么能飞的东西,就像你做的那样。"

"要想飞,你必须要了解风从哪儿来,你还需要有不同的肌肉,为你提供力量或者改变方向。你的大脑必须能够迅速地协调所有这一切,否则你就会一头摔下来,或者更糟——直接飞进捕食者的嘴巴!"

"People may have big brains, but I do not think they use them."

"Well, they have certainly not yet learned to fly, or make anything fly, as well as you do."

"To be able to fly, you have to know where the wind is coming from. You also need certain muscles to power you, and other muscles to change direction. Your brain must be able to quickly coordinate all of this or you will fall flat on your face, or worse – fly straight into the mouth of a predator!"

"听起来所有这些对我来说都太难了。"老鼠叹息道。

"我们有些种类的苍蝇可以连续飞10公里,中间不需要进食补充能量。我们甚至可以在空中读取罗盘,辨别方向。"

"All that sounds difficult to me," sighs the mouse.

"Some of us can fly for up to 10 kilometres without any food for fuel. We can even read a compass and tell direction while we're up in the air."

可以读取罗盘,辨别方向

Can read a compass and tell direction

除了南极洲以外，我们可以在任何地方生存

We have learned to live everywhere except Antarct

"太神奇了！我还以为只有帝王蝶能够做到这一点。"

"永远不要低估一只苍蝇。我们的种类很多，虽然我们身材很小，但除了南极洲以外，我们可以在任何地方生存。"

"你的生活真是丰富多彩，而我和兄弟姐妹们却不得不被关在这儿的笼子里，测试那些化妆品。看这儿，试试这支红色的口红吧。"

"Amazing! I thought only monarch butterflies are able to do that."

"Never underestimate a fly. There are so many of us and even though we are tiny, we have learned to live everywhere in the world except Antarctica."

"You lead such an interesting life, while my brothers and sisters and I have to sit here in a cage, testing cosmetics. Here, try on some of this red lipstick."

"口红？不用了，谢谢！我喜欢自己一身黑的样子。"

"好吧，我喜欢白色。我很高兴自己是一个白化病患者，我希望我的孩子将来也是白色的，但不要红嘴唇！"

……这仅仅是开始！……

"Lipstick? No thanks! I like being black all over."

"Yes, and I like being white. I am happy to be an albino, and I do hope that my children will one day be white too, without the red lips!"

... AND IT HAS ONLY JUST BEGUN!...

……这仅仅是开始！……

...AND IT HAS ONLY JUST BEGUN!...

# Did You Know?

## 你知道吗？

More than a 100 million mice and rats are killed in American laboratories because of testing. While the use of dogs and cats in experiments is strictly controlled, the use of mice in laboratories is largely unregulated.

因为做实验，超过一亿只小鼠和大鼠在美国的实验室中丧生。虽然对实验使用的狗和猫有着严格控制，但对于在实验室中使用老鼠在很大程度上是缺乏监管的。

House mice were the prime reason for taming domestic cats. Nowadays, mice are also kept as pets. The first record of a pet mouse is in the Erya, the oldest surviving Chinese encyclopaedia, which dates back to the 3rd century BC.

对付家鼠是人类驯服家猫的首要原因。如今，小鼠也被作为宠物饲养。有关宠物鼠最早的记录出现在《尔雅》中，那是现存最古老的中文百科全书，其历史可以追溯到公元前3世纪。

𝒜lbino animals have a lack of colour pigment in their skin, hair, and eyes. This condition also affects people. There are also albino plants that have lost chlorophyll pigmentation and have white flowers.

白化动物的皮肤、头发和眼睛中缺乏色素。这种情况也发生在人类中间。同样也有白化植物，它们因为缺乏叶绿素沉着而开着白色的花朵。

𝒜 mouse has a 1 000 times more neurons than a fly. The fly, however, seems to be able to process information faster than a mouse.

老鼠的神经元比苍蝇要多1000倍，然而，苍蝇处理信息的速度似乎比老鼠更快。

Flies beat their wings 220 times per second. While in flight, the fly's brain, consisting of a 100 000 neurons, can process information in a split second and enable it to change course swiftly.

苍蝇每秒钟扑打220次翅膀。在飞行过程中,苍蝇的大脑(包含有100 000个神经元)可以在一瞬间对信息进行处理,并迅速改变飞行方向。

Flies do not flap their wings, but rather create a vortex (small tornado-like air movement) to help generate lift. An internal, biological gyroscope guides their flight.

苍蝇并不扇动翅膀,而是创造一个旋涡(类似空气运动的小型龙卷风)来帮助产生升力。苍蝇通过体内的生物陀螺仪来引导它们飞行。

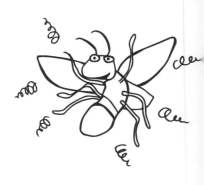

Flies' wings have sensors on them and therefore they can taste with their wings.

苍蝇的翅膀上分布着传感器，因此，它们可以用自己的翅膀品尝味道。

Flies' eyes are the fastest visual system on the planet.

苍蝇的眼睛是这个星球上速度最快的视觉系统。

# Think About It
## 想一想

Would you like to live in a cage like a mouse, even if you know you are helping people? What other ways of testing the toxicity of products can you suggest people use?

你喜欢像老鼠一样住在笼子里吗,即使你知道自己是在帮助人类?你还可以建议人们使用什么其他方法测试产品毒性呢?

你觉得苍蝇聪明吗?

Do you think a fly is smart?

Does the size of a brain matter? Or does it matter more how the brain is used?

大脑的体积大小重要吗?还是大脑的使用方法更为重要?

你能想象如果你是一名白化病患者,自己的生活会有什么不同吗?

Can you imagine how your life would be different if you were an albino?

Let's create a beta movement, an optical illusion where a series of static images on a screen creates the illusion of a smoothly flowing scene.
To find out more, have a look at:
https://en.wikipedia.org/wiki/Beta_movement
Static images do not physically change but give the appearance of motion because they are rapidly being turned on and off, faster than the eye can follow. This optical illusion is caused by the human eye's inability to process information that is displayed faster than 10 frames per second. Flies are able to process this much better than we can. Now take a look at some LED lights. Even though the lights are individually controlled, our eyes and brain perceive them as a continuous movement of light.

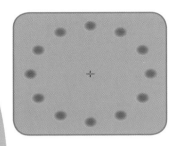

让我们来创造一个β似动现象，这是一种光学错觉，即屏幕上的一系列静态图像产生一个平滑运动场景的错觉。要了解更多信息，可以参阅：https://en.wikipedia.org/wiki/Beta_movement。静态图像没有发生物理变化，但给人一种运动的感觉，因为它们被迅速地接通和断开，其速度比眼睛注视的速度更快。这种光学错觉是由于人眼无法处理显示速度大于每秒10帧的信息而引起的。苍蝇在处理这种信息方面比我们人类强得多。现在观察一些LED灯。即使这些灯被单独控制，在我们的眼睛和大脑的感知中，它们也是光的连续运动。

# TEACHER AND PARENT GUIDE

## 学科知识
### Academic Knowledge

| | |
|---|---|
| 生物学 | 脑细胞在执行不同功能时具有不同的行为；神经元是传输电子和化学信号的细胞；毒性是物质可导致生物体损伤的程度；生物毒素包括细菌和病毒；致癌物和诱变剂；被研究最多的苍蝇是果蝇；帝王蝶及其方向感；老鼠被用作一种方便的实验室研究模型；苍蝇被认为是一个具有复杂和丰富多彩生活的多才多艺的物种，擅长使用其大脑。 |
| 化 学 | 化学毒物包括铅、汞、氢氟酸和甲醇等。 |
| 物 理 | 煤尘、石棉纤维和二氧化硅等物理毒物会干扰肺的生物学过程；β似动会产生光学错觉的效果；似动现象：光脉冲引起的似动。 |
| 工程学 | 复用：如何利用现有的东西做更多事情；通过传感器和生物学指标测量毒性；基于电生理学的工程学：了解苍蝇的视觉航向控制，设计自主飞行机器人（无人机）；四轴飞行器拥有四个旋翼，而不是像直升机那样只有一个；赖卡特探测器（发明于20世纪50年代）是检测运动的简单电路，模仿的是在苍蝇眼睛中发现的神经回路。 |
| 经济学 | 产品需要测试其毒性以获得销售许可证；在广告中使用β似动效果来吸引人们的注意力，诱使他们购买产品；拍摄影视动画时使用β似动技术。 |
| 伦理学 | 在消费品和药品测试中使用动物；利用视觉效果，使大脑认为它看到了一些其实并不存在的东西；展示观察者并没有清醒地意识到但确实存在于大脑中的信息，这种行为在大多数国家是被禁止的。 |
| 历 史 | 出生于瑞士的德国人帕拉塞尔苏斯（原名菲利普·冯·霍恩海姆）在16世纪创立了毒理学学科。 |
| 地 理 | 使用指南针寻找方向，告诉我们朝哪个方向走。 |
| 数 学 | 我们自身的运动知觉是基于视觉、前庭和本体感受的输入，推断运动速度和方向的过程而移动物体的运动知觉是在给定的某些视觉输入情况下，推断一个视觉场景中物体的速度和方向的过程。 |
| 生活方式 | 我们已经习惯了使用那些对我们的生活和健康有未知影响的化学品，并准备为此承担风险；我们每天都会受到至少3000条潜意识信息的轰炸，吸引我们的注意力，给我们造成压力，并劝说我们消费产品。 |
| 社会学 | 有毒物质用量：少量的中毒剂量可能是无害的，甚至对健康有益，而大量食用某种无害的物质（如水）也可能会中毒；我们倾向于认为人类具有优越性，只有一个极小大脑的苍蝇是微不足道的，即使它能更有效地使用它的大脑。 |
| 心理学 | 知觉作为诠释周围环境的能力，导致了格式塔心理学的发展；因为一个人的外貌而自豪；追求完美有其优点和局限性。 |
| 系统论 | 我们倾向于关注一个具有两个参数（原因和结果）的函数，并将现实简化为我们自认为理解的样子，然而现实比这复杂得多，认知更多的是模式识别的结果，而不是细节的集合。 |

# 教师与家长指南

## 情感智慧
## Emotional Intelligence

### 老 鼠

老鼠似乎接受了他的困境。起初，他表露了一点自负的感觉，因为他正在承担一项重要任务——所谓的为人类服务。但他很快就否定了自己扮演的角色，认为他的工作就是在浪费时间和金钱。更糟糕的是，老鼠感觉他是在浪费自己的生命。不过，他仍保持着幽默感。他取笑苍蝇，特别强调他微小的大脑。苍蝇的回答让老鼠质疑了研究苍蝇的实用性。通过对话，老鼠渐渐认识到了自己的局限性，同时表示出对苍蝇更多的尊重。当苍蝇即将离开时，老鼠用自嘲的语气说他会送给苍蝇一些口红。老鼠坦然接受了他作为白化病患者所具有的独特之处，并希望他的孩子能有一个更美好的未来。

### 苍 蝇

苍蝇对老鼠的困境非常关注。她挺身而出维护自己的权利，并认为老鼠没有受到很好的对待。苍蝇自信满满，并有着很强的自我意识。老鼠的取笑并没有减少她的自豪感，因为她的独特能力也获得了人类的关注，包括神经系统科学家和无人机设计师等。苍蝇花时间解释了她的特性，甚至当这让老鼠感到苦恼的时候，她仍继续在谈论自己，并在结束谈话时再一次确认了她对自己身份的自豪。

## 艺术
## The Arts

让我们观赏一部电影。《蝇王》（THE LORD OF THE FLIES）是一部基于英国作家威廉·戈尔丁（WILLIAM GOLDING）的同名小说改编的电影，戈尔丁因为这篇小说获得了1983年的诺贝尔奖。小说讲述了被困在一个无人居住的小岛上的一群男生，试图管理自己，却带来了灾难性后果的故事。和朋友们或一个成年人一起观看电影，并讨论你的感受和见解。

# TEACHER AND PARENT GUIDE

## 思维拓展
### Systems: Making the Connections

大脑有许多独特的功能，不同物种的脑容量各不相同。人们对大脑在过去400万年的发展（包括其大小）进行了详细的研究；老鼠大脑内的神经元比苍蝇多1000多倍。然而这并不意味着老鼠一定胜过了苍蝇。相反，苍蝇似乎能够更好地利用它的大脑，它对其有限的神经元的有效部署，启发人们改进无人机的设计方式和视觉感知。苍蝇似乎能够非常快地处理信息。它通过眼睛、翅膀上的传感器，脸上的触须捕获信息。它甚至还有一个内置的陀螺仪，来检测方向和确定方位。处理完所有这些信息后，苍蝇以闪电般的速度做出决策和采取行动。而人类的做法是以一种相当线性的方式部署大量的神经元，在处理所有的数据时进行详细的因果分析。苍蝇依托广泛关联的网络或神经元模式，似乎能够更系统地对此进行处理。这意味着，即使神经元的数量有限（如苍蝇仅有100 000个神经元），数学上的模式数量也是无限的。当一只苍蝇逃脱捕食者的抓捕时，其行为不是一个简单的反射，而是一种基于规划、模式识别、快速的数据处理以及迅速纠正措施而采取的行动。科学家们受到苍蝇能力的启发，认识到它为复杂系统的设计提供了大量的机会。苍蝇独特的飞行行为包含了物理和生物科学、流体动力学、复合材料结构以及复杂的非线性数学。对这方面的研究需要一个独特的、复杂的、跨学科的方法，并需要整合生物学、工程学、物理学和数学。苍蝇可能并不被视作像老鼠那样方便的实验室动物，但至少也是有趣的、复杂的昆虫，具有惊人的神经功能。

## 动手能力
### Capacity to Implement

尝试用相机拍摄一张苍蝇的照片，这可能会是一个很大的挑战！拍一张苍蝇的头部特写。现在去当地的宠物商店，拍摄老鼠的照片，同样也拍一张漂亮的头部特写（或者在互联网上寻找苍蝇和老鼠头部的照片）。将这些照片和你父亲或母亲的照片放在一起。现在想象一下，你必须要指出这三者谁的脑子最快。现在你准备好对三者的大脑进行比较并讨论哪一个更优秀了吗？

# 教师与家长指南

故事灵感来自

## 迈克尔·迪金森
### Michael Dickinson

迈克尔·迪金森于 1989 年在华盛顿大学动物学系获得博士学位。上学期间，迈克尔通过在业余时间做厨师挣学费完成了大学学业，当时他每天早上要做 200~300 顿饭。在此期间，他开始意识到一个事实，即食物的外观和气味同味道一样重要，从而产生了对感觉的兴趣。在大学里，他的研究最初集中在苍蝇翅膀上感觉细胞的生理学方面。这激发了他对苍蝇的空气动力学和控制飞行所需的电路的兴趣。《科学家》杂志（The Scientist: Magazine of Life Sciences）称迈克尔"苍蝇小子"，因为他将动物学、神经科学和流体力学联系在一起。迈克尔目前在华盛顿大学领导着迪金森实验室，并喜欢用尤克里里琴弹奏爵士乐。

更多资讯

http://www.the-scientist.com/?articles.view/articleNo/32422/title/Fly-Guy/

图书在版编目（CIP）数据

神奇的苍蝇：汉英对照 /（比）冈特·鲍利著；
（哥伦）凯瑟琳娜·巴赫绘；姚晨辉译 . -- 上海：学林
出版社，2016.6
（冈特生态童书 . 第三辑）
ISBN 978-7-5486-1072-4

Ⅰ . ①神… Ⅱ . ①冈… ②凯… ③姚… Ⅲ . ①生态环
境 – 环境保护 – 儿童读物 – 汉、英 Ⅳ . ① X171.1-49

中国版本图书馆 CIP 数据核字 (2016) 第 122656 号

© 2015 Gunter Pauli
著作权合同登记号 图字 09-2016-309 号

---

冈特生态童书

**神奇的苍蝇**

| | |
|---|---|
| 作　　　者—— | 冈特·鲍利 |
| 译　　　者—— | 姚晨辉 |
| 策　　　划—— | 匡志强 |
| 责任编辑—— | 李晓梅 |
| 装帧设计—— | 魏　来 |
| 出　　　版—— | 上海世纪出版股份有限公司 学林出版社 |
| 地　　　址： | 上海钦州南路 81 号　电话 / 传真：021-64515005 |
| 网　　　址： | www.xuelinpress.com |
| 发　　　行—— | 上海世纪出版股份有限公司发行中心 |
| | （上海福建中路 193 号 网址：www.ewen.co） |
| 印　　　刷—— | 上海丽佳制版印刷有限公司 |
| 开　　　本—— | 710×1020　1/16 |
| 印　　　张—— | 2 |
| 字　　　数—— | 5 万 |
| 版　　　次—— | 2016 年 6 月第 1 版 |
| | 2016 年 6 月第 1 次印刷 |
| 书　　　号—— | ISBN 978-7-5486-1072-4/G · 407 |
| 定　　　价—— | 10.00 元 |

（如发生印刷、装订质量问题，读者可向工厂调换）

# Work 84

# 像蜘蛛一样纺纱

## Spinning like a Spider

### Gunter Pauli

冈特·鲍利 著

凯瑟琳娜·巴赫 绘
姚晨辉 译

## 丛书编委会

主　任：贾　峰
副主任：何家振　郑立明
委　员：牛玲娟　李原原　李曙东　吴建民　彭　勇
　　　　冯　缨　靳增江

## 丛书出版委员会

主　任：段学俭
副主任：匡志强　张　蓉
成　员：叶　刚　李晓梅　魏　来　徐雅清　田振军
　　　　蔡雯奇　程　洋

特别感谢以下热心人士对译稿润色工作的支持：

姜竹青　韩　笑　贾　芳　刘　晓　张黎立　刘之杰
高　青　周依奇　彭　江　于函玉　于　哲　单　威
姚爱静　刘　洋　高　艳　孙笑非　郑莉霞　周　蕊

# 目录

| | |
|---|---|
| 像蜘蛛一样纺纱 | 4 |
| 你知道吗？ | 22 |
| 想一想 | 26 |
| 自己动手！ | 27 |
| 学科知识 | 28 |
| 情感智慧 | 29 |
| 艺术 | 29 |
| 思维拓展 | 30 |
| 动手能力 | 30 |
| 故事灵感来自 | 31 |

# Contents

| | |
|---|---|
| Spinning like a Spider | 4 |
| Did you know? | 22 |
| Think about it | 26 |
| Do it yourself! | 27 |
| Academic Knowledge | 28 |
| Emotional Intelligence | 29 |
| The Arts | 29 |
| Systems: Making the Connections | 30 |
| Capacity to Implement | 30 |
| This fable is inspired by | 31 |

一只蜘蛛正在忙着织一张巨大的网,这张网至少有一平方米大小,而且肯定是这附近最大的一张网。

一只蝙蝠绕着网飞来飞去,评头论足道:"这不仅仅是一张网,它更是一件艺术品!"

"谢谢你的夸奖。"黄金圆蛛回应道,"这确实是一件独特的作品,需要用技巧和耐心来构建。在我开始织网之前,我就对这个位置进行了研究,并应用了数学原理。"

A spider is busy spinning a huge web. It is at least a metre by a metre in size, and must be one of the biggest in the neighbourhood.

A bat swings around and remarks, "This is not just a web; it is a piece of art!"

"Thank you for the compliment," responds the golden orb spider. "It is indeed a unique piece that requires skill and patience to build. Before I started spinning, I studied this site and applied the maths."

这不仅仅是一张网,它更是一件艺术品!

This is not just a web; it is a piece of art!

即使是蛇也会被我的网缠住

Even snakes get entangled in my web

"你的网太大了，不会只捕捉到昆虫。你是否偶尔也用它抓到过鸟呢？"

"即使是蛇也会被我的网缠住。有一次，一条小蛇被我的网困住，最后把我的网弄破了。当然，我不敢吃她，只是把网修补好了。"

"既然你吐的丝那么强韧，人类难道不会对你感兴趣吗？"

"Your web is too big to only catch insects. Don't you catch birds in it once in a while as well?"

"Even snakes get entangled in my web. Once, a young one ended up in my web and destroyed it. Of course I did not dare to eat her; I only recovered my silk."

"Seeing that your silk is so strong, have people not become interested in you?"

"当然会！但我们蜘蛛不喜欢规则和约束，因此很难被驯化。"
"那人类有没有试图收取你的网？"
"哦，有的，一些人试图用我的丝做衣服，他们甚至尝试用我的丝织成手套、降落伞、渔网和绳索。"

"You bet! We spiders are quite unruly and undisciplined however, so it is difficult to domesticate us."
"Have people tried to harvest your webs yet?"
"Oh yes, some tried to make clothing out of my silk and they even tried to turn my fibres into gloves, parachutes, fishing nets, and ropes."

我们蜘蛛不喜欢规则和约束

We spiders are quite unruly and undisciplined

……我们的丝在亚洲被当作绷带

...our silk has been used in Asia as bandages

"我听说你甚至可以入药。"

"确实是这样。多年来,我们的丝在亚洲被当作绷带,用以止血。现在,英国的一些聪明人正在研究如何使用我们的丝培育神经。"

"培育神经?"

"I heard that you're even getting into medicine."

"Yes, indeed. For years our silk has been used in Asia as bandages to stop bleeding. But now some smart people in England figured out how to use our silk to grow nerves."

"Grow nerves?"

"是啊！还在培育肌肉和骨骼之间的软骨！"

"培育神经和组织……这听起来几乎太不可思议了！"

"听起来是这样，但我的丝确实有助于细胞生长，这是一个可以分享的美好礼物。"

"Yes! And to grow cartilage between muscles and bones!"

"Grow nerves and tissue ... that sounds almost too magical."

"It is, but my silk really does help cells to grow. It is a wonderful gift to be able to share."

……培育肌肉和骨骼之间的软骨!

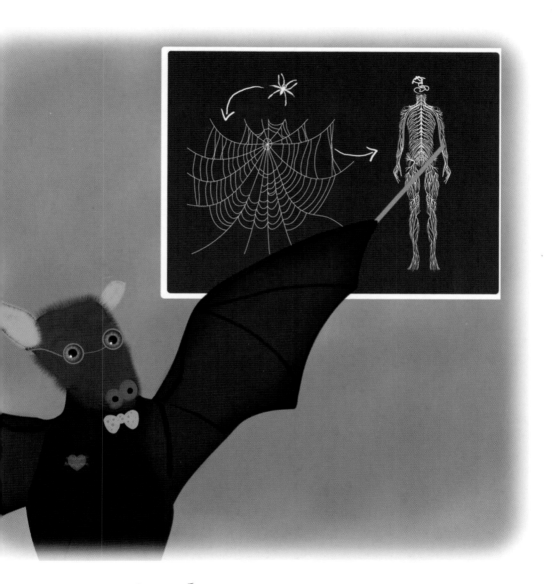

...grow cartilage between muscles and bones!

……用你的丝供应全世界?

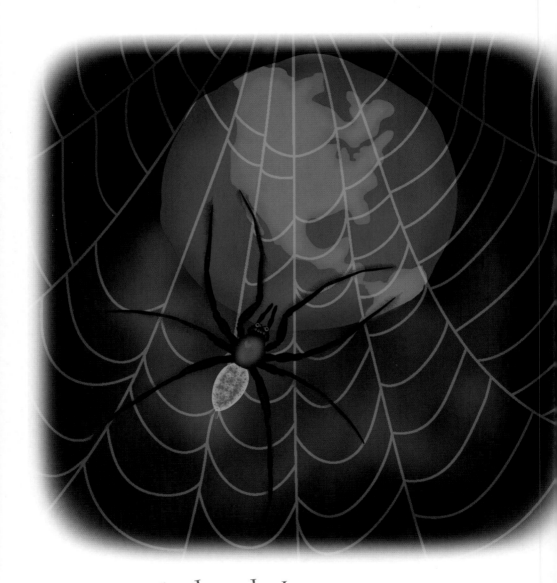

...to supply the whole world with your silk?

"这会让你非常受欢迎的,不过你如此精美的丝能够供应全世界吗?"

"想要满足全世界对丝的需求,黄金圆蛛的数量永远也不够多。"

"所以说这只不过是一个画饼充饥的梦想?"蝙蝠问道,"你最终会让太多的人失望的。"

"This will make you very popular, but will you ever be able to supply the whole world with your wonderful silk?"

"There will never be enough golden orb spiders to make all the silk the world needs."

"So is this not just a big pie-in-the-sky idea then?" asks the bat. "You will end up disappointing so many people."

"你为什么会这么说？我已经与蚕达成了协议，让他们为我的项目提供帮助。"

"他们是食草动物，你是食肉动物。你确定你们的丝是一样的吗？"

"Why do you say that? I have actually made a deal with the silkworms to help me with my project."

"They are herbivores; you are a carnivore. Are you sure you make the same silk?"

……与蚕达成了协议

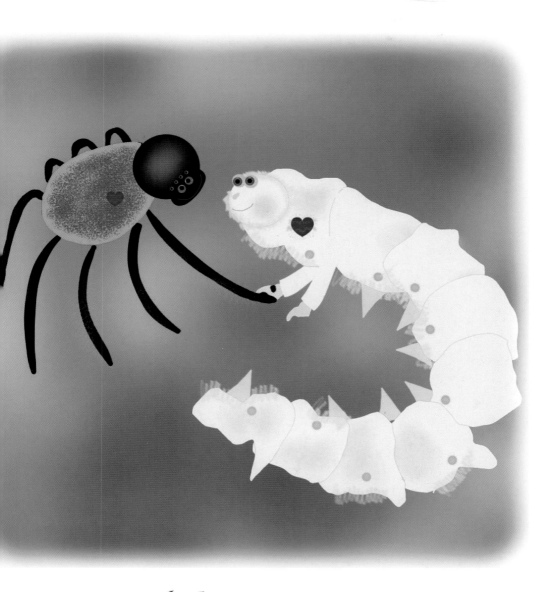

...made a deal with the silkworms

……我们可以将其转换为一种更强韧的材料

...we could convert it into a stronger material

"不，虽然我们都吐丝，但性质并不相同。然而，如果我们给柔软的丝增加压力和水分，我们可以将其转换为一种更强韧的材料。"

"有多强韧？"

"这么说吧，也许强韧到有一天你可以利用我的丝再造人体的组成部分。"

"Yes, we both make silk, but not the same kind. If we, however, add pressure and moisture to their soft silk, we could convert it into a stronger material."

"How strong?"

"Well, maybe so strong that one day you could regenerate parts of the human body from my silk."

"你的丝真的很特别!"

"没错,但是要想一个新的点子获得成功,最好能够改变游戏规则。"

"所以你没有改变和破坏任何规则,而是利用科学的奇迹造了新的规则?"

"是的,我想用这种方法让这个世界变得更美好。"

……这仅仅是开始!……

"You silk is very special indeed!"

"That's true, but if you want to be successful with a new idea, you'd better change the rules of the game."

"So you are not bending and breaking any rules, you are making new ones by using the marvels of science?"

"Exactly. And that is how I intend to make this a better world."

... AND IT HAS ONLY JUST BEGUN!...

……这仅仅是开始！……

...AND IT HAS ONLY JUST BEGUN!...

# Did You Know?

# 你知道吗？

The web of the golden orb spider could be called a "web of steel". Measuring up to one-by-one metre, it is not only the strongest and largest kind of web, but can last for several days. The silk boasts a golden colour.

黄金圆蛛的网可以被称为"钢丝网"。根据对一平方米蛛网的测量，它不仅是一种最强韧和面积最大的网，而且可以保持数天之久。黄金圆蛛的丝呈现出一种耀眼的金色。

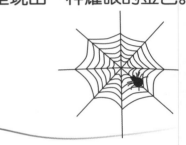

The web of the golden orb spider can trap small birds, which the spider does not eat. To avoid damage, the spider leaves silk curtains on its web, which act like a safety strip across a glass door and builds strong barrage threads around the main web.

黄金圆蛛用网可以捕获它并不吃的小型鸟类。为避免蛛网被破坏，蜘蛛在网上放置了丝帘，其作用类似于玻璃门上的安全条，围绕主网构建强韧的拦阻线。

The male golden orb spider is about 1 000 times smaller than the female. He lives in the web, and steals her food.

雄性黄金圆蛛的大小约是雌性黄金圆蛛的1000分之一，它住在网上，窃取雌性黄金圆蛛的食物。

Spider silk is much tougher than the synthetic Kevlar, which requires harsh chemicals for its production. Spider silk is made from protein and can be recycled and reconditioned indefinitely on site by the spider.

蜘蛛丝比合成芳纶坚韧得多，合成芳纶的生产需要刺激性的化学物质。蜘蛛丝是由蛋白质制成的，可以回收，而且蜘蛛能够在现场对蛛网进行随意翻新。

𝒯he Madagascar golden orb spider has very poor vision. Instead of using sight, it uses its very sensitive sense of touch to feel vibrations on the web and track down its prey.

马达加斯加黄金圆蛛的视力很差，因此它不是直接用眼睛看，而是利用非常灵敏的触觉感受蛛网的振动并追踪猎物。

𝒮ome spiders attract their prey with the odour of dead, rotting organic matter blended into their webs. Some spiders also use ant-repelling chemicals in their silk. A spider either eats its meal right away or wraps up its prey in silk and stores it for a time when food is scarce.

一些蜘蛛将死去的、腐烂的有机物融入网中，利用它们散发的气味吸引猎物。一些蜘蛛也在它们的网中使用一些引诱性的化学物质。捕获猎物后，蜘蛛要么立即将猎物吃掉，要么用蛛丝将猎物包裹并储存起来，等食物短缺时再进食。

蜘蛛织网用的建筑材料是从自己的身体中制造出来的。试想一下，如果我们可以利用自己的细胞建造住宅又会怎样！蜘蛛会吃掉全部或部分的旧网，再用可回收翻新的材料来编织新网或蛛网的一部分。

Spiders produce the building materials for their webs from their own bodies. Imagine if we could construct dwellings from our own cells! Spiders consume all or parts of their old webs and then weave new webs or parts of their webs with recycled and reconditioned material.

蛛丝，或者与其类似的蚕丝，是具有生物相容性的，可被用作软骨的替代物，促进细胞增殖，并可作为神经再生的引导材料。

Spider silk, or its analogue produced from silkworms' silk, is biocompatible, can be used as a substitute for cartilage, promote cell proliferation, and act as guiding material for nerve regrowth.

# Think About It
## 想一想

Would you farm with spiders to produce silk? Or do you have a better idea?

你会养殖蜘蛛生产蛛丝吗？或者你还有更好的想法吗？

你能想象用蛛丝来移植头发吗？如果用蛛丝做原料，如何来移植头发？

Can you imagine removing hair with silk? If silk is the raw material, how will hairs be removed?

Are you planning to improve yourself with existing rules, or will you become the best you can be by making new rules?

你是否打算在现有规则下提升自己，还是说通过制定新的规则，你就能做到最好？

你认为那些不可能实现的梦想，最终会让起初欣赏这一想法的人失望吗？

Do you think dreams that are impossible to achieve will ultimately disappoint the people who loved an idea at the outset?

Go outside and look for a spider in the garden. It will not be hard to find. Do not touch the spider with your bare hands unless your guardian is 100% sure it is not venomous and will not bite you. Try to find its web and look at the construction of the web. Find the outer corners, look where the sun is, figure out where the wind comes from, and check both sides (front and back) of the web. Assess the logic the spider must have used to build the web in that specific place. Does it make sense, and why?

走到室外，去花园中寻找一只蜘蛛。这并不会很难。但是不要直接用手触碰蜘蛛，除非你的父母100%肯定它是无毒的，而且不会咬你。试着找到它的网，观察一下网的构造。找到转角的位置，看看太阳在什么地方，确定一下风向，并检查蛛网的两边（前面和后面）。推测一下蜘蛛在这个特定位置织网肯定用到的某种逻辑，这是否有意义？为什么？

# TEACHER AND PARENT GUIDE

## 学科知识
### Academic Knowledge

| | |
|---|---|
| 生物学 | 蜘蛛是呼吸空气的节肢动物，拥有40 000多个不同品种；蜘蛛是食肉动物；在某些种类中，成千上万的蜘蛛会共用蛛网；蜘蛛一般只有不到一年的寿命，那些长寿的蜘蛛会用丝带编织越冬的巢，帮助它们抵御严冬；蛛丝具有避免细菌和真菌感染的作用；软骨是依附在骨头上的柔韧结缔组织，没有骨头那么硬和脆，但比肌肉更硬且弹性较差。 |
| 化 学 | 蛛丝由蛋白质组成；醌类物质可能是导致黄金圆蛛的网呈现金色的原因；对蜘蛛毒素的研究导致了β受体阻滞剂的发现。 |
| 物 理 | 阳光照射下的蛛网能诱捕蜜蜂，因为蜜蜂会受到模拟蓝天的紫外反射的吸引；在阴凉处，蛛网的黄色融入了背景树叶的颜色之中，充当伪装；蜘蛛会拆除其蛛网较低的部分，确保在强风吹过时蛛网不会破坏；结构化的结晶丝向非结晶丝的转换。 |
| 工程学 | 利用压力和湿度对毛虫（蚕）吐出的丝进行转化，使其与黄金圆蛛的丝具有相同功能；蜘蛛使用步足将丝从身体中抽出来；为了加快纺丝过程，蜘蛛会迎着风，利用风的力量来加速；芳纶具有阻挡子弹的能力。 |
| 经济学 | 对蜘蛛网的工作方式可以进行成本效益分析：如果没有功能性的蛛网，蜘蛛会饿死，如果蜘蛛投资不足（编织了一个小网），蛛网将无法获得足够的食物，如果蜘蛛投资过多，将导致其死亡。 |
| 伦理学 | 从学习关于自然的知识到向自然学习的转换。 |
| 历 史 | 古埃及、希腊、伊斯兰、日本和非洲的神话，印第安文化以及澳大利亚土著艺术中都有以蜘蛛为主角的故事；术语"万维网"让我们联想到蜘蛛网。 |
| 地 理 | 新几内亚部落将蜘蛛视为一种美好的东西，而不是威胁；南太平洋岛屿使用蛛网制作渔栅和渔网；所罗门群岛上的人用蛛网做的球来困住鲨鱼；制作一件披肩和斗篷需要耗费超过一百万只马达加斯加黄金圆蛛的蛛丝。 |
| 数 学 | 蜘蛛编制圆形的网，利用不同的锚点创建复杂的图案；蛛网理论或蛛网的数学：蜘蛛只用直线来创建圆形图案，编织切线而不是弧线；蜘蛛利用其遗传本能创造了一个数学奇迹。 |
| 生活方式 | 蜘蛛非常注意节约材料，它使用自己身体产生的物质来建造蛛网，并对旧的材料进行回收。 |
| 社会学 | 我们可以将自然界中物理结构的创造视为"艺术"，或者说它鼓励人们艺术性地表达自己。 |
| 心理学 | 感知取决于观点：你是一个侏儒还是一个巨人？自然界是一个创造表达的场所；蜘蛛恐惧症：对蜘蛛的病理性恐惧。 |
| 系统论 | 蜘蛛已经进化到"理解"生态系统的背景，并加以利用为自己谋利，优化了材料的使用；蜘蛛和人类之间的关系表明了相互关联性：蜘蛛为人类创造了一个生产再生医药的机会。 |

# 教师与家长指南

## 情感智慧
### Emotional Intelligence

**蝙蝠**

蝙蝠通过赞美蜘蛛开始了谈话。他观察蛛网并注意到网的大小。他抛出一系列对蜘蛛有吸引力的问题将谈话继续下去。蝙蝠见多识广,准备充分。尽管也有自己的神奇事迹可供吹嘘,他仍保持着谦虚的态度,赞美蜘蛛的独特性。他鼓励蜘蛛对所有人大胆说出其成就。蝙蝠担心是否有足够的蛛丝提供给大家,温和地告诫蜘蛛不要让大家失望。他通过询问一系列问题,了解到更详细的情况。最后,蝙蝠就蜘蛛对生态系统的重要性,以及如果具有类似蜘蛛的能力可能带来何种社会变革发表了哲学评论,结束了这次谈话。

**蜘蛛**

面对赞美,蜘蛛受宠若惊。她充满自信,毫无保留地共享了信息。通过叙述奇闻轶事,蜘蛛在她和她的蝙蝠朋友之间营造了轻松的氛围。她意识到了蜘蛛工作的价值和重要性是由人们表现出来的兴趣所驱动的,提供了各种可能性的细节情况。蜘蛛具有战略方针和预见性。她没有感到任何压力,坚信自己能够响应大家的需求,不会让他们感到失望。她认为通过帮助发明新的医疗方法和创造新的就业机会,她可以对环境做出重要贡献。这是一个关于世界是如何联系在一起,以及个体的行为如何让他人的生活和环境更美好的典范。然而,成功并不仅仅取决于技术性能,还依赖于个体为创造一个更美好的世界而改变游戏规则的能力。

## 艺术
### The Arts

让我们来制作蜘蛛和蝙蝠纸链!将纸折叠,并剪成蜘蛛或飞行蝙蝠的形状,悬挂起来作为装饰。可根据下面网站的指导进行制作:折叠和剪成蜘蛛链:HTTP://WWW.DLTK-HOLIDAYS.COM/HALLOWEEN/MPAPERCHAIN_SPIDER.HTM。折叠和剪成蝙蝠链:HTTP://WWW.DLTK-HOLIDAYS.COM/HALLOWEEN/MPAPERCHAIN_BAT.HTM。

# TEACHER AND PARENT GUIDE

## 思维拓展
### Systems: Making the Connections

　　蜘蛛是一个非凡的物种，它们的吐丝和织网技术性能值得赞赏。然而，蜘蛛真正的突出特点是它们的资源利用效率。这归功于它们的饮食习惯，它们的身体能够产生自己生存所需的所有材料。蜘蛛可以回收所有的材料，并利用它们的旧网再生新的蛛丝，来编织一个新网捕捉食物，或者建造一个庇护所度过寒冷的冬天。蜘蛛会根据场地来调整自己蛛网的建设，并会考虑到太阳的运动、月亮的位置、树冠的保护、鸟类的飞行路径以及风向等因素。看起来蜘蛛已经塑造了它们独特的生态定位，并通过侵入数千个生态位展示了它们的资源效率和适应能力。由于这种能力，目前在地球上有超过40 000种蜘蛛，而且这个数字还在不断增长。蜘蛛以一种独特的方式运用数学和几何学理论编织它们的网，它们只使用干净的水作为溶剂，它们的丝由蛋白纤维组成，完全可回收、可再生且适应性强。蜘蛛让我们深入了解作为社会系统或生态系统的一部分，如何让自己适应自然，以确保可持续性以及应对自己的需要，同时也有利于作为系统一部分的其他物种（如人类）。

## 动手能力
### Capacity to Implement

　　列出蜘蛛网所有潜在的商业应用。首先看一下技术，列出那些以受蜘蛛启发的创新作为核心商业模式的公司。然后看一下蜘蛛的工作方法——它吐丝和织网的方式。它遵循的执行原则是什么？看看它如何进行回收利用，如何通过观察风向和伪装自己来适应环境。这会让你深入了解一个行业如何通过回收利用和自适应等可持续的做法，使自己成为有机化的结构。此外还有很多。蜘蛛不仅启示我们如何解决当今的挑战，同时也向我们展示了如何通过改变游戏规则建立新的商业模式，获得成功。试着找出游戏的规则是什么——从产品到加工、到系统。

# 教师与家长指南

## 故事灵感来自

## 弗里茨·沃尔拉特
## Fritz Vollrath

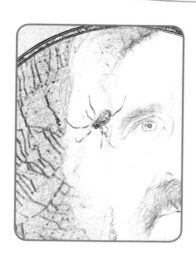

弗里茨·沃尔拉特在德国弗赖堡大学动物学系获得博士学位，他在巴拿马史密森尼热带研究中心工作时，发现了黄金圆蛛的能力。他被这种蜘蛛的表现深深打动。他研究了蛛网的工程学技巧，得知人们可以将商用的、大规模生产的毛虫（蚕）的丝转化为类似蜘蛛丝的工程材料。根据他的研究，已经成立了致力于生物材料、医用缝合线、神经再生和软骨替换技术的四家公司。

他发自内心地喜爱自然界，喜欢所有的动物，从大动物（大象）到小动物（蜘蛛）。他是拯救大象基金会的主席，每年花费大量的时间在肯尼亚的帕拉研究中心研究大象以及它们的社会和生态系统。

更多资讯

http://www.oxfordsilkgroup.com

图书在版编目（CIP）数据

像蜘蛛一样纺纱：汉英对照 /（比）冈特·鲍利著；
（哥伦）凯瑟琳娜·巴赫绘；姚晨辉译. — 上海：学林
出版社，2016.6
（冈特生态童书. 第三辑）
ISBN 978-7-5486-1071-7

Ⅰ. ①像… Ⅱ. ①冈… ②凯… ③姚… Ⅲ. ①生态环
境－环境保护－儿童读物－汉、英 Ⅳ. ① X171.1-49

中国版本图书馆 CIP 数据核字（2016）第 122655 号

---

© 2015 Gunter Pauli

著作权合同登记号 图字 09-2016-309 号

## 冈特生态童书

### 像蜘蛛一样纺纱

| | |
|---|---|
| 作　　者—— | 冈特·鲍利 |
| 译　　者—— | 姚晨辉 |
| 策　　划—— | 匡志强 |
| 责任编辑—— | 李晓梅 |
| 装帧设计—— | 魏　来 |
| 出　　版—— | 上海世纪出版股份有限公司 学林出版社<br>地　址：上海钦州南路81号　电　话/传真：021-64515005<br>网址：www.xuelinpress.com |
| 发　　行—— | 上海世纪出版股份有限公司发行中心<br>（上海福建中路193号　网址：www.ewen.co） |
| 印　　刷—— | 上海丽佳制版印刷有限公司 |
| 开　　本—— | 710×1020　1/16 |
| 印　　张—— | 2 |
| 字　　数—— | 5万 |
| 版　　次—— | 2016年6月第1版<br>2016年6月第1次印刷 |
| 书　　号—— | ISBN 978-7-5486-1071-7/G·406 |
| 定　　价—— | 10.00元 |

（如发生印刷、装订质量问题，读者可向工厂调换）

Housing
93

# 珊瑚在意

Coral Care

Gunter Pauli

冈特·鲍利 著
凯瑟琳娜·巴赫 绘
唐继荣 译

学林出版社
www.xuelinpress.com

## 丛书编委会

主　任：贾　峰

副主任：何家振　郑立明

委　员：牛玲娟　李原原　李曙东　吴建民　彭　勇
　　　　冯　缨　靳增江

## 丛书出版委员会

主　任：段学俭

副主任：匡志强　张　蓉

成　员：叶　刚　李晓梅　魏　来　徐雅清　田振军
　　　　蔡雱奇　程　洋

特别感谢以下热心人士对译稿润色工作的支持：

姜竹青　韩　笑　贾　芳　刘　晓　张黎立　刘之杰
高　青　周依奇　彭　江　于函玉　于　哲　单　威
姚爱静　刘　洋　高　艳　孙笑非　郑莉霞　周　蕊

# 目录

| | |
|---|---|
| 珊瑚在意 | 4 |
| 你知道吗？ | 22 |
| 想一想 | 26 |
| 自己动手！ | 27 |
| 学科知识 | 28 |
| 情感智慧 | 29 |
| 艺术 | 29 |
| 思维拓展 | 30 |
| 动手能力 | 30 |
| 故事灵感来自 | 31 |

# Contents

| | |
|---|---|
| Coral Care | 4 |
| Did you know? | 22 |
| Think about it | 26 |
| Do it yourself! | 27 |
| Academic Knowledge | 28 |
| Emotional Intelligence | 29 |
| The Arts | 29 |
| Systems: Making the Connections | 30 |
| Capacity to Implement | 30 |
| This fable is inspired by | 31 |

褐藻一家沿着坦桑尼亚东北部的桑给巴尔岛海岸漂浮着,他们有很多抱怨。

"如果周边的人类继续用炸药捕鱼,他们将会毁灭珊瑚。"其中一个褐藻说。

A family of brown seaweed floating along the coast of Zanzibar has a lot to complain about.

"As long as people around here fish with dynamite, they will continue to destroy the coral," says one of them.

# 褐藻一家

A family of brown seaweed

海藻和鱼类在周围有珊瑚的环境下生长得最好

Seaweed and fish grow best with coral around

"用不着你告诉我!"一只珊瑚说,"我一直试图说服这些渔民,如果他们将我们都炸掉,将没有更多的鱼了。"

"难道他们没有意识到海藻和鱼类在周围有珊瑚的环境下生长得最好?"

"是啊。而且他们需要食物,不能只靠吃褐藻为生。"珊瑚说。

"You're telling me!" says a coral. "I have been trying to convince these fishermen that if they blow up all of us, that there will be no more fish."

"Don't they realise that seaweed and fish grow best with coral around?"

"Exactly. And those people need food. They cannot live off brown seaweed alone," says the coral.

"所以他们试图吃你们,珊瑚先生,那是非常无用的。"褐藻说。"任何人想咀嚼你,都会把牙崩碎。"他补充道。

"你看,我们都必须在一起生活。只要有我们珊瑚,周围就会有鱼类和海藻。"

"And trying to eat you, Mr Coral, is quite useless," says the seaweed. "Anyone trying to chew on you will break a tooth," he adds.

"Look, we all have to work together. As long as we, the coral of the ocean, are around, there will be fish, and seaweed."

# 把牙崩碎

Break a tooth

# 六周时间生长

Six weeks to grow

"只要有你们珊瑚在周围,我们就能免遭海浪的冲击,不会被拍打得很厉害。那样,我们会生长得更快、更大,因为海里已经有我们需要的所有食物啦。"

"你们要多久才能生长成熟呢?"

"我们从100克长到1000克需要6周时间。"

"And with you coral all around, we are protected from the rough seas and being buffeted around too much. We can grow big a lot faster then, with all the food already available in the sea."

"How long before you are fully grown?"

"It takes us six weeks to grow from weighing a hundred grams to a kilogram."

"这就是说，你们在6周时间里增重10倍！真是够快的！没有其他动植物能与此相比。"

"你看到过有些妇女在养殖更多我们这样的海藻吗？她们把海藻串在水中的绳子上，而在水中我们能获得大量的食物。当然，只要有充足的食物，同时有你们保护我们免遭海浪的冲击，我想我们甚至可能比竹子长得还快。"褐藻说。

"That means you pick up ten times your weight in only six weeks! That is really fast. There is no animal or plant that can match that."

"Have you seen women planting more of us, tying seaweed twigs onto strings in the water where we can get loads of food? I think we grow even faster than bamboo, provided of course that there is an abundance of food for us — and that you protect us from the rough seas," says the seaweed.

保护我们免遭海浪的冲击

Protect us from the rough seas

# 为孩子寻找食物

Find food to feed their children

"被别人需要的感觉真好,谢谢你!但是,我们该怎样对付那些使用炸药的渔民呢?我们应该请警察来抓住他们,并把他们送进牢房吗?"

"噢,对于饥肠辘辘而且只知道通过毁灭珊瑚来捕鱼的人来说,最好的警察、最快的船和最坚固的牢房都不能阻止他们试图为孩子寻找食物。"

"It feels good to be needed, thank you. But what shall we do about fishermen using dynamite? Shall we ask the police to catch them and put them in jail?"

"Well, if people are hungry and the only way they know to fish destroys coral, then the best cops, the fastest boats, and the strongest jails won't keep them from trying to find food to feed their children."

"你说得对！穷人正遭受痛苦，没有人希望挨饿。难怪他们没有耐心。"

"你肯定是一只非常有耐心的动物。"褐藻评价道，"你需要很长很长时间才能成熟，但爆炸在瞬间就能彻底毁灭你。"

"You're right, poor people are suffering, and no one wants to go hungry. No wonder they become impatient."

"You must be a very patient animal," remarks the seaweed. "It takes you ages and ages to grow to maturity, and in less than a second an explosion can completely destroy you."

你肯定是一只非常有耐心的动物

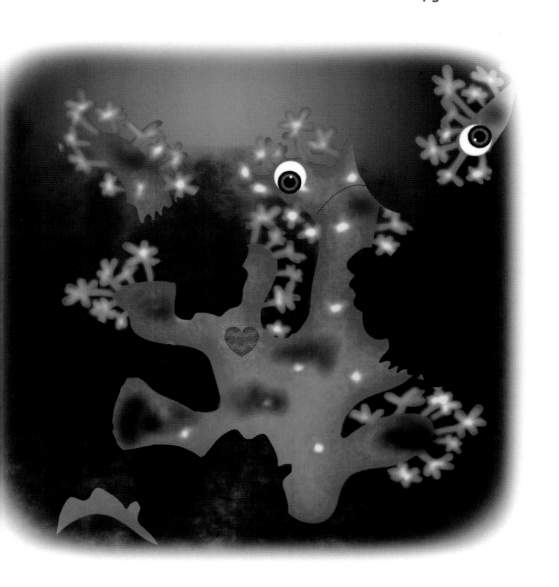

you must be a very patient animal

我们可以在你身上种蘑菇吗?

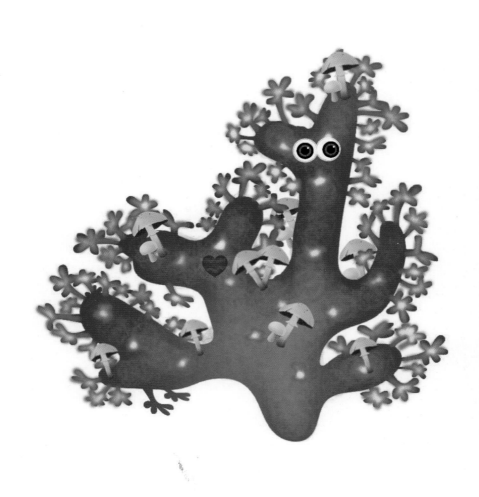

Should we grow some mushrooms on you?

"噢，我只是一个小小的珊瑚生物，也叫水螅体，需要很多我这样的个体才能创造出一块大型珊瑚礁来。虽然我可能活不长，很快死亡，但我们这个大群体存活的时间却非常长。只要我们周围的环境不被炸药、污染物或化学物品改变，我们的群体就能存活下去。"

"那就让我们设法确保在沿海生活的穷人有足够的健康食品。我们可以在你身上种蘑菇吗？"

"Well, I'm just one tiny coral creature, or polyp. It takes many of us to make up a large coral. While I may not live long and die quickly, our large colony lives for a very long time. That is as long as our environment is not changed by dynamite, pollution or chemicals."

"So let's find ways to ensure that poor people living along the coast have enough healthy food. Should we grow some mushrooms on you?"

"想尽一切办法去做吧!当人们把你夹入到饭碗中时,他们也能添加一些蘑菇!"

"这将让他们变得更聪明。"

"不仅更聪明,还更健康。你猜会怎样?在将来,甚至可以用你们海藻生产T恤!"

……这仅仅是开始!……

"By all means, do. Then when people add you to their bowls of rice, they can add some mushrooms too!"

"That will make them much smarter."

"Not only smarter, but also healthier. And guess what? In the future one will even be able to use you seaweed to make T-shirts!"

… AND IT HAS ONLY JUST BEGUN!…

……这仅仅是开始！……

...AND IT HAS ONLY JUST BEGUN!...

# Did You Know?

## 你知道吗？

In Asian cuisine, seaweed is added to soups, broths, stir-fries, and even wraps. It is used to wrap around fast foods like sushi. Seaweed is good for healthy gut flora and it also thins blood and removes heavy metals from the body.

在亚洲人的食谱中，海藻被加入到素菜汤、肉汤、炒菜甚至包装产品中，用于包裹像寿司这样的速食。海藻含有益健康的肠道菌群，也可以稀释血液，从身体中排出重金属。

Seaweed is the lungs of the ocean and an important harvester of carbon dioxide. Coral protects coastal zones from storms and erosion while providing a habitat for spawning and nursing fish.

海藻是海洋的肺，是重要的二氧化碳吸收者。珊瑚保护海岸带免遭暴风雨侵蚀，并为产卵和育卵的鱼类提供栖息地。

褐藻在日本被称为"昆布"。海藻中最重要的元素是碘，这是所有脊椎动物生存所必需的。

In Japan, brown seaweed is known as kombu. The most important element in seaweed is iodine, which is indispensable for all vertebrate life.

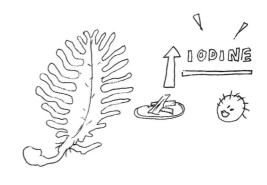

约有5亿人依赖珊瑚礁来获得食物、保护、建筑材料、收入，或开展旅游。在澳大利亚、菲律宾、伯利兹和美国，有十几个独特而重要的珊瑚礁地点被联合国教科文组织认定为"世界遗产"。

Approximately 500 million people depend on coral reefs for food, protection, building materials, income, and tourism. There are a dozen unique, important UNESCO World Heritage coral sites in Australia, the Philippines, Belize and the USA.

Coral reefs are under threat from global climate change, unsustainable fishing, and land-based pollution. About 20% of coral reefs have already been lost and 15% is under serious threat.

珊瑚礁正处在全球气候变化、不可持续的捕鱼和陆地污染的威胁之下。约 20% 的珊瑚礁已经消失，另有 15% 的珊瑚礁正受到严重的威胁。

Coral reefs support 4 000 types of fish and 800 hard coral species. That is more species per unit area than in any other marine environment. Scientists estimate that there may be as many as a million undiscovered organisms living in and around coral reefs.

珊瑚礁支持着 4000 种鱼类和 800 种珊瑚的生存，珊瑚礁单位面积上拥有的物种数量比任何其他海洋环境都多。科学家估计可能有多达上百万种未被发现的生物生活在珊瑚礁里或其周围。

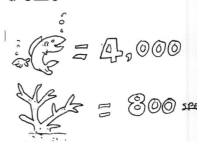

Most corals feed at night, on microscopic zooplankton, small fish, and organic particles in mucous film and strands. They use stinging cells to capture food.

大多数珊瑚在夜间进餐，取食微小浮游动物、小鱼和有机颗粒。它们用刺细胞来捕获食物。

Corals are colonial organisms, composed of hundreds to hundreds of thousands of individual animals called polyps that have stomachs that open at one end. The opening, called the mouth, is also used to clear away debris.

珊瑚属群居生物，由从数百个到数十万个的个体组成，被称为水螅体，其胃部在一端开口，也被用于清除体内残渣。

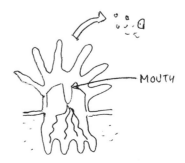

# Think About It

## 想一想

Would you use explosives to catch fish unless you were desperate?

除非处于绝境，否则你会用炸药来捕鱼吗？

你吃过含有海藻的早餐吗？

Have you ever had a breakfast that included seaweed?

How efficient is the production of food when it comes to you instead of you having to go find it?

可以直接获得而无需寻找的食物，它的生产效率如何？

当人们很饥饿并需要为他们的孩子提供食物时，你认为他们会用违法手段获取食物吗？

When people are hungry and need to feed their children, do you think they would be prepared to break the law to get some food?

# Do It Yourself!
## 自己动手！

Have you ever had the opportunity to see a live coral? Here is what you can do to protect it: The less water we use, the less polluted water ends up in the ocean, damaging and killing coral. Even if you live far from the coral reefs, be aware that fertilisers – especially non-soluble ones – flow into the ocean and damage corals. Any kind of trash pollutes water and floats down to the sea, sometimes in tiny particles. When you buy fish, or go snorkelling, or stay at a beach resort, ask people there what they are doing to protect the reefs. When we plant trees, we help protect coral, as better tree cover reduces run-off and, consequently, the amount of tiny trash particles ending up in the sea. Do your part by spreading the word about protecting coral and learning how to propagate and plant new coral.

你亲眼目睹过活的珊瑚吗？下面告诉你怎么做才能保护它。我们使用的水越少，最终流入海洋并伤害或杀死珊瑚的污水就越少。即便你住的地方远离珊瑚，也要注意各种肥料（特别是非水溶性肥料）会流入海洋伤害珊瑚。任何一种垃圾都会污染水体并流入海洋中，有时以微小颗粒的形式出现。当你买鱼、潜水或在海滨度假时，询问人们会采取什么措施来保护珊瑚。我们植树也是在帮助保护珊瑚，因为更高的植被覆盖会减少地表径流，相应减少最终流入海洋中的微小垃圾废物颗粒。你也可以尽力传播保护珊瑚的知识，并学习如何种植新的珊瑚。

# TEACHER AND PARENT GUIDE

## 学科知识
### Academic Knowledge

| | |
|---|---|
| 生物学 | 珊瑚是一个很庞大的动物家族，某些珊瑚需要8年才能成熟，生长速度最快的珊瑚每年可长15cm；红藻、绿藻与褐藻的区别；海藻是海洋之肺，产生可穿过消化道的非水溶性纤维，影响肠道菌群；珊瑚为在其中生存的物种创造生存空间，并提供全套的生态系统服务功能；巨型海藻每天生长0.5m，最终长达30~80m；海胆啃食巨型海藻，能消灭所有的巨型海藻，创造出海胆荒漠。 |
| 化 学 | 炸药由硝化甘油、硅藻土和碳酸钠制成；海带是一种褐藻，富含维生素$B_1$、镁、铁、钙、维生素$B_2$、维生素$B_5$、木脂素、碘和褐藻素；巨型海藻用于获取碳酸钠，制造炸药；虽然用海藻种蘑菇成本太高，但海藻提取物（琼脂）可用来制作微生物的培养基。 |
| 物 理 | 硝化甘油受到物理冲击就爆炸，然而当它被硅藻土吸附时，就能耐冲击；水下炸药的导火索由氧化剂供氧，不会在水中熄灭。 |
| 工程学 | 炸药主要用于开矿和道路建设；海藻生物精炼需要水，而水在实验室用来制作胶凝剂和培养基时只利用一次就被浪费了，但如果也用作灌溉就会非常高效。 |
| 经济学 | 海藻加工后的废水是很好的稻田肥料，为植物提供丰富的碘，这是它进入食物循环的途径；欧洲国家为向盐中添加碘的公司提供补贴，就是为了发展当地产业，保证碘能在食物链中循环。 |
| 伦理学 | 不要以暴制暴；采用毁灭性的方法获取食物，会在不久的将来降低你为家庭提供食物的能力；穷人是否应该因没有水和食物而失去耐心去寻找解决方案。 |
| 历 史 | 捕鱼活动在4万年前出现，开始是徒手捉，后来发展出鱼叉、网捞、鱼钩钓、陷阱诱捕等方式；一直到1940年，南非都是最大的炸药生产国。 |
| 地 理 | 在历史上，印度洋上的桑给巴尔岛是一个贸易港；菲律宾和印度尼西亚是世界上最大的海藻生产国；印度尼西亚拥有最丰富的海藻生物多样性。 |
| 数 学 | 海藻在6~8周时间里增重10倍。 |
| 生活方式 | 整日暴露在太阳下时，需要保护皮肤免受损伤。 |
| 社会学 | 炸药使用具有争议性；当犯罪的根源在于饥饿和生存需要时，压制性方式的有效性如何？ |
| 心理学 | 一个人感到受欢迎，或是被社会需要并能作出贡献的重要性。 |
| 系统论 | 珊瑚礁提供各种各样的生态系统服务，如食物来源、生存空间、孵化育幼场所、保护海岸带免遭暴风雨侵蚀、药物来源、生物多样性，以及文化根基；珊瑚礁生态系统是珊瑚、藻类和许多其他物种的共生体。 |

# 教师与家长指南

## 情感智慧
## Emotional Intelligence

**珊 瑚**

　　珊瑚正在经历一场危机,他非常清醒地知道人们迫切需要食物,而海藻不能独自提供足够的营养。他请求包括鱼类、海藻在内的所有生物共同努力。珊瑚有敏锐的观察力,对不容易理解的方面提出问询。他惊讶于海藻高水平的生产力,并认为养殖海藻可能是一个解决方案。珊瑚探究动用警察的压制性方式,但迅速懂得饥饿的紧迫性会迫使人们去做他们能做的一切以求生存。珊瑚对生命有一种沉思的态度,充分了解炸药、排放到海洋的化学物质和在水中阻碍阳光穿透的土壤颗粒所带来的挑战。最后,珊瑚提供了一些实用建议来帮助人们变得更聪明、更健康。

**褐 藻**

　　褐藻担心珊瑚的毁灭,不理解人们竟然可以毁掉自己未来的食物基地。他认识到周围有健康珊瑚的重要性,知道自己快速增重的能力,但也意识到以下事实:如果不是因为与珊瑚的共生关系,他不能如此快地生长,也不会这样高产。褐藻用简单而又清晰的逻辑进行推理,探究为什么饥饿的人在直到没有鱼或珊瑚时才停止使用炸药。褐藻接受人们需要除自己以外的更多食物的事实,并且表示单纯供应足量的食物是不够的,食物还必须健康安全。

## 艺术
## The Arts

　　是时候看照片了。你将在互联网上发现大量美丽而形态各异的珊瑚照片。选择你最喜欢的3张,与也选择了3张照片的朋友分享。从这6张照片中选出你们共同发现的最契合的共生关系,并表现出珊瑚礁与海洋中、海岸带上许多物种共同生活的3张照片。现在请第三位朋友选择自己最喜欢的3张照片,并重复上述过程。

# TEACHER AND PARENT GUIDE

## 思维拓展
### Systems: Making the Connections

　　珊瑚礁处于濒危状态，它们的衰退是由许多因素造成的：化学物品（特别是洗涤剂、漂白剂和肥料）导致珊瑚死亡；土壤侵蚀增加了水的湍流从而阻碍太阳光进入水中；虾场的创建、珊瑚被用于道路建设和旅游度假地的开发，也都促进了珊瑚礁的死亡。采用炸药不仅在一瞬间造成需要多年才能恢复的环境损害，而且也摧毁了人们赖以获得每日食物供应的基础。如果只是采取现在可用的技术来供应食物，那么不管当前生态系统能供应什么，都不足以满足每个人的基本需求。这就是为什么我们需要学会怎样与当地现有的生产方式结合，以及怎样增强生态系统的功能。我们需要利用丰富的生物多样性来增加产出，同时恢复生态系统的反馈环、乘数效应和共生关系。这里的挑战是时间。珊瑚礁的恢复需要很多年，但现在就很迫切需要食物。这促使我们要找出哪些物种生长得更快，哪种方法能更好地应对危机：这不仅要提供食物，而且要有就业机会和生态保育。海藻的种植是一个有吸引力的出发点，因为海藻是天然产品，容易种植和处理，而且在全世界都有广泛需求。目前，海藻主要被用作食物和肥料，但其应用范围明显超出这些产业。海藻种类繁多，生长迅速，有可能找到一种海藻养殖方式，使之对人类及生态系统发挥多重效益。

## 动手能力
### Capacity to Implement

　　海藻营养价值高，容易种植。列一份能从褐藻、绿藻和红藻提取的产品清单。挑战自己，找到至少20种应用方式。然后，研究在你的国家谁在种植和加工海藻，谁在进口海藻及其目的。在现有海藻商业的基础上，找出还能用海藻做些什么，这样，你可以建立起大量的海藻用途组合。就像咖啡产业那样，我们还只是看到了海藻产业的刚刚兴起。

# 教师与家长指南

## 故事灵感来自
## 凯托·姆什杰尼
## Keto Mshigeni

凯托·姆什杰尼在坦桑尼亚的高地长大,这里离海洋很远。尽管如此,他仍然发展出对海藻的浓厚兴趣。他从达累斯萨拉姆的东非大学毕业,获得植物学和地理学的学位,并辅修了教育学。他在美国的夏威夷大学获得植物科学的博士学位,并在最大的海藻生产国菲律宾开展博士后工作。他在海藻养殖上的创新方法引导印度洋上的桑给巴尔岛、奔巴岛和马菲亚岛等岛屿的沿海村庄成功地引进海藻产业,改善了农村妇女的生活。他在纳米比亚大学继续他的研究,出任学术事务方面的副校长,而且在亨蒂斯湾创建了海水养殖综合系统。目前,他是位于达累斯萨拉姆的胡贝特·凯卢吉纪念大学的校长。

更多资讯

http://www.hkmu.ac.tz/media/more/prof._keto_mshigeni_wins_aau_award_of_excelllence_in_higher_education_and_r

http://www.divenewswire.com/03/

http://www.ryandrum.com/seaweeds.htm

图书在版编目（CIP）数据

珊瑚在意：汉英对照／（比）冈特·鲍利著；
（哥伦）凯瑟琳娜·巴赫绘；唐继荣译. -- 上海：学林
出版社，2016.6
（冈特生态童书. 第三辑）
ISBN 978-7-5486-1056-4

Ⅰ．①珊… Ⅱ．①冈… ②凯… ③唐… Ⅲ．①生态环
境－环境保护－儿童读物－汉、英 Ⅳ．① X171.1-49

中国版本图书馆 CIP 数据核字 (2016) 第 125803 号
————————————————————————
ⓒ 2015 Gunter Pauli
著作权合同登记号 图字 09-2016-309 号

## 冈特生态童书
### 珊瑚在意

| | |
|---|---|
| 作　　者—— | 冈特·鲍利 |
| 译　　者—— | 唐继荣 |
| 策　　划—— | 匡志强 |
| 责任编辑—— | 程　洋 |
| 装帧设计—— | 魏　来 |
| 出　　版—— | 上海世纪出版股份有限公司 学林出版社 |
| | 地　址：上海钦州南路 81 号　　电　话／传真：021-64515005 |
| | 网　址：www.xuelinpress.com |
| 发　　行—— | 上海世纪出版股份有限公司发行中心 |
| | （上海福建中路 193 号　网址：www.ewen.co） |
| 印　　刷—— | 上海丽佳制版印刷有限公司 |
| 开　　本—— | 710×1020　1/16 |
| 印　　张—— | 2 |
| 字　　数—— | 5 万 |
| 版　　次—— | 2016 年 6 月第 1 版 |
| | 2016 年 6 月第 1 次印刷 |
| 书　　号—— | ISBN 978-7-5486-1056-4/G·391 |
| 定　　价—— | 10.00 元 |

（如发生印刷、装订质量问题，读者可向工厂调换）

# Housing 104

# 蜜蜂在家里

## Bees at Home

**Gunter Pauli**

冈特·鲍利 著

凯瑟琳娜·巴赫 绘
唐继荣 译

学林出版社
www.xuelinpress.com

## 丛书编委会

主　任：贾　峰
副主任：何家振　郑立明
委　员：牛玲娟　李原原　李曙东　吴建民　彭　勇
　　　　冯　缨　靳增江

## 丛书出版委员会

主　任：段学俭
副主任：匡志强　张　蓉
成　员：叶　刚　李晓梅　魏　来　徐雅清　田振军
　　　　蔡雩奇　程　洋

特别感谢以下热心人士对译稿润色工作的支持：

姜竹青　韩　笑　贾　芳　刘　晓　张黎立　刘之杰
高　青　周依奇　彭　江　于函玉　于　哲　单　威
姚爱静　刘　洋　高　艳　孙笑非　郑莉霞　周　蕊

# 目录

| | |
|---|---|
| 蜜蜂在家里 | 4 |
| 你知道吗？ | 22 |
| 想一想 | 26 |
| 自己动手！ | 27 |
| 学科知识 | 28 |
| 情感智慧 | 29 |
| 艺术 | 29 |
| 思维拓展 | 30 |
| 动手能力 | 30 |
| 故事灵感来自 | 31 |

# Contents

| | |
|---|---|
| Bees at Home | 4 |
| Did you know? | 22 |
| Think about it | 26 |
| Do it yourself! | 27 |
| Academic Knowledge | 28 |
| Emotional Intelligence | 29 |
| The Arts | 29 |
| Systems: Making the Connections | 30 |
| Capacity to Implement | 30 |
| This fable is inspired by | 31 |

鹦鹉产蛋的时候到了！随着森林被砍伐，树冠几乎全被毁了，她能找到一棵适合筑巢的树实属幸运。但当她回到衬有柔软树叶和羽毛的树洞里面时，鹦鹉发现一只蜂王在同样的位置建造着自己的家。

"蜂王，恕我直言，"鹦鹉说，"是我先到这儿的！"

It is time for a parrot to lay her eggs. As the forest has been cut down and the forest canopy all but destroyed, she is lucky to have found a suitable tree for her nest. Returning to the hole in the tree with soft leaves and feathers to line it, she finds that a queen bee is making herself at home in the very same spot.

"With all due respect to you, Queen," says the parrot, "I was here first!"

鹦鹉产蛋的时候到了

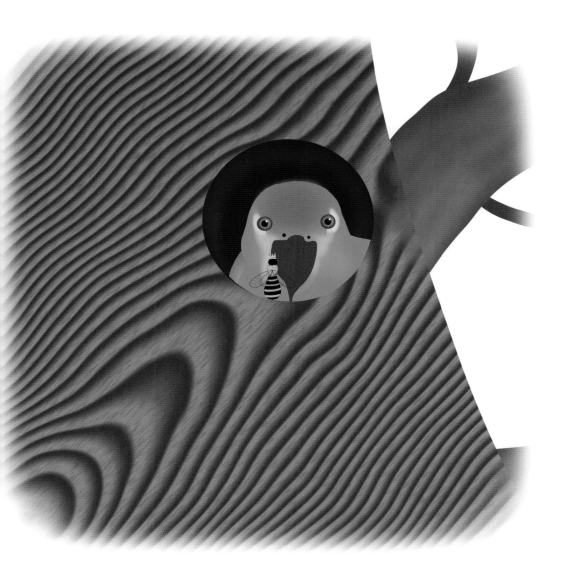

It is time for a parrot to lay her eggs

## 与蜜蜂分享筑巢地

Sharing a nesting site with bees

"远在你出现之前,我的祖先就住在这片林子里,你怎么可以宣称自己是第一个到这儿的呢?"蜂王回应道。

"好吧!那是很多代之前的事儿啦,没人记得谁真正最先到这儿的。然而,我要告诉你的是,我马上要产蛋了,我无法想象我的孩子与蜜蜂分享筑巢地的情形!"

"How can you claim to have been here first when my ancestors have been living in this forest long before you came along?" responds the bee.

"Well, that was so many generations ago that no one can remember who really was here first. What I can tell you though, is that I need to lay my eggs right now and I cannot imagine my chicks sharing a nesting site with bees!"

"那么，你为什么不能找另一棵有洞的树？我们需要取食这周围大量的花蜜，否则我们的小蜜蜂就不能生存。"蜂王争辩道。

"为了给我们的孩子提供食物，我们需要采食分布在这周围的果实和种子。我相信你会找到另一棵树放你的蜂巢。"鹦鹉说。

"Then why don't you find another tree with a hole in it? We need to feed on the nectar of flowers that grow so abundantly around here or our little ones will not survive," argues the bee.

"And we need to feed on the fruit and seeds that grow around here to provide our chicks with their food. I am sure you will find another tree for your hive," says the parrot.

我相信你会找到另一棵树放你的蜂巢

am sure you will find another tree for your hive

在这周围几乎没有树保留下来

Hardly any trees left around here

"你我都知道,在这周围几乎没什么树留下来了,找到一颗有合适树洞的树几乎是不可能的。"

"这你可不能指责我。"鹦鹉回应道,"你知道,是那些几个世纪前从欧洲来的人把我们这儿的树都砍光了。从那时起,我们就很难找到好的居住地了。"

"You know as well as I do that there are hardly any trees left around here, and that to find one with a suitable hole is just about impossible."

"As long as you don't blame me for that, " replies the parrot. " You know that it was those people from Europe who came here centuries ago who cut down all our trees. We have had a hard time finding a good place to live ever since."

"为什么你不去沙漠,住在仙人掌中?"蜜蜂提议道,"你的羽毛可以保护你免受这些尖刺的伤害,不是吗?"

"为什么你和你成千上万的小工蜂不去呢?你们能轻松地住在仙人掌中,但我尾巴太长就没那么容易。"

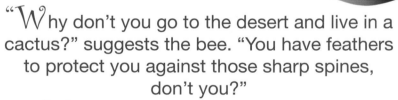

"Why don't you go to the desert and live in a cactus?" suggests the bee. "You have feathers to protect you against those sharp spines, don't you?"

"Why don't you and your thousands of little worker bees do that? You lot can easily live in a cactus. It is not so easy for me with my long tail."

你们能住在仙人掌中

you can live in a cactus

让我们尝试共同生活吧

Let's try to live together

"你知道我们在仙人掌中筑巢有多难吗?现在,让我们实际点吧。为什么我们不一起分享这棵稀有的阔叶树上这个完美的树洞呢?让我们尝试共同生活吧。"蜂王提议道。

"你真是不可理喻,蜂王!我是先来的!如果成千上万的蜜蜂在周围嗡嗡作响,从早到晚在蜂巢中进进出出,我们根本没法休息。那我们该在什么时间睡觉呢?"

"Do you have any idea how difficult it is for us to build a hive in a cactus? Now let's be practical here. Why don't we share this perfectly suitable hole in this rare hardwood tree? Let's try to live together," suggests the queen.

"You are really being unreasonable, Queen. I was here first! And with thousands of you buzzing around, entering and leaving the nest all day long, no one we will ever get any rest. When are we supposed to sleep?"

"睡觉？没时间睡觉了！作为父母，我们必须确保我们的孩子有他们的未来。如果我们能每隔一段时间打个盹，就很幸运了。"

"对啊！自从人们不考虑种植新树就砍倒我们的树木之后，生存变得困难了，我不得不飞得更远去寻找足够的食物。"

"Sleep? There is no time for sleep! As parents we have to secure a future for our little ones. We will be lucky if we manage a short nap once in a while."

"Yes, ever since people cut down our trees without worrying about planting new ones, it has been difficult to survive. I have to roam further and further to get enough food."

我们必须确保我们的孩子有他们的未来

We have to secure a future for our little ones

比与非洲蜜蜂分享要好

Better than sharing with an African bees

"与我们这些无刺蜂住在一起,至少比与非洲蜜蜂住在一起要好。他们不像我们这样安静,而且当你挡到路时他们会攻击你。我们只是个小蜂群,而且不会待很久。我们甚至能自己建一个独立的入口。"

"At least sharing with us stingless bees is better than sharing with the African bees. They are not as calm as we are and will attack you if you get in their way. We are just a small colony and we won't stay for long. We will even build ourselves a separate entrance."

"我听说过那些有攻击性的非洲杀人蜂。他们将会把我们俩都干掉,也会消灭其他所有与他们相遇的生物。我们最好设法保护自己,以及周围保留下来的生物多样性。也许,你们蜜蜂将会是好邻居……"

"那么,你愿意与我们分享你的筑巢地吗?"

……这仅仅是开始!……

"I have heard about those aggressive African killer bees. They will finish us both off, and everything else that comes in their way. It is better that we protect ourselves, and what's left of the biodiversity around here. Perhaps you bees will make good neighbours after all…"

"So, will you please share your nesting site with us then?"

... AND IT ONLY HAS JUST BEGUN!...

……这仅仅是开始！……

… AND IT HAS ONLY JUST BEGUN! …

# Did You Know?
## 你知道吗？

There are 20 000 bees species. About 500 species do not sting. Stingless bees will bite when in danger or try to get into the noses and ears of any invaders to deter them.

地球上有2万种蜜蜂，其中500种不会蛰人。无刺蜂遇险时会叮咬入侵者，或设法钻进它们的鼻子和耳朵中。

Bee biodiversity is threatened around the world, primarily as a result of the introduction of the more productive African bee that, despite its aggressive behaviour, can produce up to 100 kg of honey per beehive per year.

全世界的蜜蜂多样性正受到威胁，主要是由于引进了更高产的非洲蜜蜂所致。非洲蜜蜂虽然攻击性强，但每个蜂巢每年只能产100千克蜂蜜。

While African bees produce large amounts of honey collectively as a colony, the highest individual productivity comes from stingless bees. Melipona marginata living in swarms of about 300 bees, produce 3 litres of honey per year, which is sold at 10 times the price of standard honey.

虽然非洲蜜蜂群体能生产大量的蜂蜜，但单只蜜蜂产量最高的却是无刺蜂。一群约300只，生活于沼泽地的麦蜂（Melipona marginata）每年能生产出3升蜂蜜，这种蜂蜜的价格是标准蜂蜜价格的10倍。

Just like the parasitic bird, the cuckoo, lays its eggs in another bird's nest for the other birds to raise its chicks, there are bees that lay their eggs in cells prepared by other bee species, such as the solitary or bumble bees, and therefore are called cuckoo bees.

像杜鹃这样的巢寄生鸟类在其他鸟巢中产蛋以便让其他鸟类抚养自己幼鸟一样，也有某些种类的蜜蜂在其他蜜蜂种类（如独居蜂和大黄峰）准备的蜂巢中产卵，被称为"杜鹃蜜蜂"。

Bees change (metamorphose) like butterflies. A bee passes through four life stages: egg, larva, pupa and adult. The first three stages occur in the brood cell of the nest. Males hatch from unfertilised eggs and females from fertilised eggs.

蜜蜂像蝴蝶一样会发生身体变化，这一过程称为变态。一只蜜蜂需要经历四个生命阶段：卵、幼虫、蛹和成虫，其中前三个阶段发生在蜂巢中的抚幼室。雄蜂从未受精的卵发育而来，而雌蜂从受精卵发育而来。

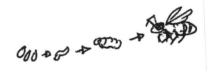

Bees keep most of their stored honey cold so that it is less likely to be eaten by other creatures that do not like cold food. In a beehive heat moves from the outside to the inside, and rises from below to the top, self-warming the honey when the need arises.

蜜蜂把大部分蜂蜜冷藏储存，这样就不会被不喜欢冷藏食物的生物吃掉。有需要时，热量会从蜂巢外部进到内部，从底部上升到顶部，从而对蜂蜜自行加温。

Bees collect water and use it to cool the hive through evaporation. They also use water to dilute honey before they consume it. A colony of 50 000 worker bees and 1 000 drones can collect 400 kg of nectar, water, and pollen per year.

蜜蜂采集水，通过水分蒸发给蜂巢降温，在吃蜂蜜前会用水稀释。一个由5万只工蜂和1000只雄蜂组成的蜂群，每年能采集400千克花蜜、水和花粉。

Just as surprising perhaps as finding out that there are stingless bees, is learning that there are flightless parrots. The New Zealand night parrot, called a kakapo in the Maori language, is a flightless, nocturnal, ground-dwelling parrot. It is endemic to New Zealand, which means that it occurs only there. It climbs trees and drops down like a parachute.

得知有不能飞行的鹦鹉存在，就像发现无刺蜂那样令人惊奇。在新西兰毛利语中被称为"kakapo"的鸮鹦鹉，就是一种不能飞行的夜行性地栖鹦鹉，也是新西兰独有的物种。它们能爬树，并像降落伞一样下坠。

# Think About It
# 想一想

In your garden, would you prefer bees that can produce honey but do not sting?

在你的花园中，你是不是更喜欢那种能产蜜不会蛰人的蜜蜂？

如果两种生物正在竞争同一个生存空间，你会建议其中一个退出争斗，还是建议他们设法一起生活呢？

If two creatures were competing for the same living space, would you advise one to give up the battle, or suggest both find a way to live there together?

Who do you blame for the fact that there are now so few trees left in certain regions?

对于在某些地区只剩下如此稀少的树木的现实，你认为该指责谁？

具有良好的睡眠是否比确保后代的未来更重要？

Is having a good sleep more of a priority in life than securing a future for one's offspring?

# Do It Yourself!
## 自己动手！

Is anyone prepared to share his or her home with someone who may be very different: as different as a bee from a parrot? Who is willing to share their bedroom with others who are desperate to find a safe place in which to grow up? Would it make any difference if you were asked to tolerate the presence of others for only a short period, a few weeks for instance? How keen are we to maintain our privacy, and under which conditions are we prepared to make an exception? Make a list of possible exceptions that may motivate people to accept sharing the privacy that seems sacrosanct to them.

会有人愿意与自己非常不同的人分享自己的家吗？即便这种差异会大到像蜜蜂与鹦鹉之间那种程度？有人愿意与其他竭尽全力寻找安全成长空间的人分享自己的卧室吗？如果你被要求忍受他人的存在，有时只要很短时间，有时却要几周，是否会对你造成影响？我们是不是很想维护自己的隐私？会在哪些情况下准备破例呢？这些特殊情况会激励人们做出牺牲去分享隐私。请列出这样一份可能情况的清单。

# TEACHER AND PARENT GUIDE

## 学科知识
### Academic Knowledge

| | |
|---|---|
| 生物学 | 蜜蜂是可以飞的昆虫，与黄蜂和蚂蚁密切相关；麦蜂建立蜂蜡隧道来保护蜂巢入口；花粉和蜂蜜储存在蜂蜡中；蜜蜂只吃植物，充足的食物会刺激蜂王产卵；蜜蜂在1.25亿年前从黄蜂中分化而来；有些鹦鹉的雄性和雌性看上去非常相似；鹦鹉寿命长达80~100年。 |
| 化　学 | 无刺蜂的蜂蜜含有高度抗菌的物质；蜂胶是蜜蜂用来密封蜂巢裂缝的一种树脂状物质；工蜂有蜡腺，在它们35天的寿命中，有10~16天在生产蜂蜡；蜡腺将糖转化为片状的蜂蜡，蜂蜡被咀嚼并与唾液混合后变成白色。 |
| 物　理 | 冬季为蜂巢加热时，蜜蜂并不是加热整个蜂巢，最温暖处位于蜂巢中心；为了给蜂巢降温，蜜蜂采集水，拍打翅膀让空气循环；蜜蜂能忍受最高50℃、最低−45℃的极端温度。 |
| 工程学 | 蜜蜂用自己生产的蜂蜡筑巢，在空间和建材上应用了经济最大化原则。 |
| 经济学 | 麦蜂蜂蜜在中美洲和非洲售价很高，当地人把它当作药品，特别是用来治疗眼疾；为了保护当地的无刺蜂，巴西曾经实施禁售，但并未取得成功，现在无刺蜂交易自由，巴西各地的无刺蜂群反而健康生长。 |
| 伦理学 | 巴西法律禁止破坏现有的野生无刺蜂群，但允许转运由蜜蜂自己在野外创建的新蜂群；野生鹦鹉是最常被捕捉和作为宠物贸易的动物，导致其野生种群的缩减。 |
| 历　史 | 马德里抄本是仅有的3部保存下来、在哥伦布发现美洲大陆以前的玛雅图书之一，可以追溯到公元900年，描写了古老的玛雅文明对蜂的利用；当亚历山大大帝将鹦鹉从印度引入希腊时，鹦鹉首次在欧洲出现；蜜蜂有1亿年的历史，而鹦鹉有4千万年。 |
| 地　理 | 中美洲的玛雅文明已饲养蜜蜂；马达加斯加拥有很多种无刺蜂。 |
| 数　学 | 蜂群优化算法被用于管理和工程学中；随着球体半径的增加，其表面积以半径平方的速度增加，而体积以半径三次方的速度增加，这意味着随着蜂群变大，其体积比表面积增加更快，即更大的蜂群以更慢的速度损失热量；蜂窝为六边形，钝角是109° 28′，锐角为70° 32′，符合工程经济学。 |
| 生活方式 | 蜜蜂越来越多地被当作宠物饲养，尤其是能在都市环境饲养的无刺蜂；野生动物贸易受到濒危物种国际贸易公约（CITES）的监管。 |
| 社会学 | 野生鹦鹉从未与猫或老鼠接触，因而未发展出任何反捕食行为，这使它们在引进这些动物的地区有很高的灭绝风险；当大量的幼蜂孵化后，蜂巢就变得过度拥挤，部分蜜蜂将成群飞离蜂巢，寻找新的家园。 |
| 心理学 | 鹦鹉心理学，指鹦鹉容易与人类亲近；鹦鹉的主人更有同情心。 |
| 系统论 | 非本地的非洲蜜蜂不会为所有开花植物授粉，所以引入非本地的蜜蜂物种不仅影响本地蜜蜂物种，还使植物多样性降低。 |

# 教师与家长指南

## 情感智慧
### Emotional Intelligence

鹦鹉

鹦鹉很有礼貌,行为举止像淑女。她平静地指出自己马上就要产蛋了。她的论点前后一致,确信会找到解决方案,不接受任何超过她的能力范围的责任。鹦鹉在争论中非常坚定,表明在仙人掌中筑巢不是她的选择,坚定地回绝了蜂王。鹦鹉重申是她最先到那儿的,因此有权继续待下去。然后她解释道,蜜蜂嗡嗡作响将让共同生活变得几乎不可能。最后,鹦鹉不再解释,开始讨论她们如何共同生存。

蜂王

蜂王的要求很直接,并且在交谈中很坚定,承认鹦鹉先发现那棵树,但强调她的祖先最先出现在那个生态系统中,应该有优先权。蜂王让鹦鹉去找另外一棵树,也意识到对鹦鹉来说这样做极为困难,于是建议鹦鹉把仙人掌作为一种选择。最后,蜂王重申共同生活的建议,因为她知道这在短时间内很有必要。当鹦鹉拒绝蜂王的提议时,蜂王占据道德高地,谈到确保后代未来的重要性。

## 艺术
### The Arts

你的活动选择是,要么用六边形做一个蜂巢模型,要么做一个鹦鹉尾巴的模型。如果你选择做蜂巢模型,就找出你将需要多少个六边形的"窗户"。如果你选择做鹦鹉尾巴的模型,就用彩色铅笔来画各种各样的羽毛,涂上明亮美丽的颜色。现在,将彩绘的羽毛剪下来,拼装在一起来模拟鹦鹉的尾巴。当然,我们不希望用真正的鹦鹉羽毛来做这个尾巴,因为买卖真正的羽毛将会鼓励非法野生动物产品贸易。

# TEACHER AND PARENT GUIDE

## 思维拓展
### Systems: Making the Connections

森林砍伐通常只与植物的毁灭联系在一起，但我们很少意识到它对许多其他生命形式具有毁灭性的级联效应，威胁这些生命的生存。然而，当森林被单一种植园取代时，野生物种就没有生存空间了。当森林被砍伐殆尽又不种植任何植物时，土地就变成了干旱的热带稀树草原。这会给所有物种带来灾难，因为土地暴露在太阳下，土壤温度升高，偶尔降雨时，也没有植被保护，结果导致大规模的水土流失。冷的雨滴无法穿透坚硬的热土，这种现象称为"煎锅"效应。森林砍伐让大自然很少有机会能再次恢复为有活力的生态系统。少数生存下来的物种，在非常有限的空间里竞争避难所。由于无法保护各种生存要素和逃脱捕食者，即便有能充分满足生存需要的营养，像鹦鹉、蜜蜂这样的物种，生存机会仍然很少。这则童话包含一个特定的哲学反思：资源有限时，我们能在多大程度上接受与他人一起生活带来的不便？特别是与我们并不认识的人，这些人或许与我们截然不同。在环境危机时期，尤其需要社会的凝聚力。它不仅发生在一个家庭或一个物种中，而且发生在所有家庭和同一生态系统的所有物种中。

## 动手能力
### Capacity to Implement

看看我们能做些什么来增加环境中的蜜蜂数量。首先，我们要种植更多的开花植物，找出哪些是本地现有的物种。选择有更多花粉的单花瓣植物，确保你种植的植物有不同颜色的花，因为黄色、白色、蓝色和紫色将比粉色、橙色和红色吸引更多的蜜蜂。选择不同花期的植物组合，这样蜜蜂全年都有食物。同时，选择会为你和家人提供水果和蔬菜的植物。草本植物也会吸引蜜蜂到你的花园来，特别是迷迭香、薰衣草、薄荷、鼠尾草和百里香。

# 教师与家长指南

## 故事灵感来自
## 德斯蒙德·图图
## Desmond Tutu

德斯蒙德·图图大主教的父亲是一位小学教师,母亲在一所盲人学校工作。德斯蒙德喜欢阅读,年轻时最喜欢的书是《伊索寓言》和莎士比亚的戏剧。他于1954年从南非大学毕业,回到自己学习过的中学教英语和历史。他于1961年被任命为牧师,于1966年获得国王学院神学硕士学位。德斯蒙德·图图是首位被任命为约翰内斯堡圣公会教长的黑人,后来短暂地出任主教。他是南非著名的圣公会教士,最著名的是他在反对南非种族隔离制度上的作用。1978年,他被任命为南非教会理事会的秘书长,成为南非黑人权利的主要发言人。

### 更多资讯

www.fromthegrapevine.com/lifestyle/5-intriguing-structures-inspired-bees

www.birdchannel.com/bird-news/bird-entertainment/bird-history.aspx

www.beeswaxco.com/bees-making-wax.php

www.wikihow.com/Attract-Honey-Bees

图书在版编目（CIP）数据

蜜蜂在家里：汉英对照／（比）冈特·鲍利著；
（哥伦）凯瑟琳娜·巴赫绘；唐继荣译．－－ 上海：学林
出版社，2016.6
 （冈特生态童书．第三辑）
 ISBN 978-7-5486-1058-8

Ⅰ．①蜜… Ⅱ．①冈… ②凯… ③唐… Ⅲ．①生态环
境－环境保护－儿童读物－汉、英 Ⅳ．① X171.1-49

中国版本图书馆 CIP 数据核字（2016）第 125806 号

---

© 2015 Gunter Pauli
著作权合同登记号 图字 09-2016-309 号

冈特生态童书
蜜蜂在家里

| | | |
|---|---|---|
| 作　　者 —— | 冈特·鲍利 | |
| 译　　者 —— | 唐继荣 | |
| 策　　划 —— | 匡志强 | |
| 责任编辑 —— | 程　洋 | |
| 装帧设计 —— | 魏　来 | |
| 出　　版 —— | 上海世纪出版股份有限公司 学林出版社 | |
| | 地址：上海钦州南路 81 号　电话／传真：021-64515005 | |
| | 网址：www.xuelinpress.com | |
| 发　　行 —— | 上海世纪出版股份有限公司发行中心 | |
| | （上海福建中路 193 号 网址：www.ewen.co） | |
| 印　　刷 —— | 上海丽佳制版印刷有限公司 | |
| 开　　本 —— | 710×1020　1/16 | |
| 印　　张 —— | 2 | |
| 字　　数 —— | 5 万 | |
| 版　　次 —— | 2016 年 6 月第 1 版 | |
| | 2016 年 6 月第 1 次印刷 | |
| 书　　号 —— | ISBN 978-7-5486-1058-8/G·393 | |
| 定　　价 —— | 10.00 元 | |

（如发生印刷、装订质量问题，读者可向工厂调换）

# Housing 80

# 手牵手
## Hand in Hand

**Gunter Pauli**

冈特·鲍利 著
凯瑟琳娜·巴赫 绘
唐继荣 译

学林出版社
www.xuelinpress.com

## 丛书编委会

主　任：贾　峰
副主任：何家振　郑立明
委　员：牛玲娟　李原原　李曙东　吴建民　彭　勇
　　　　冯　缨　靳增江

## 丛书出版委员会

主　任：段学俭
副主任：匡志强　张　蓉
成　员：叶　刚　李晓梅　魏　来　徐雅清　田振军
　　　　蔡雩奇　程　洋

**特别感谢以下热心人士对译稿润色工作的支持：**

姜竹青　韩　笑　贾　芳　刘　晓　张黎立　刘之杰
高　青　周依奇　彭　江　于函玉　于　哲　单　威
姚爱静　刘　洋　高　艳　孙笑非　郑莉霞　周　蕊

# 目录

| | |
|---|---|
| 手牵手 | 4 |
| 你知道吗? | 22 |
| 想一想 | 26 |
| 自己动手! | 27 |
| 学科知识 | 28 |
| 情感智慧 | 29 |
| 艺术 | 29 |
| 思维拓展 | 30 |
| 动手能力 | 30 |
| 故事灵感来自 | 31 |

# Contents

| | |
|---|---|
| Hand in Hand | 4 |
| Did you know? | 22 |
| Think about it | 26 |
| Do it yourself! | 27 |
| Academic Knowledge | 28 |
| Emotional Intelligence | 29 |
| The Arts | 29 |
| Systems: Making the Connections | 30 |
| Capacity to Implement | 30 |
| This fable is inspired by | 31 |

一只海獭将一个巨藻紧紧地抱在怀里,睡着了。巨藻也喜欢被海獭抱着。突然,海獭醒了,并四处张望。

A sea otter is holding a branch of kelp tightly in her arms. She is asleep. The kelp just loves to be embraced by the otter. Suddenly the sea otter wakes up and looks around.

喜欢被海獭抱着

Loves to be embraced by the otter

早上好

Good morning

"早上好,"巨藻说,"亲爱的朋友,我期待与你度过每一天,这让我保持健康和快乐。"

"噢,醒来后能听到一个可爱的声音告诉我说,我在把自己的事情做好时,也能帮助到你,这真是太好了!"海獭高兴地回答道。

"Good morning," says the kelp. "I am looking forward to another day with you, my dear friend, keeping me healthy and happy."

"Oh, it is so nice to wake up and to hear a lovely voice telling me that I am doing good for you by doing good for me," responds the otter.

"你做的事情对我的帮助可大了!如果你不把海胆、海螺、鲍鱼、海蟹甚至鱼类吃掉,我将永远无法生长成林!"

"我得承认我的胃口确实有点大。为了跟上我快节奏的生活,我每天需要吃自己体重十分之一重量的食物。"

"You are doing so much good for me. If you were not filling your belly with sea urchins, snails, abalone, crabs, and even fish, I would never be able to grow into a forest!"

"I must admit I do have a good appetite. In order to keep up with the fast pace of my life, I need to eat a quarter of my body weight every day."

我的胃口确实有点大

I do have a good appetite

我提供一个家……

I provide a home...

"是啊,你的胃口帮助我将巨浪的能量散布出去,保护海岸带免遭风暴冲击。我能为海葵、海绵和鱼类提供一个家,甚至为人类提供食物,为你提供睡觉的安乐窝。"

"Well, your appetite helps me to spread the power of powerful waves, to protect the coast from the pounding of the storms. And I can provide a home to anemones, sponges and fish. I can even give food to people and offer you a place to sleep."

"有人说你是关键物种,因为你将来自太阳的能量转换到食物中。而且,你不仅在海洋中提供能量,在陆地上也同样提供。"

"好吧,你见过由于吃了我被冲上岸的叶片上的大量虱子而变得非常肥胖的鸡吗?这些鸡几乎难以行走,因为他们真是太重了!"巨藻笑着说道。

"Some say that you are a key species because you convert energy from the sun into food. And you do not give energy only in the sea, but also on land."

"Well, have you ever seen such fat chickens, eating the tons of lice found on my beached fronds? These chickens can hardly walk because they weigh so much!" laughs the kelp.

你见过非常肥胖的鸡吗?

Have you ever seen such fat chickens?

……让他们的冰激凌有柔软口感

...make their ice cream soft

"人类把你加到他们的冰激凌和牙膏中,这是真的吗?"

"是真的呀!人类这么需要我,是不是棒极了?虽然他们已经尝试过了,但还未发现哪种化学物质能让他们的冰激凌尝起来有我这样的柔软口感。"

"Is it true that people put you in their ice cream and tooth paste?"

"Yes, isn't it wonderful that people want me this much? Even though they've tried, they have never found a chemical that can make their ice cream taste as soft on the palate as I do."

"嗯，人类喜欢我柔软的皮毛，并用它覆盖他们的皮肤。"

"那样柔软的皮毛几乎让你失去生命。你有最厚实的皮毛！人们为了得到它，将捕捉并杀死你们全家。"

"你知道，人类甚至开始计算我每平方厘米有多少毛发，但他们数不清楚。"海獭笑着说道。

"Well, people love my soft fur. They love to cover their skin with my skin."

"That softness has nearly cost you your life. You have the thickest fur ever! People will hunt and kill your whole family for it."

"You know, people once even started counting how many hairs I have per square centimeter, and they lost count," laughs the sea otter.

人类喜欢我柔软的皮毛

People love my soft fur

……利用工具来获取你们的食物

...use tools to get your food

"每当我见到你,我就想起黑猩猩。"

"算了吧,我看起来一点儿也不像黑猩猩!我觉得自己可爱多了。"

"你和黑猩猩都是可爱的哺乳动物,都能利用工具来获取食物。这就意味着你们像人类一样聪明。"

"好吧,你还能指望我破开贝壳吗?如果我试着去把贝壳咬碎,会把自己牙齿弄断的。"

"我最喜欢你的一点就是你具备良好的行为举止,用餐结束后,你会洗手和刷牙。"

"Whenever I see you I think of chimpanzees."

"Come on, I look nothing like a chimpanzee! I certainly think I am much cuter."

"Both you and the chimp are lovely creatures, both are mammals, and both use tools to get your food. That means you are as smart as humans."

"Well, how else do you expect me to crack shellfish open? I would break my teeth if I were to try and crunch them to pieces."

"What I love most about you is that you have good manners. You wash your hands and brush your teeth after you have had a meal."

"你今天给了我这么多的赞美!难怪我喜欢蜷缩在你的叶片里。"

"你这样说让我很高兴!但我的确知道,你更喜欢与你的家人手牵手一起睡觉。"

"当然啦!如果你身边有这样一群家人和朋友,难道你不喜欢和我一样手牵手吗?"

……这仅仅是开始!……

"You have so many compliments for me today! No wonder I just love to roll in your fronds."

"It is so nice of you to say that to please me, but I do know you prefer to hold hands with your family while sleeping."

"Of course, would you not like to do the same, if you were surrounded by such a wonderful troop of family and friends?"

... AND IT HAS ONLY JUST BEGUN!...

……这仅仅是开始！……

...AND IT HAS ONLY JUST BEGUN!...

# Did You Know?
## 你知道吗?

Sea otters do not have fat (blubber) to keep them warm, instead they have the densest fur of all animals in the animal kingdom, with more than 100 000 hairs per square centimeter to insulate them against the cold. That is up to 800 million hairs for an adult male sea otter.

海獭并没有脂肪来为它们保暖,但它们有动物王国中最致密的皮毛,每平方厘米有多达10万根毛发,以此来隔绝寒冷。这就是说,一只成年海獭有多达8亿根毛发。

Sea otters and kelp are both keystone species, meaning that their role and impact in their ecosystem is greater than other species.

海獭和巨藻都是关键物种,这意味着它们在其生态系统中的作用和影响力比其他物种要大。

Sea otters have a rich and varied diet, eating about 40 different marine species. Their great appetite spurs the growth of kelp forests.

海獭的食性复杂，能吃约 40 种不同的海洋物种。它们的大胃口能够刺激巨藻林的生长。

Orcas and great white sharks are predators of sea otters.

虎鲸和大白鲨是海獭的天敌。

Kelp forests are one of the most productive and dynamic ecosystems on Earth, and are recognised as a great fixer of carbon dioxide in the sea.

巨藻林是地球上最有生产力和活力的生态系统之一，并被认为是海洋中的一个大型二氧化碳吸收器。

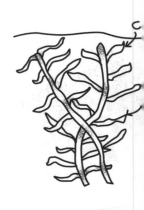

Kelp is used in ice cream, dressings, cosmetics and toothpaste, and as an ingredient in preparations that make people lose body weight. It is even used for water- and fireproofing fabric.

巨藻被用于制作冰激凌、调料、化妆品和牙膏，以及作为人类减肥制剂的组成成分，甚至被用来制作防水和防火纤维。

Sea otters do hold hands while sleeping. After every meal they feverishly clean their teeth and wash their hands and fur.

海獭在睡觉时手牵手。每次用餐结束，它们狂热地刷牙、洗手、清洗皮毛。

By 1911, when commercial fur trade was banned, the worldwide population of sea otters had dropped to perhaps only 2 000 animals. The Russian population of sea otters is now the largest in the world.

当商业性皮毛贸易于1911年被禁止时，全世界的海獭数量下降到也许只有2000只了。目前，俄罗斯的海獭数量是世界上最多的。

# Think About It

## 想一想

**C**ould you ever have imagined that there are animals that hold hands and brush their teeth, naturally?

你是否想象过有很自然地手牵手并刷牙的动物？

如果父母没有坚持让你去做的话，你能否学会刷牙、洗手？

**W**ould you have learned how to brush your teeth and wash your hands if your parents had not insisted on it for over and over again?

**H**ow would you look if you had a hundred thousand hairs on each square centimeter of your scalp?

如果你每平方厘米的头皮上有10万根头发，你看上去会怎样？

海獭的大胃口使巨藻长成巨大的海藻林，或许像热带雨林一样大。你认为这是否很有趣？

**D**o you think it is funny that the great appetite of the sea otter makes kelp grow into a huge forest, perhaps as big as the rainforest?

# Do It Yourself!
# 自己动手!

Let us make an ice cream, the simple way. A recipe may say you need milk, cream, sugar and vanilla and even a touch of salt. We like to make an ice cream the easy way, and only use a banana. It is vegan, gluten-free, dairy-free and has no added sugar. And thanks to the richness of pectin, there is no need to add alginates that usually keep the ice cream soft on the palate. I am sure that everyone who tries it will love it.

让我们用简单的方法制作一个冰激凌。食谱上可能说,你需要牛奶、奶油、糖、香草,甚至一些盐。但我们想用简单的方法来制作冰激凌,只用一根香蕉。这是一种纯素食,不含麸质或奶制品,不添加糖分,由于香蕉富含果胶,也不需添加用来保持冰激凌柔软口感的海藻酸盐。我相信每个品尝过的人都会喜欢的。

# TEACHER AND PARENT GUIDE

## 学科知识
## Academic Knowledge

| | |
|---|---|
| 生物学 | 海洋中的藻类数百万年基本保持不变；巨藻是世界上生长最快的植物，可以长达40m；海獭通过触觉发现黑暗中的猎物，从肛腺分泌出刺激性气味来自卫；被冲上岸的巨藻会吸引虱子，鸟类捕食虱子，形成食物循环；巨藻林的生物多样性像热带雨林一样高。 |
| 化 学 | 微生物发酵的海藻酸钠至今还无法与海藻生成的天然产品匹敌；巨藻富含氮、钾，是极好的肥料，广泛用于马铃薯种植；巨藻中含有纯碱和钾碱，用于玻璃制造和肥皂生产；巨藻可以提供丰富的碘。 |
| 物 理 | 冰激凌的柔软度由空气和海藻酸钠来决定；海獭的代谢率较高，以便应对在寒冷的外部环境下的热量损失；海獭的肾脏能通过一个类似于反渗透的过程从海水中提取淡水并产生浓缩尿，所以它们可以喝海水。 |
| 工程学 | 牵手对于海獭来说，类似于锚对船的作用：这是一种避免在海上漂走的方式；通过在表面创造只有十分之一毫米大小的气穴来达到柔软的效果；反渗透是一种从海水（或被污染的水）中生产饮用水的方法。 |
| 经济学 | 对天然产品的过度要求导致资源枯竭和物种灭绝，因此迫切需要进行市场监管。 |
| 伦理学 | 过度利用后，由于自然资源本身不能及时更新，导致供应减少、价格上涨，进一步刺激资源破坏，我们怎么能盲目追随市场呢？ |
| 历 史 | 海獭是印第安人图腾柱上的重要元素；海獭和海狸在19世纪的毛皮贸易中处于中心地位；在长达4000年时间里，巨藻是爱尔兰和苏格兰经济成分之一，并在1946年的马铃薯饥荒时期成为一种替代食物；在19世纪，用巨藻生产的沼气被用于工厂照明。 |
| 地 理 | 世界上最大的巨藻林沿着冷水海岸带分布；一条连绵的"巨藻公路"从日本延伸开来，向北沿着西伯利亚经过白令海峡到达阿拉斯加，向南沿着加利福尼亚海岸线直达厄瓜多尔。 |
| 数 学 | 海獭能在15秒钟内击打鲍鱼壳45次；为了成功吃到鲍鱼肉，它必须潜水4次，每次1~4分钟，其间面对冰冷的海水，它必须保持体温。 |
| 生活方式 | 降低对那些用穿动物皮毛显示财富和地位的人的社会认可度。 |
| 社会学 | 由于对动植物生存的负面影响，文化如何激励人们舍弃那些曾在过去被认为是身份地位象征的产品；手牵手一起走是一种表示亲密、友谊和信任的传统，尤其存在于非洲和阿拉伯文化中。 |
| 心理学 | 个人的行为与生命网络联系在一起，看上去孤立的事物实际上并非如此：自然界所展示的友爱与呵护会对我们的情绪产生影响。 |
| 系统论 | 生态系统的制约与平衡：海獭控制巨藻的捕食者，这样巨藻就能沿着海岸带生长成林，进而保护海岸带免遭潮汐、风暴影响。 |

# 教师与家长指南

## 情感智慧
### Emotional Intelligence

**巨 藻**

巨藻享受与海獭共处的时光。他意识到海獭家庭带来的帮助，并对此非常感激。如果没有海獭的辛劳，巨藻家庭不可能像森林那样繁盛。巨藻解释说，巨藻林能保护海岸带免遭风暴冲击。巨藻进一步分享信息，他被冲上岸的叶片吸引虫子，而虫子是鸟类重要的食物来源。巨藻显示出自信，指出即使经过多年研究，但还没有人工合成的替代物能代替他的作用。巨藻也富有同情心，充分意识到为了获得海獭皮毛而进行的捕杀导致海獭灭绝的危险。然后，谈话变为赞美，最后以对海獭喜欢手牵手的赏识而结束。

**海 獭**

海獭从观察周围环境开始一天的生活。海獭感到放心，知道自己在巨藻林中的作用。同时，他也完全清楚巨藻林在生态系统中扮演的角色。海獭并没有对为了获得皮毛而一直捕杀他的人类保持怨恨，似乎已经忘记和原谅了。海獭更愿意笑谈人类试图数他的毛发数量，而不是指责人类过去的错误。海獭被巨藻的友善和欣赏的语言所折服，于是也话语体贴。

## 艺 术
### The Arts

海藻广泛分布于全世界的海岸带。在大多数地区，它被视为杂草，应该扔进肥料堆。现在，你可以用它进行编织，将它转变为持久性材料。它在太阳下晒干后会变得坚硬、结实、耐磨，并能抵御任何天气条件。

# TEACHER AND PARENT GUIDE

## 思维拓展
### Systems: Making the Connections

　　生态系统是生命的网络。海岸带起到屏障的作用，保护陆地免受风暴潮、飓风和海啸损害。这片从海岸到海底的区域富含营养物质，因此是生物多样性的热点地区。巨藻林底层有海胆和海星栖居，海蟹和海龙也十分繁盛。巨藻的顶部是漂浮的，因为其具有气囊，使叶片能保持在水面，以便最大程度暴露在太阳下。一片陆地森林能支持3个门的动物生存，而巨藻林却至少能支持10个门的动物。依赖于巨藻的物种如此多种多样，以至于查尔斯·达尔文断言："……我不认为其他哪种森林有接近巨藻林毁灭所引起的那么多物种的消亡。"巨藻林还储存了数百万吨碳，要是这些碳被释放，会导致海洋的酸化。因此，应对二氧化碳过度释放到大气中的一种方法同样是种植更多的巨藻林。巨藻林有与热带雨林相近的生产力，但如果那些移动缓慢的鱼类和海洋无脊椎动物不受控制生长，它们将影响巨藻林的生长和生态功能的承担。

## 动手能力
### Capacity to Implement

　　我们需要了解更多有关巨藻林的信息。首先，你可以与你的朋友和家人分享你所学到的知识，即巨藻林和热带雨林同样重要。这种新的认识意味着你现在可以计划在南非和纳米比亚海岸的巨藻林中进行一次潜水之旅，即便不能真正成行，也要做这个计划。这个旅程将会让人惊叹，相当于参观亚马逊盆地。但为了享受这次旅程，你最好先学会如何潜水。

# 教师与家长指南

故事灵感来自

## 纳尔逊·曼德拉
## Nelson Mandela

纳尔逊·曼德拉通过与他爱的人和最好的朋友一起手牵手，向全世界展示他的友爱和慈悲。当他离开监狱时，他牵起妻子的手，同时举起握紧的拳头。他永远不对逮捕他的人持怨恨态度，而是以尊重的方式对待他们，甚至偶尔与他们开玩笑，了解他们的家庭情况并记得所有细节。这传递了一个团结、呵护、亲密和信任的信息。即便西方文化嘲笑手牵手的行为，纳尔逊·曼德拉清楚地表明这种非洲传统并没有错，有错的是那些认为这种传统有错的人。与朋友和家人牵手的纳尔逊·曼德拉为自己是一名非洲人而感到骄傲，并执着于维护他从父母那里学到的传统。

### 更多资讯

http://mg.co.za/article/2014-12-03-hold-hands-in-friendship-and-be-proud-to-be-an-african/

http://oceanfocus.org/focus-areas/threatened-habitats/kelp-forests/

http://animals.nationalgeographic.com/animals/mammals/sea-otter/

www.livescience.com/7042-ancient-people-kelp-highway-america-researcher.html

图书在版编目（CIP）数据

手牵手：汉英对照 ／（比）冈特·鲍利著；（哥伦）凯瑟琳娜·巴赫绘；唐继荣译. —— 上海：学林出版社，2016.6

（冈特生态童书. 第三辑）

ISBN 978-7-5486-1054-0

Ⅰ. ①手… Ⅱ. ①冈… ②凯… ③唐… Ⅲ. ①生态环境－环境保护－儿童读物－汉、英 Ⅳ. ① X171.1-49

中国版本图书馆 CIP 数据核字（2016）第 125799 号

---

© 2015 Gunter Pauli

著作权合同登记号 图字 09-2016-309 号

---

### 冈特生态童书

#### 手牵手

| | |
|---|---|
| 作　　者—— | 冈特·鲍利 |
| 译　　者—— | 唐继荣 |
| 策　　划—— | 匡志强 |
| 责任编辑—— | 程　洋 |
| 装帧设计—— | 魏　来 |
| 出　　版—— | 上海世纪出版股份有限公司 学林出版社 |
| | 地　址：上海钦州南路81号　　电话／传真：021-64515005 |
| | 网址：www.xuelinpress.com |
| 发　　行—— | 上海世纪出版股份有限公司发行中心 |
| | （上海福建中路193号 网址：www.ewen.co） |
| 印　　刷—— | 上海丽佳制版印刷有限公司 |
| 开　　本—— | 710×1020　1/16 |
| 印　　张—— | 2 |
| 字　　数—— | 5万 |
| 版　　次—— | 2016年6月第1版 |
| | 2016年6月第1次印刷 |
| 书　　号—— | ISBN 978-7-5486-1054-0/G·389 |
| 定　　价—— | 10.00元 |

（如发生印刷、装订质量问题，读者可向工厂调换）

Housing
82

# 感谢你的音乐
Thank you for the Music

Gunter Pauli

冈特·鲍利 著

凯瑟琳娜·巴赫 绘
唐继荣 译

## 丛书编委会

主　任：贾　峰

副主任：何家振　郑立明

委　员：牛玲娟　李原原　李曙东　吴建民　彭　勇
　　　　冯　缨　靳增江

## 丛书出版委员会

主　任：段学俭

副主任：匡志强　张　蓉

成　员：叶　刚　李晓梅　魏　来　徐雅清　田振军
　　　　蔡雩奇　程　洋

特别感谢以下热心人士对译稿润色工作的支持：

姜竹青　韩　笑　贾　芳　刘　晓　张黎立　刘之杰
高　青　周依奇　彭　江　于函玉　于　哲　单　威
姚爱静　刘　洋　高　艳　孙笑非　郑莉霞　周　蕊

# 目录

| | |
|---|---|
| 感谢你的音乐 | 4 |
| 你知道吗？ | 22 |
| 想一想 | 26 |
| 自己动手！ | 27 |
| 学科知识 | 28 |
| 情感智慧 | 29 |
| 艺术 | 29 |
| 思维拓展 | 30 |
| 动手能力 | 30 |
| 故事灵感来自 | 31 |

# Contents

| | |
|---|---|
| Thank you for the Music | 4 |
| Did you know? | 22 |
| Think about it | 26 |
| Do it yourself! | 27 |
| Academic Knowledge | 28 |
| Emotional Intelligence | 29 |
| The Arts | 29 |
| Systems: Making the Connections | 30 |
| Capacity to Implement | 30 |
| This fable is inspired by | 31 |

一只燕子在她正对着阳台天花板的巢中忙碌。一条正在晒太阳的小猎狗一边欣赏背景音乐，一边注视着这项泥土工程。

"你筑巢这么忙，可能没有任何空闲时间为我唱歌了。"小猎狗说。

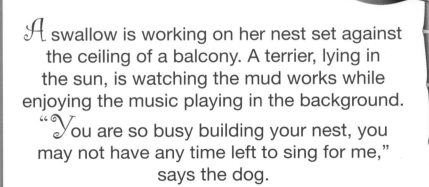

A swallow is working on her nest set against the ceiling of a balcony. A terrier, lying in the sun, is watching the mud works while enjoying the music playing in the background.

"You are so busy building your nest, you may not have any time left to sing for me," says the dog.

你可能没有任何空闲时间为我唱歌了

You may not have any time left to sing for me

我们鸟类一直会找时间来唱歌

We birds always find time to sing

"哦，不会的！我们鸟儿一直会找时间来唱我们最优美的歌曲。"燕子回答道，"但现在我必须找个安全的地方来下蛋。"

"那么，你什么时候有空，能让我再次欣赏到你甜美的歌声？"小猎狗问。

"Oh no, we birds always find time to sing our beautiful songs," responds the swallow, "but right now I have to build a safe place to lay my eggs."

"So when will you have time to let me enjoy the sweetness of your voice again?" asks the terrier.

"听着！鸟儿并不会为其他动物唱歌，我们只在自己想要唱歌的时候才唱。当太阳升起时，当月光激发我们的情感时，或当鲜花盛开且食物丰盛时——任何我们想分享生活快乐的时候，我们就唱歌。"

"你和伙伴们的歌声让我非常陶醉，你们已经激发了世界上最美好的事物。全人类历史上最受尊敬的作曲家莫扎特，就在他的一部钢琴协奏曲中加入了紫翅椋鸟的歌声。"

"Look here, birds do not organise concerts for others, we sing when we feel like it. When the sun rises, when the moon inspires us, or when the flowers are out and food is abundant. Whenever we like to share the joys of life, we just sing."

"I am so enchanted by your songs, and those of your fellows. You have inspired the best in the world. Like Wolfgang Amadeus Mozart, the most admired composer of all times, who included a starling's song in one of his piano concertos."

# 莫扎特

Wolfgang Amadeus Mozart

"是啊,那人非常著名。但那是紫翅椋鸟,不是燕子。即便这样,我们也为自己的同类激发了伟大数学家的灵感而骄傲。"

"你是指音乐家吧!"

"Oh yes, that one is famous. But that was a starling, not a swallow. Still we are very proud that one of our brothers inspired the great mathematician."

"You mean musician!"

"莫扎特是一位伟大的数学家。他的作品节奏影响人们的脑电波，帮助孩子们学习。当他们在欣赏古典音乐时，几何和代数会学得更好。"

"真不可思议！要知道，长久以来，人类认为飞行只需要羽毛和翅膀就够了。他们去尝试了，但不久后就意识到飞行不仅仅只是把身体提升到空中。唱歌也是这样，它需要的不只是控制呼吸和声带, 还有更多的方面。"

"Mozart was a great mathematician. The rhythms of his compositions affect people's brainwaves, helping children to study. They do better at algebra and geometry while enjoying classical music."

"Amazing! You know, for a long time people thought that all they needed to fly were feathers and wings. They tried but soon realised that it takes more than that to lift a body into the air. It is the same with singing. More is needed than only having control over your breath and vocal cords."

……帮助孩子们学习

... helping children to study

你们甚至启发了弗里达

You even inspired Frida

"没错,人类需要受到启发才能创造出这些带来如此多快乐的奇妙声音。"

"你们甚至启发了弗里达。我知道她的丈夫被她美妙的声音所打动,所以亲切地称她为'弗里达·福格尔桑'。福格尔桑在德语中意思是'鸟儿的歌'。"

"That's right. One needs to be inspired to create these marvellous sounds that give so much pleasure."

"You even inspired Frida. I know her husband was so inspired by her wonderful voice that he lovingly called her 'Frida Vogelsang'. Vogelsang is German for 'bird song'."

"弗里达？你是指来自瑞典流行组合阿巴合唱团的安妮·弗瑞德吗？她和三位瑞典艺术家朋友让世界上许多国家的数百万人，甚至数十亿人与他们一起歌唱。她极其欣赏我们燕子，作为一位姑娘，她连续多日观察我们幼鸟的成长。"

"但这些幼鸟并没产生任何音乐，是吧？他们只是粗声号叫来索要食物。"

"By Frida do you mean Anni-Frid from the pop group ABBA? She and her three Swedish co-artist friends made millions, even billions, of people sing along with them, in many countries around the world. She simply adores us swallows and as a girl spent days watching our chicks grow up."

"But those chicks did not make any music, did they? They were just squealing and squawking for food."

来自流行组合阿巴合唱团的安妮·弗瑞德

Anni-Frid from the pop group ABBA

他们将很快就学会像他们的父母那样歌唱

They will soon learn to sing like their parents

"嗯，我们鸟儿的首要任务是找到一位伴侣、产卵并孵化，这是不是生命的奇迹呢？然后，喂养我们的幼鸟，看他们长大——从光溜溜的留巢雏鸟成为有毛的离巢雏鸟，不久后满世界高速飞行，并捕食昆虫。他们很快就学会像他们的父母与祖父母那样歌唱。"

"哦，绝对是这样的！难怪安妮·弗瑞德看到你们燕子时会很开心，用她华丽的嗓音在全身心地歌唱。"

"Well, isn't the wonder of life for us birds first and foremost to find a mate, lay eggs and hatch them? And then to feed our chicks and watch them grow. Watch our little featherless nestlings become fledglings, soon to fly around the world, at high speed, catching insects in flight. They will soon learn to sing like their parents and grandparents."

"Oh, absolutely. No wonder that Anni-Frid felt like singing with all her heart, using her magnificent voice, when she enjoyed the sight of you all."

"我很荣幸!但你知道吗?是我爸爸最先找到这个绝佳场所的,我现在与我的家庭仍一起在这筑巢。他先找到一个好地方,然后开始歌唱,并炫耀自己擅长飞行。他通过这种方式吸引我妈妈来到这里,并与他一起度过余生。"

"是呀!能有莫扎特和阿巴合唱团的歌曲作为背景音乐,谁愿意去其他地方生活呢?"

"他们只会想待在家里载歌载舞,像我们一样成长!"

……这仅仅是开始!……

"I do feel honoured! But did you know that it was my dad who first picked this excellent place, where I am also nesting now with my family? He first picked a good spot and then started singing and showing off how good he was at flying. He did this to attract my mom to come and spend the rest of her life here with him."

"Well, with Mozart and ABBA playing in the background, no one will ever want to live anywhere else."

"They will just want to stay home and sing and dance like we did growing up!"

… AND IT HAS ONLY JUST BEGUN!…

……这仅仅是开始！……

...AND IT HAS ONLY JUST BEGUN!...

# Did You Know?

## 你知道吗？

Adult swallows and their chicks learn to recognise each other's voices. The chicks use begging calls to ask for more food from their parents. Chicks are fed up to 8 times an hour.

成年燕子以及它们的幼鸟学会辨识相互之间的叫声。幼鸟用求食声来向父母索要更多的食物。幼鸟每小时喂食可多达8次。

Swallow nests are constructed from about 1 000 mud pellets made from clay, silt and salts. The salts keep bacteria under control. The insides of nests are lined with grasses and feathers for comfort and warmth.

燕巢是用大约1000粒由黏土、泥沙和盐制作的泥丸所筑成，其中盐用来控制细菌。为了舒适和保暖，巢内还有干草和羽毛。

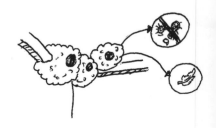

𝒮wallows are unique birds in that they mate while in the air. They return to the same nests year after year, repairing old and weakened nests. Offspring build their nests near those of their parents.

燕子在空中飞行时交配，这在鸟类中是很独特的。它们年复一年地回到同一个巢，并会修复残破的巢穴。后代会选择在父母亲的巢穴附近筑巢。

𝒜 swallow will bring up to four hundred meals a day to the nest. These meals consist of about 20 insects (compacted into an easy to eat meal), controlling the insect population.

一只燕子每天将携带食物回巢多达400次。它们的食物包括大约20种昆虫（成鸟要将这些昆虫压缩成幼鸟容易进食的状态），因而燕子能控制昆虫的种群数量。

Wolfgang Amadeus Mozart for years enjoyed the company of a pet starling. Some parts of the third movement of his 17th piano concerto, KV 453, were inspired by this bird's calls.

沃尔夫冈·阿玛多伊斯·莫扎特多年享受着一只宠物紫翅椋鸟的陪伴。他的第十七钢琴协奏曲（KV453）第三乐章的某些部分灵感来自这种鸟类的叫声。

Mozart considered music to be mathematics that you can hear. Listening to music composed by Mozart enhances mental performance in mathematics and is known as the "Mozart Effect".

莫扎特认为音乐是能聆听的数学。聆听莫扎特创作的音乐能增强在数学方面的表现，这被称为"莫扎特效应"。

如果按现在的标准向莫扎特支付作曲的版税，他将能买下整个萨尔茨堡市，那是他出生的地方。

Had Mozart been paid royalties for his compositions, as is standard practice now, he would have been able to buy the entire City of Salzburg, where he was born.

阿巴合唱团在四十多年里售出了3.8亿张唱片，而且这个数字还在增长。他们的专辑《黄金典藏精选》销量超过披头士乐队的经典专辑《佩珀军士的孤心俱乐部乐队》。

ABBA sold 380 million records over 40 years and this number is still growing. ABBA's album Gold – Greatest Hits outsold the Beatles' classic, Sgt Pepper's Lonely Hearts Club Band.

# Think About It

## 想一想

Are you inspired by bird song?

你受过鸟儿歌声的启发吗?

聆听莫扎特的音乐让你感到快乐吗?聆听阿巴合唱团的歌曲,是否让你想跳舞?

Does listening to the music of Mozart make you feel happy? And does listening to ABBA make you feel like dancing?

When you build a house for yourself, would you like it to be close to that of your parents?

当你为自己建造房子时,是否希望能离父母近一点?

你什么时候会想唱歌?是清晨醒来时、沐浴时、去学校的路上,还是回家的路上?

When do you feel like singing? Early in the morning when you wake up, in the shower, on your way to school, or on your way home?

# Do It Yourself!
# 自己动手!

Let's have a look at birdcalls. How many different birds do you know of that have beautiful or memorable calls? Make a list of your favourite birdcalls, and share it with you friends and family. You can play them some of the many recordings of birdcalls that are available on CD. Do any of these birds live near you, in cages or aviaries, or in the wild? Find out what you need to do to attract wild birds to your garden so that there will be no need to ever catch, sell or cage birds.

让我们了解下鸟类的叫声。你知道有多少种鸟类具有美妙或令人难以忘怀的叫声？列出一个你最喜欢的鸟叫声清单，并与你的朋友和家人分享。你可以播放一些鸟叫声，在光盘中有许多这样的录音。这些鸟类生活在你身边的鸟笼、鸟舍还是野外？想想你需要做些什么来吸引野鸟到你的花园中，这样就不用抓捕、出售或笼养它们了。

# TEACHER AND PARENT GUIDE

## 学科知识
### Academic Knowledge

| | |
|---|---|
| 生物学 | 雄性燕子的叫声为雌性提供了一个判断潜在伴侣生理状况的机会；燕子用唱歌来表达兴奋、求偶、与其他燕子交流，以及当捕食者到来时发出警报；当人类开发了声带并学会控制呼吸后，声音也得到演化；紫翅椋鸟能模仿人类制造声音，比如电话铃声、轿车喇叭声或莫扎特的音乐。 |
| 化 学 | 亚洲食谱中的燕窝含有氨基酸、碳水化合物、矿物盐，以及糖蛋白和唾液酸。 |
| 物 理 | 为了确定燕窝营养价值的真实性，需要采用电子显微术、能量色散X射线显微分析等多种物理化学技术。 |
| 工程学 | 家燕筑杯状巢，崖燕筑穹顶巢，这些结构没有来自下方的支撑，但却撑了成年燕子和4～6枚蛋，以及蛋孵化后要待在巢里超过3周的留巢雏。 |
| 经济学 | 燕窝位居最昂贵的动物产品之列；全球的燕窝贸易每年高达40亿美元；基于版权的收益；娱乐业在世界经济中的作用和重要性。 |
| 伦理学 | 把鸟儿关进笼子来享受它们的歌声，与帮助鸟儿在家附近筑巢让它们自由自在，但我们仍能享受歌声的区别。 |
| 历 史 | 在伊卡洛斯神话中，一个学会飞行的人飞得太靠近太阳，以致把翅膀粘在一起的蜡融化了，他跌回了地面；燕子在罗马时期被用作信使；中国食用燕窝汤已有1000多年历史。 |
| 地 理 | 燕子在除南极洲以外的所有大洲生活；欧洲和北美洲的燕子是长距离迁徙的鸟类；尼日利亚的冬季栖息场所能吸引超过100万只燕子；由于栖息地丧失，燕子变得濒危甚至有灭绝的危险；泰国、印度尼西亚和马来西亚是出口燕窝的三个主要生产国。 |
| 数 学 | 基于斐波那契代码的高度结构化的音乐对心理功能和数学学习能力的影响；数拍子、节奏、音阶、音程、节奏型、符号、和声、拍号、泛音、乐音和音名，所有这些都将音乐与数学联系在一起；毕达哥拉斯用数字来表达音符之间的音程，并从几何图形中推导出乐音；声波能被数学方程定量化描述。 |
| 生活方式 | 我们的生活变得如此忙碌，以至于我们忘记唱歌跳舞；父母和祖父母给孩子们唱歌，通常能引导他们发展出对音乐的热爱。 |
| 社会学 | 一些种类的燕子在北美和南非被称为紫崖燕；作为食虫鸟类，燕子扮演了有益的角色。 |
| 心理学 | 孩子们每次在练习乐器、演奏音乐时所锻炼的视觉和空间技能，可以强化他们心理与身体的连接；在采用抽象的符号来描述存在于脑海中和纸上的模式方面，音乐家和数学家相似；音乐能在情感、精神和身体上激发我们，帮我们记住信息和学习经验；音乐能创造一种高度集中的学习状态，此时大量信息能被大脑处理和学习。 |
| 系统论 | 音乐不只是纯粹的艺术，它也与数学有关联，并在许多人的生活中起到重要作用，唤起生活的节奏。 |

# 教师与家长指南

## 情感智慧
### Emotional Intelligence

**小猎狗**

这只小猎狗担心会失去聆听鸟类美好歌声的机会。他特别喜欢听燕子唱歌，所以特别关心这一点。他意识到音乐具有启发和激励的作用，但他对燕子将莫扎特称为数学家感到惊奇。然而，这引发了他对工程实际与生活现实进行对比的思考。小猎狗欣赏燕子在人类中营造积极心态的力量，这种力量激发人类欣赏自然。小猎狗坚持认为唱歌与长声尖叫不同，他愉快地接受音乐带来的乐趣。

**燕 子**

即便燕子下定决心把音乐作为日常生活的一部分，但也清楚地表明她在生活中的优先事项是哪些。她并不刻意取悦其他伙伴，但她准备全身心地与他们分享乐趣。她谦逊而诚实，表达出对那些卓越表现者的钦佩，希望将荣誉颁发给那些应该获得荣誉的伙伴。燕子对小猎狗提出质疑，启发他透过表面现象来思考问题。然后，在她专注于她最重要的事务——养育后代时，分享了一些有关生活和设定优先事项的智慧。燕子怀念她的父母，并对能在一个载歌载舞的快乐家庭氛围下成长满怀感激。

## 艺术
### The Arts

你会吹口哨吗？如果会，就聆听鸟儿的叫声，并模仿它们的歌声，甚至可以与你的朋友们举办一个口哨音乐会。如果不会，则可以用一块木头来制造口哨。从互联网上查找相关指导，用一些简单的工具来帮助你雕刻。

# TEACHER AND PARENT GUIDE

## 思维拓展
### Systems: Making the Connections

在地球上，声音始终是我们生活的一部分。鸟类具有通过美妙的声音来表达喜悦、呼唤伴侣或发出警报的独特能力。鸟类在城市周边筑巢，用其优美的节奏和旋律鼓舞着人们。遗憾的是，一些人将鸟类从天然栖息地中抓走并运往全世界，迫使它们在鸟笼中度过余生。然而，仍有可能在家附近创造吸引鸟类的良好环境，为它们筑巢提供条件，帮助它们在寒冬和暴风雨天下存活。音乐和舞蹈一直是人类文化和传统中非常重要的组成部分，甚至比我们的说话能力还重要。然而，由于缺乏足够的时间和精力在家播放音乐或唱歌跳舞，拜金主义的生活方式常常妨碍我们享受音乐和舞蹈的快乐。技术也扮演了重要的角色，让人们只有一种享受音乐的形式，即限于在电子设备上听音乐或看音乐视频。我们意识到音乐和舞蹈对我们思想、情绪和人际关系的影响，它们能激发创造力，增强情感上的联系。此外，还存在"莫扎特效应"：结构化的音乐改善我们的心理表现，特别是在代数、几何以及时空技能上。在人生旅程中，音乐会陪伴我们。不管我们的肩上或心里有什么负担，每当我们在黎明时分聆听到第一只鸟儿的叫声，正如同我们聆听凯特·斯蒂文斯演唱的歌曲《破晓》那样，我们都能以微笑来开始新的一天。

## 动手能力
### Capacity to Implement

有多少音乐家给你带来过真正的快乐？我们可以列出一些世界著名的音乐家，但你了解在当地演出的艺术家吗？在周围打听一下，你的老师们中是否有优秀的音乐家。或者，你有让你享受音乐的家庭成员吗？你的任务是在社区中发现音乐家和歌唱家，也就是那些能让你享受音乐，或给你带来幸福的人。你愿意邀请他们来举行一场演唱会吗？

# 教师与家长指南

故事灵感来自
## 安妮-弗瑞德·罗伊斯（弗里达·林斯塔）
## Anni-Frid Reuss (Frida Lyngstad)

安妮-弗瑞德·罗伊斯（弗里达·林斯塔）出生于挪威，后移居瑞典。她在13岁时开始表演，并发展出对包括格伦·米勒、杜克·艾灵顿和考特·贝西的所有曲目在内的各种爵士乐的热爱。她于1963年组建了自己的乐队，并于1967年录制了她的第一首歌曲。后来她加入阿巴合唱团，成为现代史上最成功的音乐组合之一的成员。她坚定地致力于环境问题，是瑞典保护机构"自然脚步"（the Natural Step）的董事会成员，也是非政府组织"环境艺术家"的主席。

她喜欢花时间去瑞士高山和地中海沿岸，帮助健康领域的前沿企业取得成功，并投身于慈善工作。

更多资讯

http://thekidshouldseethis.com/post/how-do-cliff-swallows-build-their-mud-pellet-nests

www.electrummagazine.com/2013/06/mozart-and-mathematics/

https://m.youtube.com/watch?v=kZYw7VvB44s

www.ams.org/samplings/math-and-music

## 图书在版编目（CIP）数据

感谢你的音乐：汉英对照 /（比）冈特·鲍利著；
（哥伦）凯瑟琳娜·巴赫绘；唐继荣译. -- 上海：学林
出版社，2016.6
（冈特生态童书. 第三辑）
ISBN 978-7-5486-1055-7

Ⅰ. ①感… Ⅱ. ①冈… ②凯… ③唐… Ⅲ. ①生态环
境－环境保护－儿童读物－汉、英 Ⅳ. ① X171.1-49

中国版本图书馆 CIP 数据核字（2016）第 125804 号

---

ⓒ 2015 Gunter Pauli
著作权合同登记号 图字 09-2016-309 号

## 冈特生态童书
### 感谢你的音乐

| | |
|---|---|
| 作　　者—— | 冈特·鲍利 |
| 译　　者—— | 唐继荣 |
| 策　　划—— | 匡志强 |
| 责任编辑—— | 程　洋 |
| 装帧设计—— | 魏　来 |
| 出　　版—— | 上海世纪出版股份有限公司 学林出版社 |
| | 地　址：上海钦州南路 81 号　电话／传真：021-64515005 |
| | 网址：www.xuelinpress.com |
| 发　　行—— | 上海世纪出版股份有限公司发行中心 |
| | （上海福建中路 193 号　网址：www.ewen.co） |
| 印　　刷—— | 上海丽佳制版印刷有限公司 |
| 开　　本—— | 710×1020　1/16 |
| 印　　张—— | 2 |
| 字　　数—— | 5 万 |
| 版　　次—— | 2016 年 6 月第 1 版 |
| | 2016 年 6 月第 1 次印刷 |
| 书　　号—— | ISBN 978-7-5486-1055-7/G·390 |
| 定　　价—— | 10.00 元 |

（如发生印刷、装订质量问题，读者可向工厂调换）

# 蜜蜂的数学

## The Mathematics of Honey

Gunter Pauli

冈特·鲍利 著

凯瑟琳娜·巴赫 绘

唐继荣 译

## 丛书编委会

主 任：贾 峰

副主任：何家振　郑立明

委 员：牛玲娟　李原原　李曙东　吴建民　彭 勇
　　　　冯 缨　靳增江

## 丛书出版委员会

主 任：段学俭

副主任：匡志强　张 蓉

成 员：叶 刚　李晓梅　魏 来　徐雅清　田振军
　　　　蔡雩奇　程 洋

特别感谢以下热心人士对译稿润色工作的支持：

姜竹青　韩 笑　贾 芳　刘 晓　张黎立　刘之杰
高 青　周依奇　彭 江　于函玉　于 哲　单 威
姚爱静　刘 洋　高 艳　孙笑非　郑莉霞　周 蕊

# 目录

| | |
|---|---|
| 蜜蜂的数学 | 4 |
| 你知道吗？ | 22 |
| 想一想 | 26 |
| 自己动手！ | 27 |
| 学科知识 | 28 |
| 情感智慧 | 29 |
| 艺术 | 29 |
| 思维拓展 | 30 |
| 动手能力 | 30 |
| 故事灵感来自 | 31 |

# Contents

| | |
|---|---|
| The Mathematics of Honey | 4 |
| Did you know? | 22 |
| Think about it | 26 |
| Do it yourself! | 27 |
| Academic Knowledge | 28 |
| Emotional Intelligence | 29 |
| The Arts | 29 |
| Systems: Making the Connections | 30 |
| Capacity to Implement | 30 |
| This fable is inspired by | 31 |

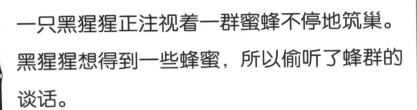

一只黑猩猩正注视着一群蜜蜂不停地筑巢。黑猩猩想得到一些蜂蜜,所以偷听了蜂群的谈话。

"你们吃了太多的蜂蜜,但却没有生产足够的蜂蜡。"蜂王斥责道。

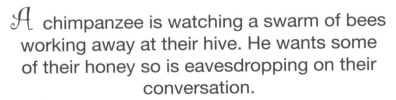

A chimpanzee is watching a swarm of bees working away at their hive. He wants some of their honey so is eavesdropping on their conversation.

"You are eating too much honey and not making enough wax," scolds the queen bee.

你们吃了太多的蜂蜜

you are eating too much honey

更多蜜蜂扇动他们的翅膀

More bees to flap their wings

"抱歉，陛下。"一只工蜂回答道，"我们正竭尽全力挤压出蜡片，用来建造蜂巢。"

"为了让温度保持在35摄氏度，我们还需要更多蜜蜂扇动他们的翅膀。"蜂王坚持道，"这是我们能让蜂蜡保持坚固，但仍具可塑性以用于筑巢的唯一方式。"

"I apologise, Your Majesty," responds one of the worker bees. "We are doing the best we can to squeeze out flakes of wax to make the honeycomb."

"We also need more bees to flap their wings to keep the temperature at exactly 35 degrees Celsius," insists the queen. "This is the only way we can keep the wax firm but still workable enough to build the hive."

"那么，安排更多的蜜蜂进到蜂巢里来吧，越多越好！"

黑猩猩偷看到蜂巢里有无数形状完美的六边形小窗口。这些蜜蜂不欢迎他的入侵，因此他在被蜜蜂蛰伤前迅速离开了。

"每间蜂室的室壁必须以正好120度的夹角相交，这样我们才能建造出这种六边形的室壁。"一只工蜂喊道。

"So let's get more bees into the hive, The more the merrier!"

The chimpanzee peeks inside and sees perfectly formed, six-sided clusters of little windows. The bees do not like his intrusion, so he quickly moves away, before he gets stung.

"The walls of every cell must meet at exactly 120 degrees so that we can build this wall of hexagons," shouts a worker bee.

正好120度

Exactly 120 degrees

永远不要把室壁造得比头发丝还粗

Never make the wall thicker than a hair

"我们还必须节约材料！"另一只工蜂说，"永远不要把室壁造得比头发丝还粗。"

"如果我们正确地建造好蜂巢，那么我们采集的花蜜将永远不会滴出，我们就有足够的花蜜来养活我们的孩子了。"那只工蜂说。

"And we must save on materials!" says another. "Never make the wall thicker than a hair's width."

"If we build the hive properly, the nectar we collected will never drip out and we can have enough to feed all our babies," says the worker.

"这就是我们要把蜂巢的蜂室倾斜的原因，这样储存的蜂蜜才不会滴出来。"

"当我们只用六边形的窗口时，我们能在蜂巢中储存更多的蜂蜜。如果我们用三角形或四边形的窗口，那就不能像这样储存蜂蜜了。"

"That's why we tilt the cells of our honeycomb so that the stored honey will not drip out."

"We can keep more honey in our hive when we only use six-sided windows. It would never work if we were to have triangular or square-shaped windows."

把蜂房倾斜

Tilt our honeycomb

# 正方形或长方形的砖瓦

Squares or rectangular tiles

"为什么人类总是用正方形或长方形的砖瓦、墙壁和窗户呢？"一只蜜蜂好奇地问。
"因为他们从未真正地认真对待数学。"

"Why are people always making use of squares or rectangular tiles, walls, and windows?" wonders one bee.
"Because they never really took maths seriously."

"但人类擅长研究数学,他们甚至为计算机开发出能为你运算一切的程序。"

"擅长?"一只工蜂惊讶地问,"他们知道的数学是线性的,而自然界中的绝大部分事物都不遵循这样的规律。我们,以及我们的六边形窗口,就是这样一个例外。"

"But people are good at studying maths. They even developed programs for computers that calculate everything for you."

"Good?" asks one worker in dismay. "Most maths they know is linear and hardly anything in nature follows that logic. We, with our six-sided windows, are one of the exceptions."

能为你运算一切的计算机

Computers that calculate everything for you

# 气泡、泡沫和水晶怎样形成

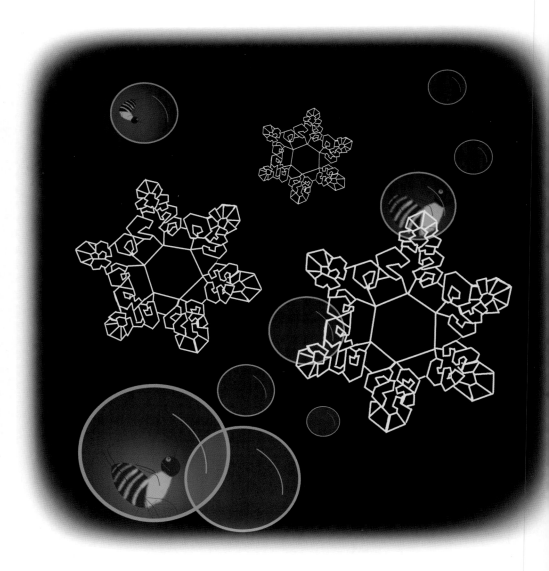

How bubbles, foam, and crystals form

"那么,当人类连气泡、泡沫和水晶怎样形成,以及植物细胞为什么从不线性生长都不懂时,为什么他们还想去研究基因?"

"这是因为太多人不知情。或许他们中的一些人喜欢那种方式,满足于他们所知道的一切。"

"So why do people want to play around with genes, when they don't even know how bubbles, foam, and crystals form, and why plant cells never grow in a straight line?"

"It is because so many people are uninformed. Perhaps some of them prefer it that way and want to stay in the comfort zone of what they know."

"因此,这些智人更应该被称为'不智之人'!"

"好吧,让我们相信他们的孩子将会做得更好一些吧。"蜂王叹息道。

……这仅仅是开始!……

"So these Homo sapiens should rather be called Homo non sapiens!"

"Well, let's trust that their children will do better," sighs the queen.

… AND IT HAS ONLY JUST BEGUN!…

……这仅仅是开始！……

...AND IT HAS ONLY JUST BEGUN!...

# Did You Know?

## 你知道吗？

Honeybees produce their own antitoxins to break down more than 100 kinds of pesticides and chemical treatments against mites. When commercial bees are fed high fructose corn syrup (HFCS), they are no longer able to do this.

蜜蜂产生自己的抗毒素，能分解超过100种农药和抗螨虫的化学物质。当商用蜜蜂被喂食高果糖浆（HFCS）时，它们将不再具备这种能力。

Bees visit 600 flower sites and pollinate over 40 crops every season. Commercial bees' ability to pollinate is negatively affected when fed HFCS containing minute amounts of chemicals (20 parts per billion), which is well below permitted concentrations.

蜜蜂每个季度访问600处开花点，为40多种农作物传粉。当喂食的高果糖浆含有微量人造化学物质（20ppb）时，即便远低于允许浓度，蜜蜂的授粉能力仍然会受到负面影响。

Neonicotinoids are the most toxic pesticides for bees. Pesticides are applied to plants, seeds, and on virtually all forms of genetically modified corn, but also to cauliflower, broccoli, apples, and pears.

烟碱类物质对蜜蜂来说是最毒的农药。农药应用于一般植物、种子、所有形式的转基因玉米,还有花椰菜、西兰花、苹果和梨。

Bees that constantly explore new environments for food have a different form of genetic brain activity than those that live in hives. The chemical signals from the brain are similar to those in thrillseeking people.

与那些生活在蜂巢中的蜜蜂相比,为获取食物而经常探索新环境的蜜蜂有一种不同的大脑遗传活动。来自它们大脑中的化学信号与寻求刺激的人类大脑情况相似。

𝒮tingless bees are not docile. They will still attack: Male bees collide in midair with intruders and females bite their enemies until one of them drops dead. These tactics are used to defend themselves and to invade other hives to spread their genes.

无刺蜂的蜜蜂并不温顺，它们仍然会攻击：雄峰在半空中撞击，雌蜂叮咬入侵者，直至敌人跌落地面死亡。这种战术策略用于保护自己，以及入侵其他蜂巢以传播基因。

𝐵ees can choose to collect highprotein pollen or sugarfilled nectar. Nectar is a rich sugar source and the material for making honey, while pollen is high-protein food for bee larvae.

蜜蜂能选择收集高蛋白的花粉，或富含糖分的花蜜。花蜜是糖类丰富的来源，也是制造蜂蜜的原料，而花粉是蜜蜂幼虫的高蛋白食物。

黑猩猩生活在非洲中部地区的茂密森林中，能使用5种不同的自制工具来取食蜂巢中的蜂蜜，而这些蜂巢位于树上20米高处，或深达地下1米。

Chimpanzees living in dense forests in Central Africa use as many as five different self-made tools to get honey from beehives located up to 20 m up a tree and up to 1 m underground.

3000万年来，蜜蜂一直采用集体决策的方式。在多达1万只蜜蜂的蜂群里，有300~500只蜜蜂加入寻找新巢址的队伍，区域面积大到30平方千米。

Bees have been taking collective decisions for 30 million years. In a swarm of 10 000 bees there are 300-500 bees that join the search for new nest sites over areas large as 30 square kilometres.

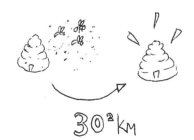

# Think About It

## 想一想

If bees pollinate apple and broccoli flowers and the insecticides used to protect these crops kill bees, should we continue using insecticides?

如果蜜蜂为苹果和西兰花的花传粉，而用于保护这些农作物的杀虫剂会杀死蜜蜂，我们还应该继续使用杀虫剂吗？

蜜蜂是聪明的建筑师，高效利用空间和建筑材料。我们人类建筑师是否应该学习蜂巢的工程学和数学呢？

Bees are smart architects, using space and building materials efficiently. Should architects study the engineering and mathematics of beehives?

Bees are not only architects; they are also very efficient in working together in large numbers. Could managers of companies learn anything interesting from them?

蜜蜂不仅是建筑师，而且在群体共同工作时也非常有效率。公司里的经理们能否从他们那里学到一些有趣的东西？

除了六边形之外，蜜蜂从来不在他们的建筑中采用直线，那么为什么人类常用直线和 90 度夹角呢？

If bees never use straight lines in their buildings, except for the hexagonal tile, why do people mostly make use of straight lines and 90° angles?

# Do It Yourself!
# 自己动手！

Triangles, squares, and hexagons are some of the only geometric shapes that can be used in tiling. Cut out some cardboard tiles, at least 20 of each shape, and try to cover a sheet of paper with them. Calculate how many tiles you will need of each of these geometrical shapes to cover a sheet of paper. Which shapes were the easiest to make and which ones were easiest to put together? After doing this on a flat, two-dimensional surface, see if you can glue the tiles together to make a three-dimensional beehive shape.

三角形、正方形和六边形是仅有的能用于砖瓦的几何形状。剪出一些纸板，每种形状至少20块，设法用它们来盖住一张纸。对于这些形状的纸板，计算每种形状需要多少块才能盖住一张纸。哪种形状最容易剪，哪种形状最容易拼在一起？在平坦的二维表面完成这项工作后，你能否把这些纸板用胶水粘在一起，用来制作一个三维的蜂巢形状。

# TEACHER AND PARENT GUIDE

## 学科知识
### Academic Knowledge

| | |
|---|---|
| 生物学 | 蜜蜂依赖被储存的花粉越冬，使冬季成为净化身体的"节食期"；蜜蜂通过摇摆运动进行通讯交流；雄峰只进行交配，并不采集花蜜或传粉，当食物缺乏时常被从蜂巢中赶出来；蜂王寿命长达5年；蜜蜂身体完全被毛覆盖，甚至眼睛上也是如此，以此最大可能地采集花粉。 |
| 化学 | 烟碱类物质被广泛用作杀虫剂，这也是蜜蜂减少的原因之一；对蜜蜂来说，农药降解产物通常比农药原来的化学物质毒性更大；蜂蜜含有天然的防腐剂，因而长久不变质。 |
| 物理 | 圆形比六边形的面积与周长之比更大，但当你用圆形砖瓦时，空间会浪费；蜜蜂扇动翅膀来为蜂巢加热，使已消化的花蜜中的液体蒸发来生成蜂蜜；蜜蜂也可扇动翅膀来为蜂巢降温；碳环结构中六边形是最稳定的结构，而石墨由类似蜂巢的六边形碳原子网络结构组成。 |
| 工程学 | 蜜蜂建造蜂巢时，从制作粗糙的圆形小室开始，因为这对材料的利用最充分，然后将这些材料弯曲成六边形的排列；幼蜂分泌圆珠笔尖大小的蜡片，其他的蜂则生产厚度不超过0.2毫米的圆柱形腔室。 |
| 经济学 | 商用蜜蜂在授粉方面的效率不如野蜂，这证明野生昆虫不能被替代；据估计，昆虫授粉对世界经济的价值贡献每年超过2000亿美元。 |
| 伦理学 | 一些人缺乏学习能力，而另一些人却拒绝学习；某些科学家希望能批准进行遗传操作，将人类遗传物质与合成化学物质结合，但他们并未意识到这样做将对生态系统造成冲击。 |
| 历史 | 埃及人从4500年前便养蜂，而在埃及神话中，蜜蜂由太阳神"拉"的眼泪转化而来；公元四世纪的几何学者帕普斯提出，蜜蜂选择六边形为蜂室的形状，是因为这样只需使用最少量的蜂蜡。 |
| 地理 | 除了在植物只能自花授粉或依赖鸟类传粉的南极洲外，其他所有大洲都有蜜蜂分布。 |
| 数学 | 蜜蜂需要3~4千克蜂蜜才能生产出1千克的蜂蜡；为了生产一茶匙的蜂蜜，一只蜜蜂必须往返花丛150次，而为了生产1000升的蜂蜜，需要造访花丛400万次。 |
| 生活方式 | 虽然一些人接受转基因食品，但许多人并不容忍这种食物消费，而是更愿意接受他们本地的文化、传统和生态系统。 |
| 社会学 | 蜜蜂采取集体决策方式已长达3000万年，这有助于它们的生存；蜜蜂被称为社会性昆虫，因为它们生活在群体中，相互依赖；"嘴唇抹了蜂蜜"是描述口才好的一种表达方式。 |
| 心理学 | 持之以恒：耗费多年精力证明六边形是最有效的储存模式。 |
| 系统论 | 为了可持续性，我们必须用一个系统来替代另一个系统，而不是用一件产品来替代另一件产品，或者一个过程替代另一个过程。 |

# 教师与家长指南

## 情感智慧
### Emotional Intelligence

蜜蜂

蜜蜂很热情，行为举止彬彬有礼，并且勇于认错，竭尽所能把事情做得最好。他们做了很多努力，并忠诚于集体。他们的工作精度很高，体现了他们的非凡技能，以及掌控环境的能力。他们欢迎其他成员加入他们的群体，并分享他们的知识和目标：建设蜂巢、找到食物。蜜蜂保卫蜂巢和后代，决心攻击任何入侵者。这证明体型不是最重要的方面，蜜蜂能依赖他们庞大的个体数量来对付所有的敌人。蜜蜂执着于资源效率，不纵容浪费，并懂得对几何学的机智利用将提高他们的效率。蜜蜂将他们的数学与人类的数学进行对比，得到的结论是：人类肯定是无知的。然而，他们希望人类的下一代能做得更好。

## 艺术
### The Arts

找一些蜜蜂的特写图片，研究它们以及蜂巢六边形的窗口形状。想象一下你能用哪些方式将你所看到的整合进一件艺术作品。你可以从生物学家安娜·瓦尔马（Anna Varma）那里得到启发，她拍摄了一些令人惊奇的蜜蜂及其生活条件的图片。现在画一幅包含六边形与蜜蜂一家的图像，你得花时间才能让你的绘画反映出蜜蜂的忙碌。

# TEACHER AND PARENT GUIDE

## 思维拓展
### Systems: Making the Connections

　　蜜蜂是生态系统中的重要角色,是生态系统健康非常重要的指示器。蜜蜂已经被许多文化所赞美,如玛雅、中国、埃及和希腊文明。蜜蜂在社会组织机构、资源效率和物理建筑结构上具有令人印象深刻的指挥能力,它还有非凡的免疫系统,从来不生病。蜂蜜、蜂蜡和蜂王浆总是被用于治疗。最近蜜蜂开始遭受睫毛上螨类的侵害,但很快就弄清楚这是由蜂巢中人造化学物质积累引起的生理应激所致。即便蜜蜂拥有自己的抗毒素,但随着时间的推移,微量但多种多样的人造化学物质,以及食物从天然(花蜜和花粉)到高果糖浆的转变,导致世界上80%的蜜蜂种群崩溃。野蜂没有能力为上千公顷的植物授粉,而喂食高果糖浆的商用蜜蜂缺乏建立抗病群体的营养物质,这导致作物的生产力迅速降低。蜜蜂生态系统服务功能的下降不能只归咎于某种人造化学物质或某种食物类型,它是在过去数十年里出现的多种因素综合影响的结果,因而不可能迅速解决。虽然兴旺繁盛的蜜蜂群体已经激发政策制定者和管理专家的灵感,但蜜蜂当前的状况迫使我们寻找让整个自然回归演化路径的方式和手段。尽管有机种植是改善当前状况的重要一步,我们不得不承认,生态系统中不断积累的成千上万种化学分子,使得该过程变得复杂而缓慢。这是一个典型案例,说明我们不能简单地改变或禁止某种人造化学物质,而是要改变整个系统。

## 动手能力
### Capacity to Implement

　　你是否近距离看过蜂巢?蜜蜂生活在任何有花并气候允许的地方,这给我们绝好的机会,去热带、温带以及寒冷气候下观察蜜蜂的灵活性和适应性,以及它们如何生产蜂蜜、蜂王浆。问问自己:我能做什么来帮助我所在地区的蜜蜂兴旺繁盛?现在,想想你能做什么来让蜜蜂更健康,当然也包括增加它们的生产力。关键是找到帮助蜜蜂在大自然中发挥其预期功能的方式。做一些探究,每次针对一个蜂巢,致力于实现你的想法。

# 教师与家长指南

## 故事灵感来自
## 巴克明斯特·富勒
## Buckminster Fuller

巴克明斯特·富勒出生于美国马萨诸塞州米尔顿市。他在学习几何学上有困难，但他高兴地用在树林中发现的材料自制工具。他推广了由德国工程师瓦尔特·鲍尔斯费尔德设计的穹顶（一种半球形的结构），这些穹顶具有连续的张力和不连续的压力。在半个世纪里，富勒有许多设计和发明，被授予许多专利。他认为社会很快将依赖于可再生能源。他致力于建筑学、工程学和设计学上的能源和材料的效率问题，希望将科学原理应用于解决人类的问题。他设计了1967年蒙特利尔世界博览会的"生物圈"项目，主要由六边形小格和六个五边形小格组成。蒙特利尔市和加拿大政府后来将这座建筑物转变为一所介绍水资源和可持续发展的博物馆。

### 更多资讯

http://inhabitat.com/6-buzzworthy-backyard-beehive-designs/

http://mathworld.wolfram.com/HoneycombConjecture.html

http://proof.nationalgeographic.com/2015/05/13/for-a-biologist-turned-photographer-a-beehive-becomes-a-living-lab/

图书在版编目（CIP）数据

蜜蜂的数学：汉英对照／（比）冈特·鲍利著；
（哥伦）凯瑟琳娜·巴赫绘；唐继荣译. －－ 上海：学林
出版社，2016.6
（冈特生态童书. 第三辑）
ISBN 978－7－5486－1057－1

Ⅰ. ①蜜… Ⅱ. ①冈… ②凯… ③唐… Ⅲ. ①生态环
境－环境保护－儿童读物－汉、英 Ⅳ. ① X171.1－49

中国版本图书馆 CIP 数据核字 (2016) 第 125805 号

ⓒ 2015 Gunter Pauli
著作权合同登记号 图字 09－2016－309 号

# 冈特生态童书
## 蜜蜂的数学

| | |
|---|---|
| 作　　者—— | 冈特·鲍利 |
| 译　　者—— | 唐继荣 |
| 策　　划—— | 匡志强 |
| 责任编辑—— | 程　洋 |
| 装帧设计—— | 魏　来 |
| 出　　版—— | 上海世纪出版股份有限公司 学林出版社 |
| | 地　址：上海钦州南路 81 号　电话／传真：021-64515005 |
| | 网址：www.xuelinpress.com |
| 发　　行—— | 上海世纪出版股份有限公司发行中心 |
| | （上海福建中路 193 号 网址：www.ewen.co） |
| 印　　刷—— | 上海丽佳制版印刷有限公司 |
| 开　　本—— | 710×1020　1/16 |
| 印　　张—— | 2 |
| 字　　数—— | 5 万 |
| 版　　次—— | 2016 年 6 月第 1 版 |
| | 2016 年 6 月第 1 次印刷 |
| 书　　号—— | ISBN 978－7－5486－1057－1/G · 392 |
| 定　　价—— | 10.00 元 |

（如发生印刷、装订质量问题，读者可向工厂调换）

# Energy 108

# 伴日航行

## Sailing with the Sun

**Gunter Pauli**

冈特·鲍利 著
凯瑟琳娜·巴赫 绘
高　芳　李原原 译

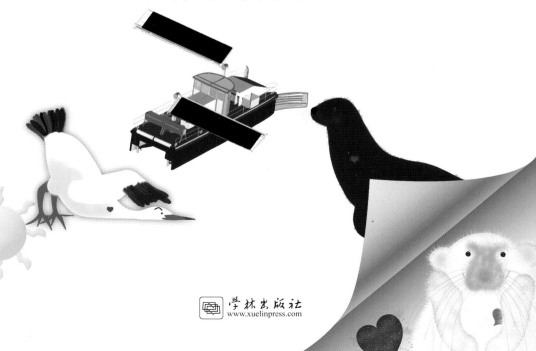

学林出版社
www.xuelinpress.com

## 丛书编委会

主　任：贾　峰
副主任：何家振　郑立明
委　员：牛玲娟　李原原　李曙东　吴建民　彭　勇
　　　　冯　缨　靳增江

## 丛书出版委员会

主　任：段学俭
副主任：匡志强　张　蓉
成　员：叶　刚　李晓梅　魏　来　徐雅清　田振军
　　　　蔡雩奇　程　洋

特别感谢以下热心人士对译稿润色工作的支持：

姜竹青　韩　笑　贾　芳　刘　晓　张黎立　刘之杰
高　青　周依奇　彭　江　于函玉　于　哲　单　威
姚爱静　刘　洋　高　艳　孙笑非　郑莉霞　周　蕊

# 目录

| | |
|---|---|
| 伴日航行 | 4 |
| 你知道吗？ | 22 |
| 想一想 | 26 |
| 自己动手！ | 27 |
| 学科知识 | 28 |
| 情感智慧 | 29 |
| 艺术 | 29 |
| 思维拓展 | 30 |
| 动手能力 | 30 |
| 故事灵感来自 | 31 |

# Contents

| | |
|---|---|
| Sailing with the Sun | 4 |
| Did you know? | 22 |
| Think about it | 26 |
| Do it yourself! | 27 |
| Academic Knowledge | 28 |
| Emotional Intelligence | 29 |
| The Arts | 29 |
| Systems: Making the Connections | 30 |
| Capacity to Implement | 30 |
| This fable is inspired by | 31 |

一只海狮在看一艘船启航。一群海鸥正准备加入这次航行。

"这艘双体船有四个帆,比普通的多两个。而且那些帆不是普通的帆。"海狮说。"显然不是普通的船,这一定是艘昂贵的船。"

A sea lion is watching a boat set sail. A flock of seagulls are getting ready to join the voyage.

"This catamaran has four sails. That is two more than usual. And those are no ordinary sails," remarks the sea lion. "This is clearly no ordinary vessel. It must be an expensive boat."

这一定是艘昂贵的船

It must be an expensive boat

但是你看，它有太阳能电池

But look, it has solar cells

"但是你看,它有太阳能电池。"海鸥评论道。"我从这上面可以看到,这艘船甚至利用螺旋桨引起的湍流产生动力。"他补充说。

"这是怎样起作用的呢?"海狮很好奇。

"But look, it has solar cells," observes the seagull. "The boat even gets power from the turbulence created by a propeller. I can see that from up here," he adds.

"How does that work?" wonders the sea lion.

"嗯，就像那些混合动力汽车利用刹车产生动力一样，这艘船有一对螺旋桨，可以把推进力和水流所产生的多余尾流全部利用起来。"

"这是那种'绿色'船只中的一艘吗？"海狮问道。

"Well, just like those hybrid cars get power from using their brakes, this boat has a double propeller that catches all the excess wake from propulsion and the current."

"Is this then one of those 'green' boats?" asks the sea lion.

这是那种"绿色"船只中的一艘吗?

Is this one of those "green" boats?

它被称为"蓝色"船只

This is called a "blue" boat

"不,它被称为'蓝色'船只。毕竟,水是蓝色的,天空是蓝色的,从外太空看地球也是蓝色的。"

"这听起来很浪漫,但我们得面对现实。一艘有四个帆、太阳能电池以及双螺旋桨的船肯定更贵吧?三种能源替代了化石燃料,那些沉重、肮脏的东西才便宜!"海狮惊呼道。

"No, this is called a 'blue' boat. After all, the water is blue, the sky is blue, and the earth viewed from outer space is also blue."

"That sounds very romantic, but let's be realistic here. A boat with four sails, solar cells and double screws must surely be more expensive? Three sources of power instead of just fuel, that heavy, dirty stuff that is supposedly cheap!" exclaims the sea lion.

"嗯,这种双体船不使用内燃机而使用电动引擎。在无风的时候,帆被展平,内置的太阳能电池就会从早到晚工作。"

"打断一下,太阳能电池不是只在白天工作吗?我从没听说太阳能可以在晚上工作。"

"Well, this catamaran does not use a combustion engine but rather an electric engine. And when the wind is calm, the sails are flattened and the built-in solar cells work day and night."

"Excuse me, but don't solar cells only work during the day? I've never heard of solar power working in the dark."

太阳能电池不是只在白天工作吗?

Don't solar cells only work during the day?

我真的不知道

I really don't know

"真的吗?你认为外面漆黑一片、天空晴朗时会发生什么呢?"海鸥问道。

"我真的不知道。"海狮回答。

"Really? What do you think happens when something is black and it's dark outside, with clear skies?" asks the seagull.

"I really don't know," replies the sea lion.

"嗯,试想一下。如果白天变热,那么夜晚它会变……?"

"变冷吗?"

"是的。如果你捕捞到很多新鲜的鱼,你会用冷水做什么呢?"

"Well, think about it. If it gets hot during the day, so during the night it will get …?"

"Cold?"

"Indeed. And what can you do with cold water if you have lots of freshly caught fish?"

你会用冷水做什么呢?

What can you do with cold water?

# 新鲜的鱼需要冷藏

Freshly caught fish need to be kept cool

"嗯，我猜想新捕捞到的鱼是不是需要冷藏来保持新鲜和美味？"

"完全正确。"海鸥回答，"我们需要帆捕捉风和阳光来发电，需要用冷水保持鱼的新鲜，以及用螺旋桨的湍流来产生额外的电力。这是一艘绝妙的双体船。"

"Well, I suppose freshly caught fish need to be kept cool to stay fresh and tasty?"

"Exactly," replies the seagull. "We need sails to catch the wind, sunlight to make electricity, cool water to keep the fish fresh, and the turbulence of the propeller to generate additional power. This is a wonderful catamaran."

"并且,
如果鱼被冷鲜保
存,可以保持味道鲜美。
嗯……我最好还是和你一起加入
这次航行,而不是直到你返航了
还在岸边徘徊。"
……这仅仅是开始!……

"And, if fish is kept cold and fresh, it remains so tasty. Mm ... I may as well join you on this voyage instead of lingering on the shore until you return."
… AND IT HAS ONLY JUST BEGUN!…

……这仅仅是开始!……

...AND IT HAS ONLY JUST BEGUN!...

你知道吗？

一艘帆船同时被风推动和拉动着向前。它应用的空气动力学原理与风在飞机机翼旁的流动是一样的。

𝒜 sailboat is both pushed and pulled forward by the wind. It works on the same aerodynamic principle as that of the wind flow over the wings of an aircraft.

19世纪90年代发电制动被发明，它首次应用于丰田和本田的汽车制造，现已被应用于电气铁路。

𝒢enerating power by braking, as first used in cars manufactured by Toyota and Honda, was already invented during the 1890s and applied to the electric railway.

双体船有两个船体，在6000年前被泰米尔人首次使用。"双体"一词来源于泰米尔语"kattumaram"。

The catamaran has two hulls and was first used by the Tamils 6 000 years ago. The term "catamaran" is derived from the Tamil word *kattumaram*.

当只有一个电源时，电源故障的风险很大。然而，当有三种不同的电力来源，且有一种是可再生能源时，连续获得电力的机会就会很高。

When there is only one power source, the risk of power failure is high. However, when there are three different sources of power, and renewable ones at that, the chance of continuous access to power is high.

方尖碑是高高的四面锥形的纪念碑，公元前200年在古埃及建造时，由双体船在尼罗河上运载。

The obelisks, the tall, four-sided, tapered monuments originally made in Ancient Egypt, were transported on the Nile in catamarans in 200 BC.

如今世界上有400万艘渔船，使用桨和帆的只有180万艘，而其他船只——超过50%的船都有一个引擎。印度尼西亚有70万艘渔船，其中一半没有引擎。

There are 4 million fishing boats in the world today. 1,8 million of these use only oars or sails, whereas the others – just over 50% of them – have an engine. Indonesia alone has 700 000 fishing boats, half of which are without an engine.

In 3 000 BC, the Egyptians made cotton sails to use the wind to assist propulsion by oars. The combination of sails and manpower (to pull the oars) allowed the Egyptians to sail the Mediterranean. China introduced sails at about the same time.

在公元前3000年，埃及人制作棉质的帆利用风来协助桨的推进。风帆和人力的组合（拉桨）使埃及人在地中海航行成为可能。中国大约在同一时期引进了帆。

Black ice, also called clear ice, (a thin layer of ice that allows the black road or surface underneath it to show through) is generated on surfaces when the open night sky radiates heat. This ice is produced even when the ambient temperature is not freezing.

黑冰，也称为透明冰（一层薄薄的冰，我们能看见在它下面的黑色道路或表面），当空旷的夜空散发热量时，黑冰在表层生成，甚至在环境温度不足以冻结时也能产生。

# Think About It
# 想一想

Do you think solar panels can work at night?

你认为太阳能电池板能在晚上工作吗?

绿色产品尤其是绿色能源比传统产品和能源更昂贵吗?

Are green products in general and green energy in particular more expensive than conventional products or power?

Should the government subsidise green products and power to make them cheaper and more competitive?

政府应该补贴绿色产品和能源使它们更便宜、更有竞争力吗?

石油的价格是使用这种燃料的唯一成本吗?还是存在其他隐性成本?

Is the price of petroleum the only cost involved in using this kind of fuel? Or are there any hidden costs?

## 自己动手!

Have you ever gone sailing? If you live too far from the sea or a lake, study sailing with a miniature boat in the sink or bathtub. Pay special attention to how the boat manages to sail against the wind. This is an interesting technique that everyone should master: Using the force against us to go forward. This does not only apply to sailing, but is also a good life lesson.

你曾经航海过吗？如果你住的地方离海或湖太远，那就在水槽或浴缸里研究一下小型船的航行吧，特别注意如何操控船逆风航行。这是一个很有趣的技术，每个人都应该掌握：使用反作用力推动我们前进。这不仅适用于航行，也是一条很好的生活经验。

# TEACHER AND PARENT GUIDE

## 学科知识
### Academic Knowledge

| 生物学 | 海狮和海豹的区别。 |
|---|---|
| 化 学 | 沸石和熔盐类的化学物质吸收和释放热量比较快，在低温下储存可以避免不必要的化学反应。 |
| 物 理 | 当帆和风的方向相同时，它推着船前进，当逆风航行时，帆拉动船，把空气引向船的后部并加速；逆着风的帆受到一个反向的力（牛顿第三定律）；帆的一面高压，另一面低压，压差使帆获得一个提升力，正如空气流过机翼上下的方式；辐射冷却靠的是将热量辐射到外部空间。 |
| 工程学 | 再生制动是一种能量回收机制，在一辆车减速时，它的动能被转化为可以立即使用或储存的能量，这种技术最近被用于混合动力汽车；宇宙船在真空中穿梭，靠辐射散发多余的热量；屋顶降温系统结合了高光学反射率和高红外发射率，白天减少太阳的热传递，晚上靠辐射增加散热；智能电网使我们能协调多种能源连续供电。 |
| 经济学 | 我们面临的一个问题是绿色经济作为一种潜在的理想生活方式，维护起来是非常昂贵的；所谓廉价化石燃料的隐性成本：尽管最初看起来很便宜，但它涉及许多环境成本和对我们健康造成的危害，这些都是我们所说的转移成本，即成本由社会承担；船靠三种可再生能源获取动力更加昂贵，但由于不需要化石燃料，也是值得的。 |
| 伦理学 | 我们怎样才能维持绿色经济？它是如此昂贵的可持续绿色生活方式以至于只有富人才能承担得起。 |
| 历 史 | 波利尼西亚人以及德拉威人穿越太平洋航行时首次使用双体船；1886年，弗兰克·斯普拉格在美国首次使用再生制动。 |
| 地 理 | 从外太空看到的地球是蓝色的，伴有一些白色、绿色和棕色斑点；在大海中的航行告诉我们，两点之间最短的旅行方式不在一条直线上，首先因为地球是圆的，其次因为水流和风使船选择沿着阻力最小的路径航行。 |
| 数 学 | 安德烈·柯尔莫哥洛夫的湍流能量定律；翼帆的几何设计提供了比传统帆更大的升力；翼或帆顶部和底部表面之间的曲面是不对称的，这能帮助产生空气动力。 |
| 生活方式 | 现代社会根据需要采用开关简单地开关电能，而大自然的能量系统要更复杂，但仍然可以控制；人们倾向于简单和传统的生活方式，但那是不可持续的，而可持续性的生活方式被认为是复杂的。 |
| 社会学 | 陪同他人航行的乐趣。 |
| 心理学 | 不要相信一些事情，即使它被解释得很好也能举出例子，利用逆着我们的力来前进。 |
| 系统论 | 用多种能源产生能量，减少能源需求，使用可再生能源，有可能让所有人类活动自给自足。 |

# 教师与家长指南

## 情感智慧
### Emotional Intelligence

**海 狮**

海狮正在观察分析新的帆船。他注意到太阳能电池内置在坚硬的帆里,从而得出结论:这想必是一艘昂贵的船。他意识到自己的无知,但很好奇并有勇气提问、探索,并试着去了解,得出结论:绿色或蓝色技术听起来浪漫,但一定很昂贵。他并不相信太阳能可以在晚上工作,但也承认,他并不真正了解。他聆听海鸥的解释,并激发了加入海鸥航行的动力,享受海鸥的陪伴和美味的食物。

**海 鸥**

海鸥准备着解释一切。他观察和分析船所使用的不同能源的区别,相当了解一切是如何运转的。他有清晰的观点并愿意分享意见。他意识到自己的一些见解不容易理解时就进行解释,通过提问来帮助海狮看透本质。海狮竭力使用常识时,海鸥坚持尝试,并通过提示来建立海狮的自信,否则他可能会自卑。这创造了一个二者的组合,海鸥也成功激励了海狮加入他的钓鱼之旅。

## 艺术
### The Arts

你如何表达湍流?要求学习者每次使用不同风格的图画表达湍流。每个人的图画风格不同,湍流的表达也会不同。现在看看这些不同的图画,选择其中最成功表述湍流的。画幅图来说明如何能够获取湍流的能量而不至于被损失掉。

# TEACHER AND PARENT GUIDE

## 思维拓展
### Systems: Making the Connections

  现代社会大多数家庭家里都通电，但却不知道电是从哪来的。现在电力已经能够进行远距离传输，而能量的产生已经成为经济进步和发展的动力，但当前的系统是不可持续的。碳排放达到一定水平，就会使空气中集聚大量的有害颗粒物，导致气候变化，增加健康风险。

  向可再生能源发展的转换进展十分缓慢，而且大多数未经测试。太阳能电池板放置在屋顶，风车放置于山脊，消化器里产沼气，这些能源混合后输送到现有的输电网，对传统能源，即煤炭、石油、核能和天然气进行补充。

  船的设计集成了多种可再生能源，如风能、太阳能、水能，它考虑到满足多种需求：运输和流动、冷却、照明、定位、安全，等等。所产生的能量大部分被直接利用，如用风力推动船的前进，用冷藏保持鱼的新鲜。如果风力首先用来产生电动引擎的能量，再推动船前进，那么就只剩下不到一半的能量能够被利用了。

  当把可再生能源和能源效率与几何学、流体动力学和材料学相结合来减少阻力时，虽然系统变得更加复杂，但性能却得到了改善。如果我们利用当地可获得的水流和风，那么可持续性发展就能取得巨大的成就。

## 动手能力
### Capacity to Implement

  机翼或帆的形状决定了飞机如何飞行，或者一艘船如何航行。选择你所喜欢的三种类型的帆或者机翼，学习如何把它们画出来。然后试着解释它们如何影响一艘船或一架飞机的空气动力。不需要了解特定、精确的几何知识，看看你能不能画出空气在船帆或机翼流动的方式，尝试解释其中的原理。为什么船帆或机翼能够工作呢？

# 教师与家长指南

故事灵感来自

## 安德烈·柯尔莫哥洛夫
## Andrey Kolmogorov

　　安德烈·柯尔莫哥洛夫生于坦波夫，距离莫斯科500多千米。他五岁时就已经开始学数学了。他毕业于莫斯科国立大学和莫斯科门捷列夫化学技术研究所，后来成为一位大学教授。他专注研究湍流，后来专注于复杂性算法理论。1956年，他出版了一本关于概率论基础的书。他不仅是一位大学教授，也积极参与天才儿童教育学的发展。他生前是俄罗斯科学院院士，获得过众多奖项和荣誉，包括列宁奖和沃尔夫奖。

更多资讯

　　https://terrytao.wordpress.com/2014/05/15/kolmogorovs-power-law-for-turbulence/

　　http://www.skybrary.aero/index.php/Aerofoil

# 图书在版编目（CIP）数据

伴日航行：汉英对照／（比）冈特·鲍利著；
（哥伦）凯瑟琳娜·巴赫绘；高芳，李原原译．－－上海：
学林出版社，2016.6
（冈特生态童书．第三辑）
ISBN 978-7-5486-1068-7

Ⅰ．①伴… Ⅱ．①冈… ②凯… ③高… ④李… Ⅲ．
①生态环境－环境保护－儿童读物－汉、英 Ⅳ.
① X171.1-49

中国版本图书馆 CIP 数据核字（2016）第 126072 号

————————————————————————————

© 2015 Gunter Pauli
著作权合同登记号 图字 09-2016-309 号

## 冈特生态童书
### 伴日航行

| | | |
|---|---|---|
| 作　　者—— | 冈特·鲍利 | |
| 译　　者—— | 高　芳　李原原 | |
| 策　　划—— | 匡志强 | |
| 责任编辑—— | 程　洋 | |
| 装帧设计—— | 魏　来 | |
| 出　　版—— | 上海世纪出版股份有限公司 学林出版社 | |
| | 地　址：上海钦州南路81号　电话／传真：021-64515005 | |
| | 网址：www.xuelinpress.com | |
| 发　　行—— | 上海世纪出版股份有限公司发行中心 | |
| | （上海福建中路193号 网址：www.ewen.co） | |
| 印　　刷—— | 上海丽佳制版印刷有限公司 | |
| 开　　本—— | 710×1020　1/16 | |
| 印　　张—— | 2 | |
| 字　　数—— | 5万 | |
| 版　　次—— | 2016年6月第1版 | |
| | 2016年6月第1次印刷 | |
| 书　　号—— | ISBN 978-7-5486-1068-7/G·403 | |
| 定　　价—— | 10.00 元 | |

（如发生印刷、装订质量问题，读者可向工厂调换）

Energy
73

# 旗帜飘飘

Fluttering Flags

Gunter Pauli

冈特·鲍利 著
凯瑟琳娜·巴赫 绘
高 芳 李原原 译

学林出版社
www.xuelinpress.com

## 丛书编委会

主　任：贾　峰
副主任：何家振　郑立明
委　员：牛玲娟　李原原　李曙东　吴建民　彭　勇
　　　　冯　缨　靳增江

## 丛书出版委员会

主　任：段学俭
副主任：匡志强　张　蓉
成　员：叶　刚　李晓梅　魏　来　徐雅清　田振军
　　　　蔡雩奇　程　洋

特别感谢以下热心人士对译稿润色工作的支持：

姜竹青　韩　笑　贾　芳　刘　晓　张黎立　刘之杰
高　青　周依奇　彭　江　于函玉　于　哲　单　威
姚爱静　刘　洋　高　艳　孙笑非　郑莉霞　周　蕊

# 目录

| | |
|---|---|
| 旗帜飘飘 | 4 |
| 你知道吗？ | 22 |
| 想一想 | 26 |
| 自己动手！ | 27 |
| 学科知识 | 28 |
| 情感智慧 | 29 |
| 艺术 | 29 |
| 思维拓展 | 30 |
| 动手能力 | 30 |
| 故事灵感来自 | 31 |

# Contents

| | |
|---|---|
| Fluttering Flags | 4 |
| Did you know? | 22 |
| Think about it | 26 |
| Do it yourself! | 27 |
| Academic Knowledge | 28 |
| Emotional Intelligence | 29 |
| The Arts | 29 |
| Systems: Making the Connections | 30 |
| Capacity to Implement | 30 |
| This fable is inspired by | 31 |

蝙蝠倒挂在他最喜欢的树上,享受着喜马拉雅山脉的清风。他望着地平线,注意到在附近的山脊上,人们正在建造风力涡轮机来发电。
"我从来都弄不明白为什么风能在这里会如此受欢迎。"他对猫头鹰说。"我们该不会真的要牺牲山谷的美丽,就只为了发点电吧。"

A bat is hanging from his favourite tree, enjoying the fresh breeze of the Himalayas. He looks over the horizon and notices that on a mountain ridge nearby people are building wind turbines to generate electricity.
"I've never understood why wind energy is so popular here," he says to the owl. "We should really never have to sacrifice the beauty of our valley just to make some electricity."

享受着喜马拉雅山脉的清风

Enjoying the fresh breeze of the Himalayas

# 风力涡轮机的噪声

Noise of the wind turbines

"噢，风力涡轮机的噪声在我脑袋周围咆哮，真的让我很难受。"猫头鹰补充道，"幸运的是，这些新的风车转得比过去慢了。至少我们不会难受得想把头砍下来！"

"顺便问一下，你知道水的密度比空气大得多吗？"蝙蝠问。

"Oh, and the noise of the wind turbines snarling around my head really bothers me," adds the owl. "Fortunately, these new windmills turn slower than in the past. So at least we will not get our heads chopped off!"

"By the way, did you know that water is much denser than air?" asks the bat.

"那又怎样？"猫头鹰回答。

"嗯，速度是每小时15千米的水流，要比速度超过每小时300千米的飓风拥有更多的能量呢。"

"真的吗？那人们为什么不试着用冰川融化后丰富的水流产生能量呢？"猫头鹰问道。

"So?" replies the owl.
"Well, water flowing at 15 kilometres per hour has more power than a hurricane blowing at more than 300 kilometres per hour."
"Really? So why are people not trying to make energy out of the abundance of flowing water we get from the melting glaciers?" asks the owl.

为什么不用水流产生能量呢？

Why not make energy out of flowing water?

学会了如何建造巨大的水坝

Taught how to build huge dams

"问题是,人们仅仅学会了如何建造巨大的水坝来发电。不幸的是,这些水坝淹没森林和村庄,毁灭所有生物的生命,其实这么做是完全没有必要的。如果人们早知道会是这样……"

"嗯,木已成舟。不如再告诉我些关于水坝的事吧!人们不再需要它们了吗?"猫头鹰问道。

"The problem is that people were only taught how to build huge dams to generate electricity. Unfortunately, these dams can flood forests and villages, ruining the lives of all living things, without really having to do so. If only they had known…"

"Well, what's done is done. Rather tell me more about these dams. Are they not needed anymore?" asks the owl.

"在过去,这是人们能想出的最好的解决方法了。但是今天,你可以把涡轮机放到水管中发电。"蝙蝠说。

"那是怎么做到的呢?"猫头鹰问道。

"水流通过管道时,涡轮就会旋转,并把由此产生的能量传递到发电机中,从而产生电力。涡轮机垂直地放置在管道中,沿着管道旋转。"

"In the old days it was the best solution they could come up with. But today you can put turbines in a water pipe to generate electricity," says the bat.

"So how does that work?" asks the owl.

"When water flows through the pipes, the turbines spin and send the energy created by doing that to a generator, which makes electricity. The turbines are placed vertically in the pipe and turn along with the pipe."

把涡轮机放到水管中

Put turbines in a water pipe

把涡轮机放到管道中为电灯提供电力

Turbines into pipes to power the lights

"这么说,把它们垂直而不是水平放置,就能安装更多的涡轮机?"猫头鹰问道。

"是啊!"蝙蝠回答。

"好简单啊!为什么人们不用这个方法取代建更多的水坝呢?"猫头鹰惊呼道。

"哦,好多年前人们就这么做了。中国的家庭还把小涡轮机放到小管道中为电灯提供电力。现在,甚至整个城市都开始利用水流发电。这种方法正在流行起来。"

"So instead of placing them horizontally, they fit in more turbines by placing them vertically?" asks the owl.

"Yes!" answers the bat.

"So simple! Why don't people do this instead of building more dams?" exclaims the owl.

"Oh, it has been done for years. In Chinese homes they even put tiny turbines into small pipes to power the lights. Even whole cities are starting to produce power by using the flow of water. They are catching on."

"真是聪明。我听说制造这些大型风力涡轮机需要大量的磁铁和稀土金属，更不用说还需要那些巨大的塔架了。"

"嗯，这些涡轮机沿用了从中世纪传下来的利用旋转产生能量的思想。但我认为这个思想时间上可能比旋转木马还要早。旋转产生过多的摩擦，而只要有摩擦存在，一切最终都将归于结束。"

"That's smart. I hear these big wind turbines need lots of magnets and rare earth metals to work, not to mention those huge masts."

"Well, these turbines follow the old logic of the Middle Ages by making use of rotation. But I think the time has come to go beyond the concept of a merry-go-round. They cause too much friction and whenever there is friction, things eventually break."

需要大量的磁铁和稀土金属

Need lots of magnets and rare earth metals

# 让东西随风摆动

Let things flutter in the wind

"那么，有没有更好的解决方案呢？"猫头鹰很好奇。

"当然有！那就是让东西随风摆动。"

"就像用来祈祷的旗帜和船帆那样在风中摆动？"猫头鹰问道。

"So, is there a better solution then?" wonders the owl.

"Absolutely! You just let things flutter in the wind."

"Flutter in the wind like prayer flags and boat sails do?" asks the owl.

"完全正确，"蝙蝠热情地说，"你可以把一块磁铁放在桅杆上，再将另一块捆在固定旗子或船帆的绳子上。当旗子或船帆在风中飘扬时，将会产生电流。"

"太好了！"猫头鹰惊呼道。

"这个山谷祈祷的人越多，产生的电也就越多。"

……这仅仅是开始！……

"Exactly," says the bat enthusiastically. "You can just put a magnet on the mast and another one on the rope that holds the flag or sail in place and as it flutters in the wind, it will generate electricity."

"Wonderful!" exclaims the owl. "The more prayers offered in this valley, the more electricity will be generated for its people."

… AND IT HAS ONLY JUST BEGUN!…

……这仅仅是开始！……

...AND IT HAS ONLY JUST BEGUN!...

# Did You Know?
## 你知道吗？

Because the blades of a wind turbine are very long – longer than the wingspan of a Boeing 747 – they have the potential to generate a lot of noise.

因为风力涡轮机的叶片很长，比波音747的机翼还长，所以它们可能产生很大的噪声。

每年，全美国会有十亿只鸟撞到建筑物的窗户上，而汽车和卡车又会杀死另外十亿只鸟。小叶片的风车表面积小，却比具有大表面积的大叶片风车更容易杀死鸟类。

Every year, throughout the United States of America, a billion birds collide against the windows of buildings. Cars and trucks kill about another billion birds. Windmills with small blades, and therefore smaller surface areas, kill more birds than windmills with very big blades and large surface areas.

Modern hydro turbines can convert as much as 90% of available energy into electricity. Only 2 400 of the 80 000 existing dams in the United States are used to generate power.

现代水力涡轮机可以把高达90%的可用能源转换成电能。在美国现有的80 000个水坝中只有2400个用于发电。

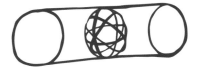

The Kingdom of Bhutan produces electricity from hydropower alone and 75% of this electricity is exported. This makes up 40% of all exports and 25% of the country's GDP.

不丹王国只靠水力发电,并且75%的电力用于出口,占全国总出口的40%和GDP的25%。

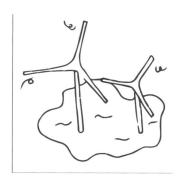

水磨自希腊和罗马时代就开始使用，转动磨盘使锤子发力来粉碎矿石。第一架风车于9世纪建于伊朗，12世纪才传入欧洲。

The watermill has already been used since Greek and Roman times to turn millstones and to power hammers to crush ore. The first windmill was constructed in the 9th century in Iran. Windmills were introduced to Europe in the 12th century.

藏传佛教的经幡（用来祈祷的旗帜）上四种表示尊贵的象征分别为：龙、金翅鸟（神话中的类鸟生物）、老虎和雪狮子。旗帜有五种颜色：蓝色、白色、红色、绿色和黄色，分别象征着天空、祥云、火焰、江河和大地。有人死了，就会升起108面经幡。

Tibetan Buddhist prayer flags display symbols of the four dignities: the dragon, the garuda (a mythical bird-like creature), the tiger, and the snow lion. The flags come in five colours: blue, white, red, green, and yellow, symbolising the sky, cloud, fire, water, and earth, respectively. When someone dies, 108 prayer flags are raised.

The flow of water, from flushing toilets to kitchen taps, can be guided through tiny turbines fitted into a pipe as small as 2 mm in diameter to generate electricity.

抽水马桶和厨房水龙头里的水流都可以通过一个安装在直径2毫米管道中的小涡轮机来发电。

Similar to how a balloon sticks to the wall after being rubbed against clothing, the fluttering motion of a flag generates energy because of the ability of charge to transfer between two materials.

类似于一个气球与衣服摩擦后就可以粘在墙上,一面旗帜的摆动也能产生能量,因为电荷具有在两种材料之间转移的能力。

# Think About It

## 想一想

Would you like your valley to be full of windmills, or would you prefer to have water flowing through underground pipes to generate power?

你是喜欢让你的山谷布满风车,还是更喜欢让水流通过地下管道来发电呢?

如果旗帜可以发电,你认为它能够满足所有人日常生活中所需要的电力吗?

If flags can generate power, do you think that would be enough to provide all the electricity people need in their daily lives?

If water is so much denser than wind, why do we invest more in wind power than in hydropower?

如果水比风的密度大很多,那为什么我们更多地投资风力而不是水力呢?

如果水流通过小管道可以发电,有没有可能在你的水龙头上安装一个传感器呢?它可以告诉你洗手的水是不是太热。

If the flow of water through small pipes can generate power, would it be possible to install a sensor in your tap, which can tell if the water is too hot to wash your hands?

# Do It Yourself!
## 自己动手!

Research the 10 main renewable energy sources of the future by consulting engineers about it. Then ask children under the age of 10 what they think the best ways are to generate electricity. Before you ask the children, tell them about the whale powering its own heart, leaves fluttering in the wind to aid trees in their absorption of $CO_2$, and the trout that swims against the current. Make a list of all the possible solutions put forward by the children and debate the differences. Who will be shaping the future? What will the energy sources of the future be?

通过咨询相关工程师，研究10种未来主要的可再生能源。问问10岁以下的孩子，他们所认为的最好的发电方式是什么。在问孩子们之前，告诉他们鲸用自己的心脏来提供动力，叶子在风中摆动来帮助树木吸收二氧化碳，以及鲑鱼逆流游泳。列出孩子们提出的所有可能的解决方案并且讨论其差异。谁将会塑造未来？未来的能源将会是什么？

# TEACHER AND PARENT GUIDE

## 学科知识
## Academic Knowledge

| | |
|---|---|
| 生物学 | 树叶摆动可促进树木吸收二氧化碳，一棵树上在树冠顶部飘动的树叶比下面静止的树叶温度低2～4℃；树叶飘动的结果是使二氧化碳的传递加倍。 |
| 化 学 | 钕和钐这两种稀土金属可提供永久性的磁场。 |
| 物 理 | 很多物理元件的质量、体积和密度都是可以测量的；水不能被压缩，但空气可以；密度随压力或温度变化而变化；理想气体定律可以用来计算干空气、水蒸气或两者混合物的密度；阿基米德原理：浸入液体中的物体受到向上的浮力，浮力的大小等于物体所排开的液体受到的重力；声音在水中的传播速度比在空气中快，这是由于水的密度更高，允许更多的粒子碰撞；热在水中的损耗速度比在空气中快20倍，这是由于水的密度较高，热传递更快。 |
| 工程学 | 永久磁场的产生；1克每立方厘米($g/cm^3$)和1千克每立方米($kg/m^3$)是常见的密度单位，$1g/cm^3$等于$1000kg/m^3$；工程师们用钕磁铁材料取代了风力涡轮机组中的齿轮箱；滚珠轴承降低了摩擦；风力涡轮机的输出已经从过去的200千瓦/转提高成现在的7.5兆瓦/转。 |
| 经济学 | 水电比核能便宜50%，是化石燃料成本的40%，天然气成本的25%。 |
| 伦理学 | 人们宁愿沿着山脊线建水坝或风车而牺牲山谷的美丽，也不愿将水通过带有涡轮机的管道来保护山谷；不丹王国保护了52%的土地和生物多样性，但却没有保护河流，它们都被水坝利用了。 |
| 历 史 | 阿基米德通过排开水的多少计算黄金的体积；风车于9世纪在伊朗发明，12世纪由十字军带到欧洲，同时期的中国发明了风驱动的提水设备。 |
| 地 理 | 喜马拉雅山脉来自梵文"喜马"(雪)和"拉雅"(住宅)，包括了世界十大山峰中的9个；喜马拉雅山脉源于印度板块和亚欧板块的碰撞，跨越五个国家(印度、尼泊尔、不丹、中国和巴基斯坦)；喜马拉雅山脉是冰雪的第三大沉积地，仅次于南极和北极；挪威99%的能源来自水力发电。 |
| 数 学 | 密度等于质量除以体积，水的密度是空气的784倍。 |
| 生活方式 | 经幡是永久性的，而生命在演化、终结，并被新的生命所取代；孩子们喜欢旋转木马，它是一种在圆形的旋转平台上放置很多座位的游乐设施。 |
| 社会学 | 经幡是用来促进和平、同情、力量和智慧的。祈祷者不是向神祷告，而是让风带着美好的愿望传播祝福和同情心。 |
| 心理学 | 摆动是一种混乱或兴奋的状态。 |
| 系统论 | 能量无处不在。问题不是缺乏能量，而是我们不知道周围所有的能量来源。 |

# 教师与家长指南

## 情感智慧
### Emotional Intelligence

**蝙蝠**

蝙蝠为风车可能毁了他的美丽山谷而感到紧张。尽管了解了这个问题可能的解决方案,他仍在哀叹人们的无知,不懂得水比风的力量更强。蝙蝠指出,政策制定者和金融家坚持利用传统的水电资源和容忍环境破坏是非常愚昧的。蝙蝠不仅表示担忧,还公开分享他对创新的认识并指出这些创新已经成功实施的案例。蝙蝠热情地解释了每个人如何能实现这些创意,并鼓励合理搭配和团体意识。

**猫头鹰**

猫头鹰抱怨风车产生的噪声。他对创新很好奇,因为他渴望找到良好的解决方案。猫头鹰表示沮丧,因为有很明显的现成的机会人们却没有利用。他希望去探索其他方法,寻找更好的解决方案。然而他很担心蝙蝠的方法,那需要使用大量的稀土金属来制造很多磁铁。通过模仿旗帜摆动来发电的想法,吸引了他全部的注意力,因为这结合了功能(电)与精神(祈祷和平)的需求,所以他很感激他的朋友蝙蝠。

## 艺术
### The Arts

很多人利用风吹过树木和水流经森林的声音来帮助他们放松。找到这些声音的录音并想象这些来源中可能产生的能量形式。让艺术激发科学吧!

# TEACHER AND PARENT GUIDE

## 思维拓展
### Systems: Making the Connections

我们周围存在很多的能量，它无时无刻无处不在。问题是，通常人们尤其是工程师们并没有注意到这一点，很少有人受过训练去寻找现存的小型本地能源。每个人都更关注大型、集中发电的方法，这种方法产生的电要通过覆盖数千公里的大型电网来传输。树木为在当地如何产生电力提供了一个很好的启发。通过模仿巧妙而简单的树叶摆动产生的电，可以与小型风力涡轮机的发电能力相媲美。风车的圆周运动、齿轮箱和涡轮都会产生摩擦和衰变，因此需要长期维护。可以用一种没有齿轮箱或涡轮的动力系统来取代它。这种动力系统产生的电足以满足当地的需求。大自然拥有用简单方法解决复杂问题的能力，利用本地分散的电力系统，比传统系统需要的基础设施、资本投资和后期维护更少。而方便易得的天然能源，只需要风力和水力就可运作。大到一个城市，小到一个家，每一幢建筑、每一所房子、每一间办公室都有水流入，但没人再次利用水的流出。如果工程师不能抓住这个机会利用可再生能源创造更多的能源，而是继续依靠不可再生能源的话，全球的电力需求将无法得到满足。相反，如果意识到了这个机会，我们便可以扭转局势，走向可持续发展。

## 动手能力
### Capacity to Implement

我们不断地寻找新能源，但必须意识到能源就在我们周围，无时无刻无处不在。当不同的两种材料接触时，就会有电荷转移。如果让它们之间保持一定距离，就会产生电压，形成电流。因此,发电机可以将环境中的任意机械能转化为电能。尝试通过两种材料摩擦来发电，用一个LED小灯把它们连接起来。也许需要一点时间来产生足够的摩擦力，使产生的电量能点亮LED灯。当灯开始闪烁时，拍个照片和我们一起分享吧！

# 教师与家长指南

## 故事灵感来自

## 肖恩·弗拉伊内
## Shawn Frayne

肖恩·弗拉伊内毕业于麻省理工学院，并获得物理学学位。2006年，他加入救援队帮助重建海地。他注意到当地人依赖煤油或柴油发电机，试图建一个便宜的风力发电机。他意识到风力涡轮机设计中正在取消齿轮箱，但涡轮技术仍然效率低下，尤其是使用范围很小时，因为所有技术都是为大规模使用而设计的。这推动了第一个无涡轮风力发电机——风箱的发明，其效率比燃气涡轮发动机高10～30倍。肖恩已经有了多项创新，并在香港的实验室组建了一支创意团队，通过黑线鳕（Haddock Invention）发明平台，在绿色包装、太阳能水消毒和能量储存方面进行了开创性工作。此外，他被《发现》杂志评为"40岁以下的20个最强大脑"之一，被《企业家》杂志评为"30岁以下30人"之一。

更多资讯

http://www.engineeringtoolbox.com/density-air-d-680.html

http://techxplore.com/news/2014-09-fluttering-flags-harvest-power.html

http://www.rexresearch.com/frayne/frayne.htm

www.lookingglassfactory.com

www.haddockinvention.com

图书在版编目（CIP）数据

旗帜飘飘：汉英对照 /（比）冈特·鲍利著；
（哥伦）凯瑟琳娜·巴赫绘 ；高芳，李原原译 . —— 上海：
学林出版社，2016.6
（冈特生态童书 . 第三辑）
ISBN 978-7-5486-1064-9

Ⅰ . ①旗… Ⅱ . ①冈… ②凯… ③高… ④李… Ⅲ .
①生态环境－环境保护－儿童读物－汉、英 Ⅳ .
① X171.1-49

中国版本图书馆 CIP 数据核字（2016）第 126070 号

———————————————————————————

© 2015 Gunter Pauli
著作权合同登记号 图字 09-2016-309 号

冈特生态童书

**旗帜飘飘**

| 作　　者—— | 冈特·鲍利 |
|---|---|
| 译　　者—— | 高　芳　李原原 |
| 策　　划—— | 匡志强 |
| 责任编辑—— | 程　洋 |
| 装帧设计—— | 魏　来 |
| 出　　版—— | 上海世纪出版股份有限公司 学林出版社 |
|  | 地　址：上海钦州南路 81 号　　电　话 / 传真：021-64515005 |
|  | 网址：www.xuelinpress.com |
| 发　　行—— | 上海世纪出版股份有限公司发行中心 |
|  | （上海福建中路 193 号　网址：www.ewen.co） |
| 印　　刷—— | 上海丽佳制版印刷有限公司 |
| 开　　本—— | 710×1020　1/16 |
| 印　　张—— | 2 |
| 字　　数—— | 5 万 |
| 版　　次—— | 2016 年 6 月第 1 版 |
|  | 2016 年 6 月第 1 次印刷 |
| 书　　号—— | ISBN 978-7-5486-1064-9/G·399 |
| 定　　价—— | 10.00 元 |

（如发生印刷、装订质量问题，读者可向工厂调换）

# Energy
## 87

# 跳蚤和虱子
## Fleas and Lice

### Gunter Pauli

冈特·鲍利 著

凯瑟琳娜·巴赫 绘
高　芳　李原原 译

学林出版社
www.xuelinpress.com

## 丛书编委会

主　任：贾　峰
副主任：何家振　郑立明
委　员：牛玲娟　李原原　李曙东　吴建民　彭　勇
　　　　冯　缨　靳增江

## 丛书出版委员会

主　任：段学俭
副主任：匡志强　张　蓉
成　员：叶　刚　李晓梅　魏　来　徐雅清　田振军
　　　　蔡雩奇　程　洋

特别感谢以下热心人士对译稿润色工作的支持：

姜竹青　韩　笑　贾　芳　刘　晓　张黎立　刘之杰
高　青　周依奇　彭　江　于函玉　于　哲　单　威
姚爱静　刘　洋　高　艳　孙笑非　郑莉霞　周　蕊

# 目录

| | |
|---|---|
| 跳蚤和虱子 | 4 |
| 你知道吗？ | 22 |
| 想一想 | 26 |
| 自己动手！ | 27 |
| 学科知识 | 28 |
| 情感智慧 | 29 |
| 艺术 | 29 |
| 思维拓展 | 30 |
| 动手能力 | 30 |
| 故事灵感来自 | 31 |

# Contents

| | |
|---|---|
| Fleas and Lice | 4 |
| Did you know? | 22 |
| Think about it | 26 |
| Do it yourself! | 27 |
| Academic Knowledge | 28 |
| Emotional Intelligence | 29 |
| The Arts | 29 |
| Systems: Making the Connections | 30 |
| Capacity to Implement | 30 |
| This fable is inspired by | 31 |

跳蚤和虱子在一个人的头发上相遇了。他们听说彼此很长时间了，但还从未见过面。

"多美好的一天！"跳蚤打破沉默说，"我是跳蚤家庭的父亲。你好啊！"

"你好，很高兴见到你！我一直期待这一时刻。"虱子说，并自我介绍，"我是虱子家庭的母亲。"

A flea and a louse come across each other in the hair growing on a person's head. They have known about each other for a long time but have never met in person.

"Good day," says the flea, breaking the silence. "I am the father of the flea family. How do you do?"

"Hello, how lovely to meet you! I've been waiting for this moment," says the louse and then she introduces herself: "I'm the mother of the lice family."

# 跳蚤和虱子

A flea and a louse

我们享受着同样的生活空间

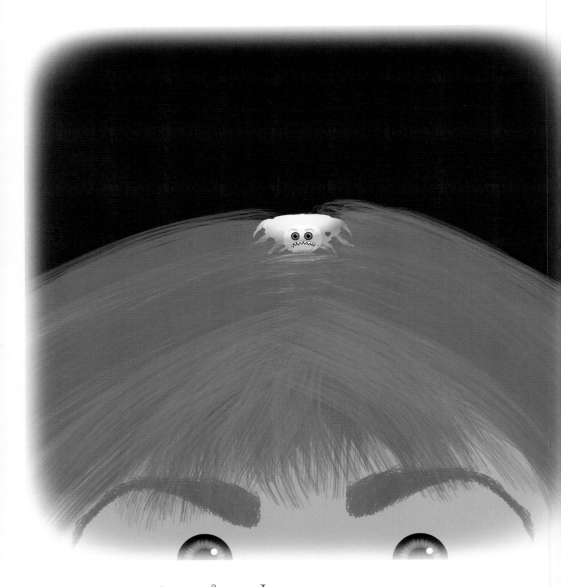

We both enjoy the same living space

"我一直在想我们何时会见面,因为我们享受着同样的生活空间。"

"的确如此,尤其是我们都爱咬寄主的皮肤,如果可能,甚至吸点他的血!"

"我一直在想,为什么你们叫一个虱子为louse,但两个或两个以上的虱子为lice?"

"I've been wondering when we would meet, as we both enjoy the same living space."

"Indeed, especially because we both love to have a bite of the skin of our host and, if possible, even a little bit of his blood!"

"I've always wondered why you call one louse a louse but two or more lice?"

"这个我帮不了你,因为我也不知道。我倒是一直想知道,你是如何跳得那么高又那么远的?"

"实际上,我们没有跳。我们的膝盖上有个弹簧,释放的时候就像弓把箭弹出去一样。可以说我们的身体结构更像一把弩。"

"你是世界上最好的跳跃者!"

"I can't help you there, as I have no clue. What I have been wondering is how you can jump so high and so far."

"Actually, we don't jump as such. We have springs in our knees that are released, like an arrow from a bow. You could say we are built like a crossbow."

"You are the best jumpers in the world!"

世界上最好的跳跃者!

Best jumpers in the world!

沫蝉才是真正的冠军

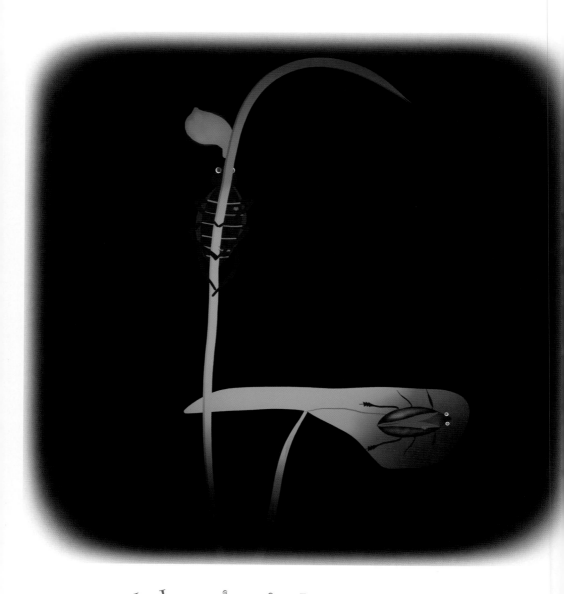

Real champion is the froghopper

"是的，如果人类有我们的弹跳力，他们就能向前飞跃90米。但沫蝉才是真正的冠军，那种小虫子可以产生相当于其体重400倍的力量，而我们只能产生100倍，蚂蚱甚至连10倍都不到。"

"这真让人佩服啊。"

"Well yes, if people had our jumping power, they would be able to leap 90 metres forward. But the real champion is the froghopper. That little bug can apply a force of 400 times its body weight, while we can do only about a 100 times. A grasshopper can't even do 10 times."

"That is still very impressive."

"谢谢,但是你知道我们以动物为食,而沫蝉只吃植物。这可能是力量差异的原因。"

"我们也喜欢吃血液和皮肤,但我们却不能像你一样会跳。"

"不,也许不是那么回事,我听说你在数百万年前就已经寄居在黑猩猩和大猩猩身上了。"

"Thank you, but you know we feed on animals, whereas the froghopper only eats plants. That may be the reason for the difference in force."

"We prefer to eat blood and skin as well, but we cannot jump like you do."

"No, perhaps not, but I'm told that you were already living on chimpanzees and gorillas millions of years ago."

只吃植物

Only eat plants

人类甚至把我放在马戏团里!

They have even put me in a circus!

"哦，是的，而且现在如果有机会，我们也会把人类作为食物的来源，无论他们喜欢与否。我想一下，我们在人类开始穿衣服时就向他们靠近，衣服里面很容易躲藏，尤其是他们不常洗澡或者洗衣服的时候。"

"你不太受人欢迎，对吗？但我已经变得很受欢迎了，"跳蚤说，"人类甚至把我放在马戏团里!"

"Oh yes, and now, if we get the chance, we will use humans as a food source too, whether they like it or not. I imagine we moved to people when early man started wearing clothing, and it was so much easier for us to hide in there, especially if they do not wash themselves or their clothes regularly."

"You are not popular with them, are you? I have become popular though," says the flea. "They have even put me in a circus!"

"我记得的！对于我们俩来说，不幸的是，人类发明了真空吸尘器，这使我们的生活非常艰难。这就是如今在周围很少见到你们的原因。"

"是的，这还不算什么。当空气太干燥时，我们会生活得更艰难，"跳蚤说，"那样我们就没有办法生存下去了。"

"I remember that! Unfortunately for both of us, people have invented the vacuum cleaner and that makes life very difficult. That is why there are now less of you around."

"Yes, and that's not all. Life also gets hard for us when the air is too dry," says the flea. "Then we have no way to survive."

真空吸尘器

Vacuum cleaner

17

我们能够自我复活

We are able to revive ourselves

"即使人类想用水淹,也无法摆脱我们。我们在水中十二小时后,可能看起来像是死了,其实我们能够自我复活,生活依旧。"虱子说。

"我不怕水,"跳蚤回答,"但我受不了他们把肥皂放入水里。那样我就失去了保护我不受细菌威胁的蜡,然后就会死掉。"

"Not even trying to drown us can get rid of us. After twelve hours under water, we may come out looking like death, but then we are able to revive ourselves and continue our lives," says the louse.

"I'm not afraid of water either," replies the flea. "But I can't stand it when they put soap in the water. Then I lose the wax that protects me so well from bacteria, and I die."

"我受不了我的寄主吃大蒜或者喝苹果醋。这让他闻起来很糟糕,我会失去食欲。这真的让我想换一个寄主。"

"对我来说很容易,我只需要把自己弹起来穿过房间就行,但可怜的你必须等到孩子们相互拥抱时才能摆脱那个散发着臭味的人!"

……这仅仅是开始!……

"And I can't stand it when my hosts eat garlic or drink apple cider vinegar. It makes them smell so bad that I lose my appetite. That really makes me want to find another host."

"It's easy for me to do that, as I just catapult myself across the room, but poor you, you have to wait until the kids are hugging each other before you can move away from a stinky one!"

… AND IT HAS ONLY JUST BEGUN!…

……这仅仅是开始！……

...AND IT HAS ONLY JUST BEGUN!...

# Did You Know?
## 你知道吗？

Fleas and lice are insects that have been feeding on people since the beginning of time.

跳蚤和虱子生来就是寄生在人类身上的昆虫。

It is assumed that the body louse evolved from the head louse about 100 000 years ago, when people started to wear clothing and lost their body hair.

据推测，体虱是由大约100 000年前的头虱进化而来，那时人类刚开始穿衣服并失去体毛。

虱子寄生在除了蝙蝠、鲸和海豚之外的所有鸟类和哺乳动物身上，甚至寄生在鱼类身上，比如养殖的鲑鱼。

Lice live on all birds and mammals except bats, whales, and dolphins and even infest fish like farmed salmon.

虱子孵化出来时是父母的微缩，称为若虫。跳蚤则是从卵孵化成幼虫，然后吐丝后结茧并在其中化蛹，最后成为成虫。

Lice hatch as miniature versions of their parents, known as nymphs. Fleas hatch from eggs, develop into larvae, and weave cocoons with their silk, in which the pupae transform before they emerge as adult fleas.

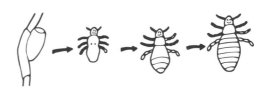

母兔临近分娩时，寄生在它身上的跳蚤会收到化学信号，开始产卵。当小兔子（称为兔子幼崽）出生时，刚孵化的跳蚤就会转移到它们身上。

When a female rabbit is close to giving birth, fleas living on her receive a chemical trigger to start the production of eggs. As soon as the little rabbits (called kits, kittens or bunnies) are born, the newly hatched fleas will make their way towards them.

尽管人们讨厌跳蚤，跳蚤市场却是非常受欢迎的旅游景点。第一个跳蚤市场是1860年前后在巴黎举办的。当时贫民窟被拆除，商人们开始在街上售卖他们的商品。

Although people detest fleas, flea markets are very popular tourist attractions. The first flea market, called the marché aux puces (market of fleas), was held in Paris in around 1860, after merchants from the demolished slums started selling their wares on the street.

跳蚤可以被驯化和教导。自19世纪30年代以来，跳蚤马戏团在英格兰蓬勃发展。现在仅存的唯一真正的跳蚤马戏团会在一年一度的慕尼黑啤酒节进行演出。

Fleas can be domesticated and taught tricks. The flea circus has thrived in England since the 1830s. The only remaining true flea circus now performs at the annual Oktoberfest in Munich.

跳蚤表演时脖子上戴个小金项圈，与黄金皮带相连。在一些跳蚤马戏团中根本就没有跳蚤在表演，演员的工作是让人们相信确实是有跳蚤在表演。

Performing fleas wear tiny gold collars around their necks and are attached to gold wire leashes. In some flea circuses there are no fleas at all. It is the job of the performer to convince people that there are indeed fleas performing.

# Think About It

## 想一想

If fleas and lice have been living along mammals for millions of years, will we ever find a way to get rid of them? Or should we welcome them as part of our family?

如果跳蚤和虱子已经和哺乳动物一起生活了数百万年,那么我们能否找到一个方法来摆脱他们呢?或者我们应该欢迎他们成为我们中的一员吗?

你是喜欢一个马戏团真的有跳蚤在表演,还是宁愿有一个小丑让你相信有跳蚤在表演呢?

Would you enjoy a circus where fleas are really performing tricks? Or would you rather have a clown make you believe that there is a flea performing?

How easy would it be to make a gold leash for a flea? Do you think it will ever get out of it alive?

为跳蚤制作一条黄金皮带容易吗?你认为他能活着摆脱这条皮带吗?

如果你用真空吸尘器清扫宠物睡觉的地方,你可以清理掉跳蚤和他们的卵,虱子和他们的幼虱。这样就足以摆脱这些寄生虫了吗?

If you vacuum the areas where your pets sleep, you will pick up fleas and their eggs, and lice and their nits and nymphs. Would that be enough to get rid of these parasites?

# Do It Yourself!
## 自己动手!

Let's put together a flea remedy that smells pleasant and keeps pests at bay.
You will need the following:
A sachet made from a natural and breathable fabric.
(You can use an old scarf.)
A handful of cedar chips.
Two teaspoons of dried lavender buds.
The peel of one lemon.
Place the cedar chips, lavender buds, and lemon peel in the sachet or scarf and sew it up. Place the bag next to where you pet sleeps. This will keep your pet (and you) free from fleas!
You will need to refresh the content every two months.
http://everydayroots.com/flea-remedies

让我们一起做一个跳蚤治疗袋吧,它气味宜人,还能让害虫陷入困境。你需要以下物品:一个透气的天然织物袋(或一条旧围巾),少量的雪松芯片,两勺干薰衣草花蕾,柠檬皮。把雪松芯片、薰衣草花蕾和柠檬皮装入袋子或围巾中缝起来。把它放在你的宠物睡觉地方的旁边。这会让你和你的宠物摆脱跳蚤!你需要每两个月更新一次里面的东西。

# TEACHER AND PARENT GUIDE

## 学科知识
## Academic Knowledge

| | |
|---|---|
| 生物学 | 寄生虫对健康的影响；头发对身体的作用；寄生虫摄取的营养来自血液和死亡的体细胞；不完全变态昆虫（如虱子）的若虫对应于完全变态昆虫（如跳蚤）的幼虫和蛹；食草动物和食肉动物的区别；跳蚤有坚硬的外壳。 |
| 化学 | 使用化学物质杀死寄生虫；肥皂能溶解蜡和脂质；大蒜中的化学物质能够产生一种强烈的气味；如何制造醋。 |
| 物理 | 弹簧弹力与肌肉力量的区别；肥皂的物理效应；亲水蜡和防护涂料的功能；干燥剂。 |
| 工程学 | 美洲和非洲原住居民所使用的弓和箭的功能；弓箭和弩在技术、性能上的区别。 |
| 经济学 | 成本因素：寄生虫造成生产力损失，以及化学品污染和后续健康问题所引发的成本。 |
| 伦理学 | 奴役动物以娱乐人类所带来的伦理问题；以谦虚的态度接受如下事实：总有某人在某一点上比你优秀。 |
| 历史 | 寄生虫的进化：从寄生在大猩猩和黑猩猩身上到寄生在人类身上。 |
| 地理 | 跳蚤和虱子在世界各地随处可见。 |
| 数学 | 如何计算你从一个地方跳跃移动到另一个地方所需的体力；如何计算克服重力所做的功。 |
| 生活方式 | 服装的产生；人体卫生、洗衣服和洗澡的重要性；在家里使用吸尘器对卫生和健康的影响；人们第一次见面时问候形式的差异；从古希腊时代起，薰衣草就被作为驱虫剂。 |
| 社会学 | 对气味的不同感知和欣赏：一个人觉得愉悦的气味可能另一个人却排斥；当孩子们玩耍时，他们彼此频繁接触，这使寄生虫在他们之间转移比在成年人之间更容易。 |
| 心理学 | 自我认可的重要性；如何表达热情；如何询问你不懂的东西，尽管那会显得你很无知。 |
| 系统论 | 物种如何为适应新环境而演化、变异和转化：一些物种，例如被认为是害虫的跳蚤和虱子，如何仍能成为生态系统的组成部分，并在其中找到自己的容身之地，即使人类还未能很好地了解其功能。 |

# 教师与家长指南

## 情感智慧
## Emotional Intelligence

跳蚤

跳蚤礼貌恭敬地对虱子进行自我介绍。他很体贴地问一些问题以便更好地理解虱子,并与她建立友好关系。他还饶有兴趣地倾听虱子的问题,并花时间来解释她所不理解的事情。即便当虱子表示对他良好弹跳力的尊重时,他还谦虚地告诉她有比自己做得更好的,并解释了为什么会出现这种情况。跳蚤在接受虱子赞美后表示友好,并试图指出虱子的特点。跳蚤有自知之明,知道自己掌握马戏表演技巧而受欢迎。当虱子突出表现自己的独特优势(能在水下生存超过12小时)时,他知道自己也能做到,但并没有夸耀自己,而是分享了恐惧。跳蚤很同情虱子,因为他很容易改变寄主,而虱子这样做却需要更多的耐心。

虱子

虱子在介绍性问候时不是很正式,并且首先分享她和跳蚤的共同点,也就是他们取食的寄主相同。她谦虚地承认不是一切都知道。她通过问题试图寻找共同点,表现出钦佩和尊重。当跳蚤希望突出虱子的一个特点时,她发现好像没有什么是与众不同的。跳蚤谈到自己的声望时,虱子温柔地告诉他,吸尘器的发明是周围跳蚤数量减少的原因。然后虱子指出自己的一个优点和一个弱点,使这两只昆虫产生了共鸣。

## 艺术
## The Arts

你要去参加一个脱口秀节目。站起来,假装有跳蚤在你的手上。让跳蚤跳跃、翻筋斗、再翻一个筋斗。让所有观众相信真的有跳蚤在你的手上。假装跳蚤跳到第一排的一位女士身上——也许是你的老师,或假装在她的头发里"找到"一只跳蚤。扮演虚拟的跳蚤马戏团主持人真是搞笑啊。

# TEACHER AND PARENT GUIDE

## 思维拓展
### Systems: Making the Connections

> 人类对跳蚤和虱子以及其他害虫的反应是试图消灭它们。然而，自人类存在开始（甚至在此之前），跳蚤和虱子就已经存在了。数千年后，问题不在于我们能否找到一种新的化学物质最终消除所有害虫，而是如果我们可以设计自己的栖息地，跳蚤和虱子就无法在其中找到容身之地了。尽管我们可能还没有认识到虱子和跳蚤的生态价值，但必须承认，摧毁我们不喜欢或不理解的事物可能会影响社会的可持续发展。如果虱子和跳蚤让我们不舒服，那么我们应该改变它们的生存条件。首先定期用真空吸尘器清理房子，特别是在狗和猫睡觉的地方。然后，不要把跳蚤、虱子及其幼虫和卵放置在垃圾箱中，它们会繁殖，所以要立即处置掉，确保它们不会再次进入房子。接下来，为这些寄生虫创造不适条件，在我们的日常饮食中添加大蒜和苹果醋。或者你可以使用植物，它们的气味对人类来说是愉悦的，而跳蚤和虱子却很讨厌。自古希腊和罗马时代以来，薰衣草就被用来控制跳蚤和虱子，它在阿拉伯文化中也很普及。这个根除和排除的区别可以由国际象棋和围棋的游戏来说明：国际象棋游戏的目的是把王将死，而围棋游戏的目的是占据所有的空间让对手无处可去。

## 动手能力
### Capacity to Implement

> 在网上或报纸上看一看，你会发现如果学校有跳蚤和虱子爆发，学校会被迫要关闭一段时间。找出当局对有关问题所采取的措施。提出一个你能想到的策略，首先要扭转危机，其次要避免问题再次发生。现在列出解决问题所需的成本，对如何使用这些钱，你是否有建议。描绘一下，如果采用了不同的方法，新的业务活动应该怎样开展。

# 教师与家长指南

故事灵感来自

## 蒂姆·科克里尔
## Tim Cockerill

蒂姆·科克里尔是一名昆虫学家，也是一个马戏团演员，毕业于英国剑桥大学，获博士学位。他专门从事生物多样性和生态系统科学研究，研究生命之间的相互作用。他攻读动物学博士学位期间关注研究婆罗洲的热带雨林，那是地球上生物多样性最丰富的地区之一。由于棕榈种植园的栽培，这些热带雨林正在经受威胁。在几个月的野外研究中，他发现了大量的新昆虫物种，现在这些物种已经被永久收录在伦敦自然历史博物馆的《昆虫学集》中。蒂姆也是一个经验丰富的演员。他是一个吞火魔术师，也是一位研究跳蚤马戏团的专家。

### 更多资讯

http://timcockerill.com/main

http://news.bbc.co.uk/2/hi/science/nature/3110719.stm

https://www.newscientist.com/article/mg21628961-900-fleadom-or-death-reviving-the-glorious-flea-circus/

图书在版编目（CIP）数据

跳蚤和虱子：汉英对照／（比）冈特·鲍利著；
（哥伦）凯瑟琳娜·巴赫绘；高芳，李原原译．－－上海：
学林出版社，2016.6
（冈特生态童书．第三辑）
ISBN 978-7-5486-1065-6

Ⅰ．①跳… Ⅱ．①冈… ②凯… ③高… ④李… Ⅲ．
①生态环境－环境保护－儿童读物－汉、英 Ⅳ．
① X171.1-49

中国版本图书馆 CIP 数据核字（2016）第 126069 号

——————————————————————————

ⓒ 2015 Gunter Pauli
著作权合同登记号 图字 09-2016-309 号

## 冈特生态童书
### 跳蚤和虱子

| | | |
|---|---|---|
| 作　　　者—— | 冈特·鲍利 | |
| 译　　　者—— | 高　芳　李原原 | |
| 策　　　划—— | 匡志强 | |
| 责任编辑—— | 程　洋 | |
| 装帧设计—— | 魏　来 | |
| 出　　　版—— | 上海世纪出版股份有限公司 学林出版社 | |
| | 地　址：上海钦州南路 81 号　　电　话／传真：021-64515005 | |
| | 网址：www.xuelinpress.com | |
| 发　　　行—— | 上海世纪出版股份有限公司发行中心 | |
| | （上海福建中路 193 号　网址：www.ewen.co） | |
| 印　　　刷—— | 上海丽佳制版印刷有限公司 | |
| 开　　　本—— | 710×1020　1/16 | |
| 印　　　张—— | 2 | |
| 字　　　数—— | 5 万 | |
| 版　　　次—— | 2016 年 6 月第 1 版 | |
| | 2016 年 6 月第 1 次印刷 | |
| 书　　　号—— | ISBN 978-7-5486-1065-6/G·400 | |
| 定　　　价—— | 10.00 元 | |

（如发生印刷、装订质量问题，读者可向工厂调换）

# Energy 101

# 利用太阳能的蓝龙
## The Solar Blue Dragon

**Gunter Pauli**

冈特·鲍利 著
凯瑟琳娜·巴赫 绘
高 芳 李原原 译

学林出版社
www.xuelinpress.com

## 丛书编委会

主　任：贾　峰
副主任：何家振　郑立明
委　员：牛玲娟　李原原　李曙东　吴建民　彭　勇
　　　　冯　缨　靳增江

## 丛书出版委员会

主　任：段学俭
副主任：匡志强　张　蓉
成　员：叶　刚　李晓梅　魏　来　徐雅清　田振军
　　　　蔡雩奇　程　洋

特别感谢以下热心人士对译稿润色工作的支持：

姜竹青　韩　笑　贾　芳　刘　晓　张黎立　刘之杰
高　青　周依奇　彭　江　于函玉　于　哲　单　威
姚爱静　刘　洋　高　艳　孙笑非　郑莉霞　周　蕊

# 目录

| | |
|---|---|
| 利用太阳能的蓝龙 | 4 |
| 你知道吗? | 22 |
| 想一想 | 26 |
| 自己动手! | 27 |
| 学科知识 | 28 |
| 情感智慧 | 29 |
| 艺术 | 29 |
| 思维拓展 | 30 |
| 动手能力 | 30 |
| 故事灵感来自 | 31 |

# Contents

| | |
|---|---|
| The Solar Blue Dragon | 4 |
| Did you know? | 22 |
| Think about it | 26 |
| Do it yourself! | 27 |
| Academic Knowledge | 28 |
| Emotional Intelligence | 29 |
| The Arts | 29 |
| Systems: Making the Connections | 30 |
| Capacity to Implement | 30 |
| This fable is inspired by | 31 |

一群蓝瓶僧帽水母沿着海岸游动时,遇到了一条孤独的沙丁鱼。

A school of bluebottles is floating along the coast when they meet a lonely sardine.

蓝瓶僧帽水母遇到了一条孤独的沙丁鱼

Bluebottles meet a lonely sardine

过度捕捞导致我们快要灭绝

Overfishing has really wiped us out

"哎呀,很少在周围见到你。"僧帽水母说。
"是的,过度捕捞导致我们快要灭绝了。你一定很高兴渔民们帮你摆脱了最大的食物竞争对手。"沙丁鱼回答。

"Oh, you are one of the few still around," says the bluebottle.
"Yes, overfishing has really wiped us out. You must be happy that the fishermen got rid of your biggest competitor for food," replies the sardine.

"我知道。但还有可怕的带须虾虎鱼,他真的很喜欢吃我们!"
"你应该也害怕蓝龙吧!"

"I know. But there is still that dreaded bearded goby fish and he really likes to eat us!"
"You should also be scared of the blue dragon!"

# 可怕的带须虾虎鱼

Dreaded bearded goby fish

这个蓝龙会喷火焰

This blue dragon spews fire

"又是龙!"僧帽水母喊道,"这些天,人人都在谈论龙。"
"这个蓝龙会喷火焰。"
"哦,是的。我太了解这一切了!他利用我们制造他的火!"

"Not dragons again!" shouts the bluebottle. "These days, everyone talks about dragons."
"This blue dragon spews fire."
"Oh yes, I know that all too well! He makes his fire from us!"

"你一定是在开玩笑!蓝龙怎么会用蓝瓶僧帽水母来制造火呢?"

"好吧,你知道我们含有可以燃烧的毒素。"

"我当然知道,被你刺到会很痛苦。"

"You must be joking! How does the blue dragon make fire from a bluebottle?"

"Well, you know that we have this toxin that burns."

"Of course I do; that's what makes your sting so painful."

被你刺到会很痛苦

That's what makes your sting so painful

像火一样燃烧

It really burns like fire

"没错。所以,蓝龙吃掉我们,把我们所有的毒液放进一个特殊的小口袋,当他受到威胁时再吐出来。这些毒液会像火一样燃烧。"

"那蓝龙是如何逃脱捕食者的呢?"

"Exactly. So, the blue dragon eats us, puts all our venom into a special little pocket, and spits it out when he is threatened. And it really burns like fire."

"How does this blue dragon get away with it?"

"正如他的名字所示——他是蓝色的。他漂浮在水面上，蓝色的肚子朝上，这使鸟类在俯瞰蓝色的海洋时，很难看到他。而且他有闪亮的灰色后背，下面的鱼从底部向上看时，也看不到他。"

"但你知道吗？他其实是一个软体动物，只是在行为上更像一条龙。"

"As the name says – he is blue. He floats on the water with his blue tummy upwards, which makes him hard for birds to see when they look down on the blue ocean. And his shiny grey back that is seen from the bottom makes him invisible to fish below."

"But did you know that he is in fact a mollusc and only behaves like a dragon?"

正如他的名字所示——他是蓝色的

As the name says - he is blue

他确实是一个没有壳的贝类

He is a shellfish without a shell

"软体动物，也就是说，他是个贝类动物？可是他都没有壳。"

"我知道这很让人费解，但他确实是一个没有壳的贝类。"

"这还不算什么，有的蓝龙是靠太阳能生存的。"

"不是吧，你一定是在开玩笑！一条利用太阳能的龙吗？"

"A mollusc, in other words a shellfish? But he doesn't even have a shell."

"I know it's confusing. He's a shellfish without a shell."

"And that is only the beginning. There are blue dragons that live on solar energy."

"No, you must be kidding me! A solar dragon?"

"虽然不可思议,但那却是事实。他让微小的藻类生长在他的身体里,供给他所需的糖,这样他可以靠阳光生存,而不用吃东西。我们僧帽水母,只不过是他的甜点。"

"对我来说,他现在听起来才像一条真正的龙!"

……这仅仅是开始!……

"Incredible but true. He has these tiny algae growing in his body that give him sugars so that he can live on sunlight instead of real food. We, the bluebottles, are only his dessert."

"Now that sounds like a real dragon to me!"

… AND IT HAS ONLY JUST BEGUN!…

……这仅仅是开始！……

...AND IT HAS ONLY JUST BEGUN!...

# Did You Know?

## 你知道吗？

蓝龙有很多名字：海燕子、蓝色天使以及蓝色海蛞蝓。

The blue dragon has many names: sea swallow, blue angel, and blue sea slug.

蓝龙靠吞咽空气并将其保存在胃旁边的一个袋内来漂浮，这个空气口袋让蓝龙能够上下浮动。

The blue dragon floats by swallowing air and keeping it in a pouch that is next to its stomach. The pocket of air enables the dragon to float upside down.

小小的蓝龙不产生任何毒液，但它从猎物那获取毒液，集中储存在触角上，这样更致命。

The tiny blue dragon does not make any venom, but it takes venom from its prey, concentrates it, and stores it in its tentacles, making it more lethal.

蓝龙是一个积极的捕食者，以比自己大得多的生物（包括僧帽水母）为食，用它50米长的触须吞噬毒素。

The blue dragon is an aggressive predator that feeds on organisms that are much larger than itself, including the Portuguese man of war, devouring its toxin laced 50 m long tentacles.

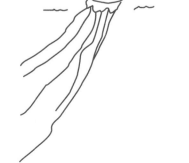

蓝龙舌头的形状就像锯子，这就是它可以很快切断猎物的原因。

The blue dragon has a tongue that is shaped like a saw. That is why it can cut off a piece of its prey very quickly.

蓝龙可以将阳光转化为糖。它们比人类更早利用太阳能进行光合作用。它们体内生长着微小的藻类，这些藻类在受保护的环境中繁殖旺盛。

Blue dragons can convert sunlight into sugars. They are ahead of humans in harnessing solar energy through photosynthesis. Some of the species capture and farm microscopic algae inside their bodies, where they flourish in a protected environment.

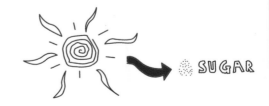

沙丁鱼只吃浮游生物，这使它们所吃的东西富含营养且安全。

Sardines eat only plankton. That makes them rich in nutrients and safe to eat.

沙丁鱼没有武器来保护自己免受捕食。它们的防御机制是1000万条以上的鱼一起成群游泳。这会迷惑捕食者，让它们傻傻地认为自己看到的生物有5千米长、1千米宽。

Sardines have no weapons to protect themselves against predators. Their defence mechanism is to swim together in schools of up to 10 million fish. This confuses predators who is fooled into thinking they are seeing a creature that is 5 km long and 1 km wide.

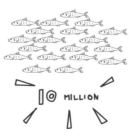

# Think About It

## 想一想

Can you imagine that a small creature of only few centimetres long could take on a large one with 50 m long tentacles?

你能想象，一个只有几厘米长的小生物可以长着50米长的触手吗？

如果你漂浮失败，你会怎样保护自己返回岸上？

If you were floating belly up, how would you protect your back?

Do you believe that firespewing dragons really exist in the water?

你相信水里真的存在喷火的龙吗？

想想钓鱼：我们应该留一些鱼在水中给海狮和鹈鹕吃吗？

Think about fishing: Should we leave some fish in the water for the sea lions and the pelicans to eat?

Explore history and look for examples where someone small took on someone big, and won. Take note of how they achieved the impossible to win the battle. Now look at Nature and find tiny predators that take on big prey, and succeed. Again, discover ways in which it is possible for a tiny predator to beat the odds and win. When you compare the examples from history and nature, draw some conclusions on how we can achieve the seemingly impossible.

探索历史，寻找以小胜大的例子，注意他们如何在这场不可能赢的战斗中取胜。现在看看大自然，寻找小的捕食者战胜大的猎物的例子。然后，看看小的捕食者可能打赢的方法及其几率。对比了历史和自然的例子后，得出对如何实现看似不可能事情的一些结论。

# TEACHER AND PARENT GUIDE

## 学科知识
### Academic Knowledge

| | |
|---|---|
| 生物学 | 死的蓝龙和僧帽水母仍然可以投刺;像蓝龙一样,海龙偶尔也会同类相食;沙丁鱼在食物链中级别很低,因此在有汞和多氯联苯污染物的环境下非常少见;水母能够麻醉食肉动物的中枢神经系统和大脑。 |
| 化 学 | 蓝龙从僧帽水母中积累刺丝囊及其毒素;沙丁鱼富含维生素B12(支持神经系统),还有磷、钙、钾、硒、ω3脂肪酸和维生素D,能增加钙的吸收,对骨骼健康至关重要。 |
| 物 理 | 运动神经末梢调节鱼群的运动;海豚与沙丁鱼一起工作,把它们当成"诱饵球";水母拥有一个复杂的视觉系统,为其在沼泽里导航;水母的外层感光器官帮助它区分明暗。 |
| 工程学 | 如何使物体在水中隐形;沙丁鱼油用于制造油漆和清漆;通过研究鱼群和鸟群的运动,交通管制员应该掌握汽车在繁忙的高速公路上如何移动会增加交通事故风险。 |
| 经济学 | 生产1千克三文鱼鱼肉需要3千克饲料鱼,如果人们更愿意吃三文鱼,我们的需求就会增加三倍,这会在提升营养质量的同时提高资源生产率,它是实现蓝色经济的一个例子——以更低的成本得到更好质量的产品。 |
| 伦理学 | 与野生捕捞的鱼相比,人工养殖的鱼仅仅提供给我们三分之一的蛋白质,我们如何证明用野生鱼类作为养殖鱼的饲料是合理的呢? |
| 历 史 | 1777年在库克船长驾驶"决心"号进行的第二次太平洋航行里,蓝龙被科学家首次记录。拿破仑·波拿巴推广沙丁鱼是为了生产沙丁鱼罐头,史上第一个鱼罐头就是为了满足他所管理的市民和士兵的需求。 |
| 地 理 | 5月至7月之间,沙丁鱼在非洲南部海岸很常见,引起了最大的海洋食肉动物(如海豚、大白鲨、铜鲨鱼、角塘鹅)和许多海鸟的聚集;蓝龙出没在南非海岸、莫桑比克、欧洲南部和澳大利亚;水母和僧帽水母用阳光作指南针为其导航。 |
| 数 学 | 大数定律是指在随机试验中,每次出现的结果不同,但是大量重复试验出现的结果的平均值却几乎接近于某个确定的值;"数字安全"理论就是每个单独的鱼在鱼群里被抓住的可能性变得相对较低。 |
| 生活方式 | 可持续的食物选择:植物性食物需要更少的能源和生产土地空间,产生较低的碳排放,只需本地养殖的植物即可。 |
| 社会学 | 用"挤得像沙丁鱼"来形容非常满的公共汽车或火车上的人;沙丁鱼成群行动,作为一种防御机制。 |
| 心理学 | 要有信心战胜远远大于自己的猎物,小的可以战胜大的。 |
| 系统论 | 当沙丁鱼被过度捕捞,哺乳的雌性海狮只能吃营养很少的食物,因此只能提供给幼崽更少的奶;沙丁鱼最近的数量增加了一倍,因为食肉动物物种,包括鲨鱼、金枪鱼、鳕鱼和海豚也被过度捕捞。 |

# 教师与家长指南

## 情感智慧
### Emotional Intelligence

蓝瓶僧帽水母

蓝瓶僧帽水母很自信，但对孤独的沙丁鱼没有同情心。知道沙丁鱼受到过度捕捞的影响后，僧帽水母感到不安，提到"龙"这个词时，他感到心慌。僧帽水母厌倦了一遍又一遍地讨论相同的问题。不过，他还是告诉沙丁鱼蓝龙的毒液是从猎物的毒液中得来的。僧帽水母难以说服沙丁鱼，这些微小的动物可以猎食比他们大的动物。僧帽水母花时间来解释蓝龙如何成功地在水中移动而不被注意。他对沙丁鱼的解释很感兴趣。

沙丁鱼

沙丁鱼艰难地忍受孤独，这是过度捕捞的结果，他已接受命运。沙丁鱼想要警告僧帽水母防备蓝龙，但得到了一个令他惊讶的答案。虽然在改变自己的生活条件上什么都没做,但他惊讶地了解到很多关于蓝龙的新事物。沙丁鱼折服于僧帽水母解释一切的能力，也折服于蓝龙独特的不引起注意的移动能力。当沙丁鱼得知蓝龙甚至可以利用太阳能时，他表示，如果这种生物真的存在，那就是他想象的龙所能做到的。

## 艺术
### The Arts

找找裸鳃类动物的照片,指出那些你觉得最不可思议的形状和颜色。看起来真的很惊人!把它们与著名的漫画和卡通人物的图片对比，有什么相似的地方吗？哪些作家和艺术家的灵感是来自蓝龙？

# TEACHER AND PARENT GUIDE

## 思维拓展
### Systems: Making the Connections

　　沙丁鱼和其他鱼类的过度捕捞不仅导致那些几十年来依赖于一种鱼持续供应的企业出现财务崩溃，而且也剥夺了其他海洋生物生存所需的营养。我们必须意识到，大量的野生沙丁鱼被用作养殖鱼类（如挪威三文鱼）的饲料，它们已被严重商业化。今天的消费者更喜欢三文鱼，并且在市场销售的三文鱼几乎都是养殖的。连那些在智利和挪威海岸捕捞到的三文鱼，也是从养鱼场逃出来的。

　　我们必须发展可持续的渔业和农业。捕一种鱼来养活其他鱼，这种方式很难满足世界上不断增长的人口对不断地生产更多的需求。我们不能期望陆地和海洋生产更多东西，而是应该更多利用现有的东西。这意味着，我们不要吃运往世界各地的高成本的三文鱼，而应该吃更多的像沙丁鱼一样在本地就能捕获到的鱼。沙丁鱼被认为是世界上最有营养的动物产品，为什么我们要用密集养殖的三文鱼取代这些富含营养的深海鱼呢？据发现，供应只有中高层收入者能买得起的昂贵鱼类（如三文鱼和鳟鱼），企业更有利可图。这使40%的人口买不起昂贵的鱼，只能找更便宜的替代品。

## 动手能力
### Capacity to Implement

　　你家里吃三文鱼吗？算算每千克三文鱼的成本，与沙丁鱼比较一下。和预期一样，三文鱼更加昂贵，因为生产1千克三文鱼肉需要消耗3千克的沙丁鱼。想办法告诉人们，用吃三文鱼的同样价格，可以吃两倍份量的沙丁鱼或鲱鱼，还能为海狮和海鸟在海洋中留下足够的食物。为什么海狮和海鸟的茁壮成长对你很重要？同你的朋友和家人分享所有你获取到的信息，提出令人信服的观点。他们可能会对数据感到很无聊，告诉他们令人难以置信的蓝龙来激发他们的兴趣吧！

# 教师与家长指南

故事灵感来自

## 伊丽莎白·曼-博尔杰塞
## Elisabeth Mann-Borgese

伊利莎白是罗马俱乐部成员，并教她的德国牧羊犬弹钢琴。

更多资讯

www.investopedia.com/terms/l/lawoflargenumbers.asp

www.dtmag.com/Stories/What%20About/04-06-whats_that.htm

www.seaslugforum.net/general.htm

The Drama of the Oceans by Elisabeth Mann-Borgese (ISBN-10: 0810903377)

http://en.wikipedia.org/wiki/Glaucus_atlanticus

http://salient.org.nz/columns/animal-of-the-week-the-blue-dragon

www.seaslugforum.net

http://en.rocketnews24.com/2012/09/13/blue-dragon-sea-slug-reminds-us-that-the-ocean-is-filled-with-strange-creatures-that-look-like-pokemon/

伊丽莎白·曼–博尔杰塞生于慕尼黑，是德国作家托马斯·曼最小的女儿。她搬到美国后，以编辑和作家为职业，最终晋升为《大英百科全书》执行秘书。事业中途，她将兴趣转移到海洋，于1970年组织了第一次海洋法会议，主题是"海洋和平"。她促使各国政府订立了《联合国海洋法公约》。在59岁，她成为达尔豪斯大学教授，在那里教学直到81岁。她的书《海洋的戏剧》是第一部关于过度捕捞和公海污染危机的作品。

# 图书在版编目（CIP）数据

利用太阳能的蓝龙：汉英对照／（比）冈特·鲍利著；（哥伦）凯瑟琳娜·巴赫绘；高芳，李原原译. ——上海：学林出版社，2016.6
（冈特生态童书. 第三辑）
ISBN 978-7-5486-1067-0

Ⅰ. ①利… Ⅱ. ①冈… ②凯… ③高… ④李… Ⅲ. ①生态环境－环境保护－儿童读物－汉、英 Ⅳ. ① X171.1-49

中国版本图书馆 CIP 数据核字（2016）第 126071 号

---

© 2015 Gunter Pauli
著作权合同登记号 图字 09-2016-309 号

**冈特生态童书**
**利用太阳能的蓝龙**

| | |
|---|---|
| 作　　者—— | 冈特·鲍利 |
| 译　　者—— | 高　芳　李原原 |
| 策　　划—— | 匡志强 |
| 责任编辑—— | 程　洋 |
| 装帧设计—— | 魏　来 |
| 出　　版—— | 上海世纪出版股份有限公司 学林出版社 |
| | 地　址：上海钦州南路 81 号　电话／传真：021-64515005 |
| | 网址：www.xuelinpress.com |
| 发　　行—— | 上海世纪出版股份有限公司发行中心 |
| | （上海福建中路 193 号　网址：www.ewen.co） |
| 印　　刷—— | 上海丽佳制版印刷有限公司 |
| 开　　本—— | 710×1020　1/16 |
| 印　　张—— | 2 |
| 字　　数—— | 5 万 |
| 版　　次—— | 2016 年 6 月第 1 版 |
| | 2016 年 6 月第 1 次印刷 |
| 书　　号—— | ISBN 978-7-5486-1067-0/G·402 |
| 定　　价—— | 10.00 元 |

（如发生印刷、装订质量问题，读者可向工厂调换）

# Energy 100

# 阳光下的雪绒花
## Edelweiss in the Sun

**Gunter Pauli**

冈特·鲍利 著
凯瑟琳娜·巴赫 绘
高 芳 李原原 译

## 丛书编委会

主　任：贾　峰

副主任：何家振　郑立明

委　员：牛玲娟　李原原　李曙东　吴建民　彭　勇
　　　　冯　缨　靳增江

## 丛书出版委员会

主　任：段学俭

副主任：匡志强　张　蓉

成　员：叶　刚　李晓梅　魏　来　徐雅清　田振军
　　　　蔡雩奇　程　洋

特别感谢以下热心人士对译稿润色工作的支持：

姜竹青　韩　笑　贾　芳　刘　晓　张黎立　刘之杰
高　青　周依奇　彭　江　于函玉　于　哲　单　威
姚爱静　刘　洋　高　艳　孙笑非　郑莉霞　周　蕊

# 目录

| | |
|---|---|
| 阳光下的雪绒花 | 4 |
| 你知道吗? | 22 |
| 想一想 | 26 |
| 自己动手! | 27 |
| 学科知识 | 28 |
| 情感智慧 | 29 |
| 艺术 | 29 |
| 思维拓展 | 30 |
| 动手能力 | 30 |
| 故事灵感来自 | 31 |

# Contents

| | |
|---|---|
| Edelweiss in the Sun | 4 |
| Did you know? | 22 |
| Think about it | 26 |
| Do it yourself! | 27 |
| Academic Knowledge | 28 |
| Emotional Intelligence | 29 |
| The Arts | 29 |
| Systems: Making the Connections | 30 |
| Capacity to Implement | 30 |
| This fable is inspired by | 31 |

在奥地利阿尔卑斯山上，一朵纯白色的雪绒花正享受着明媚的阳光，它的旁边生长着一株美丽的深蓝色的阿尔卑斯山龙胆草。

"我一直羡慕你毛绒绒的、白色的花瓣。"龙胆草说。

A pure white edelweiss is enjoying the bright sun in the Austrian Alps, where it is growing next to an Alpine gentian, with its beautiful deep blue colour. "I have always admired you for your plush, white petals," says the gentian.

在奥地利阿尔卑斯山上享受着明媚的阳光

Enjoying the bright sun in the Austrian Alps

为我写了一首优美的歌曲

Wrote a beautiful song about me

"谢谢。我的洁白让世界各地的许多人都成了我的朋友。我的奥地利朋友把我印到在欧洲各地流通的硬币上,我的美国朋友甚至为我写了一首优美的歌曲。"

"我不明白你为什么从不会被太阳晒伤。"龙胆草说。

"Thank you. My whiteness has made me many friends around the world. My fellow Austrians put me on a coin used all around Europe and my American friends even wrote a beautiful song about me."

"I have a hard time understanding how you never get sunburn," says the gentian.

"太阳根本无法影响我。"

"你厚厚的花瓣能阻挡阳光吗?"

"不能,我们根本阻挡不了任何东西,而是允许阳光穿透。"

"The sun doesn't affect me at all."

"Do your thick petals block the rays?"

"Not at all; we don't block anything. We let the sun come through."

你厚厚的花瓣能阻挡阳光吗?

Do your thick petals block the rays?

# 天空中的一个大洞

A big hole in the sky

"但这不是很危险吗?我听说人类使用特殊的面霜来保护他们的皮肤。"

"人类的确应该保护他们自己,但也应该保护环境。这样天空中就不会有一个让所有强烈的紫外线穿过的大洞了。"

"But isn't that dangerous? I hear people apply special creams to protect their skins."

"People should protect themselves, but they should also protect the environment. Then there wouldn't be a big hole in the sky that lets all these harsh sunrays through."

"天空中有一个洞？这会威胁着数百万人的健康，你是怎样在这样的山脉中生存下来的？"

"嗯，很难解释清楚，但是高山上的空气要比山谷下面更稀薄。当空气很稀薄时，天气很干燥，阳光会更容易穿过。"

"A hole in the sky? How can you survive here in the mountains with something like that, which puts the health of millions in danger?"

"Well, it's difficult to explain, but here in the mountains, when you go up very high, there is less air than down in the valley. When the air is thin, it is dry and easier for the sun to get through."

空气比山谷下面更稀薄

Less air than down in the valley

我只是让它进来

I simply let it come in

"但你还没有告诉我整个故事。你怎么从来不会被晒伤?你必须告诉我你的秘密。"龙胆草强调。

"没什么秘密。我并没有隔离阳光,只是让它进来。"

"然后呢?"

"But you are not telling me the full story. How come you never get surnburnt? You have to tell me your secret," insists the gentian.

"It's no secret. Instead of keeping the sun out, I simply let it come in."

"And then?"

"光被分散成微小的射线,这样当它到达我的身体组织时,就不会对我造成伤害了。"

"这就和武术一样!你用敌人的力量对付他自己。"

"The light is scattered into tiny little rays, so that by the time it gets to my tissues, it's all gone."

"That's what you do in martial art! You use your enemy's energy against himself."

你用敌人的力量对付他自己

Use your enemy's energy against himself

我花瓣上的纤细绒毛

Tiny hairs on my petals

"有点像，但不全是那样。不是隔离我们的朋友——太阳，而是让阳光进入我的身体，然后我把能量输送到花瓣上成千上万的纤细绒毛上。"

"真聪明！这比人类为了保护自己免受太阳光照射而在窗户、地毯、汽车油漆甚至皮肤上使用的毒素污染少多了。"龙胆草说。

"Kind of, but not really. Instead of keeping our friend, the sun, out, I let the sun in, but I channel the power to thousands of tiny hairs on my petals."

"Smart, indeed! It is so much less polluting than the toxins humans use on windows, in carpets, for car paints, and even on their skin to protect themselves from the sun," says the gentian.

"但我认为你应该分享更多藏在你那些美丽的蓝色花朵下面的小秘密……"

"好吧,我的秘密在我的根里。任何人的胃有问题时,告诉我就行了。"龙胆草眨眨眼说。

……这仅仅是开始!……

"But I think you should be sharing more of the little secrets you're hiding underneath those beautiful blue flowers of yours …"

"Well, I keep them guarded in my roots, but whenever anyone has a stomach problem, just let me know," says the gentian, and winks.

… AND IT HAS ONLY JUST BEGUN!…

……这仅仅是开始!……

...AND IT HAS ONLY JUST BEGUN!...

# Did You Know?
## 你知道吗?

The edelweiss originated in Asia and settled in Europe during the glacial periods. In Europe, it grows in the Jura, the Pyrenees, the Alps, and the Carpathians.

雪绒花源于亚洲,冰川时期定居在欧洲。欧洲的侏罗山、比利牛斯山、阿尔卑斯山和喀尔巴阡山都有雪绒花。

"Edelweiss" is German for "noble white". Its Latin name, *Leontopodium*, refers to how the shape and the woolly appearance of the flower makes it looks like a lion's paw. There are 30 different kinds of edelweiss.

"雪绒花"是德语,意思是"切花紫罗兰"。拉丁名是"*Leontopodium*",意思是花的形状和毛绒绒的外观使它看起来像一个狮子的爪子。雪绒花有30个不同的品种。

雪绒花是向日葵家族中的一员，它是一种多年生草本开花植物，生长在高达3000米的石灰石岩石上。浓密的绒毛能保护它抵御寒冷、干旱和紫外线的照射。

The edelweiss, a member of the sunflower family, is a perennial flower that lives on limestone rocks up to 3 000 metres high. Its dense hair protects it against the cold, drought, and ultraviolet radiation.

雪绒花是三个国家的国家象征：奥地利、瑞士和罗马尼亚。瑞士从1878年开始保护雪绒花，后来欧洲和亚洲的伊朗、印度、印度尼西亚、马来西亚和蒙古等国也加入进来。

The edelweiss is the national symbol of three countries: Austria, Switzerland, and Romania. Switzerland started to protect the flower in 1878 and was later joined by European and Asian nations like Iran, India, Indonesia, Malaysia, and Mongolia.

雪绒花的根可以抵御强大的山风，它的绒毛形成浓密的毯子防止其汁液受冻。

The roots of the edelweiss can resist strong mountain winds and their dense blanket of tiny hairs prevents their sap from freezing.

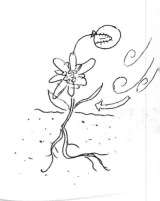

积雪覆盖的减少以及岩石和黑土对太阳能的吸收，导致山地植被没有像其他地方那样定植并危及雪绒花的生存。阿尔卑斯山阻止了南部的植被不断向北生长，这种生长会导致雪绒花的灭绝。

The decrease of the snow cover and the absorption of solar energy by rocks and dark soils have led to mountain vegetation not becoming established elsewhere and endanger the survival of the edelweiss. The Alps block southern vegetation from growing further north, leading to their extinction.

《雪绒花》是音乐剧《音乐之声》中的一首歌。1959年，《音乐之声》作为一个百老汇音乐剧出品，并紧接着在1965年被改编成电影。歌里的雪绒花在奥地利是忠诚的象征。这首流行歌曲常常被认为是奥地利民歌，尽管它其实是一首百老汇歌曲。

The song "Edelweiss" is a show tune of the musical The Sound of Music. This was first as a Broadway musical production in 1959 and then in 1965 the film adaptation followed. The edelweiss flower in the song was used as a symbol for loyalty to Austria. This popular tune is often thought to be an Austrian folk song, even though it is a Broadway composition.

龙胆草生长于世界各地，被用作中草药已经有几个世纪了，它被用来治疗消化不良、高血压、寄生蠕虫，甚至疟疾等疾病。它是巴赫花疗法（一种天然药物的组合）的核心成分。

The gentian grows worldwide and has been used for centuries as an herbal remedy to treat indigestion, hypertension, parasitic worms, and even malaria. It is a core ingredient of Bach flower remedies, a collection of natural medicine.

# Think About It

Do you like the idea that a powerful force can become harmless because its strength dissipates into thousands of little forces?

当一股强大的力量被分散成成千上万的小力量时,它会变得无害,你喜欢这个想法吗?

当你不得不抵御寒冷、大风和刺目的阳光,而数百万人赞叹你的美丽时,你感觉如何?

How would you feel if you had to defy the cold, wind and harsh sun, and are then celebrated for your beauty by millions around you?

With whom would you be prepared to share your secrets?

你愿意和谁一起分享你的秘密?

你更愿意使用防晒霜、遮阳伞,还是某种类似雪绒花所采用的方式来保护自己免受太阳辐射?

How do you prefer to protect yourself against the sun: with sunscreen, a parasol, or protection like the one the edelweiss developed?

# Do It Yourself! 自己动手！

How do you protect yourself against the sun? What kinds of sunscreens are popular? While we must protect ourselves against excessive exposure to the sun, we also need to expose ourselves regularly to the sun in order to produce vitamin D. We need to find a balance between avoiding and welcoming the sun on our skin. That balance is a dynamic one as not getting enough sun could cause your bones to become brittle. So, ask your friends and relatives who of them are protecting their skins all the time? And tell them that by doing that, they may be reducing their exposure to the sun, which is necessary to help absorb enough calcium into their bones.

如何保护自己免受太阳照射？什么样的防晒霜最流行？我们必须避免过度暴露在阳光下，同时，为了生成维生素D，我们也需要经常晒晒太阳。在晒和不晒之间，我们需要找到一个平衡。这个平衡是动态的，因为没有得到足够的阳光照射可能会导致你的骨骼变得脆弱。所以，问问你的朋友和亲人，他们每时每刻都在保护皮肤吗？告诉他们这样做可能会减少在阳光下的暴露，而晒太阳对于帮助骨骼吸收足够的钙是很有必要的。

# TEACHER AND PARENT GUIDE

## 学科知识
## Academic Knowledge

| | |
|---|---|
| 生物学 | 雪绒花是一种多年生草本植物；雪绒花很慷慨，它供养29种昆虫科，但只有少数昆虫为其授粉；一个授粉的雪绒花每个季度只生成一颗种子，成熟期是20～30天；龙胆草有400种，是两年生或多年生常绿植物。 |
| 化 学 | 雪绒花生长在碱性环境中；雪绒花与昆虫共生，因为来自植物花蜜的氨基酸对于为其授粉的昆虫的生存是不可或缺的；龙胆草生长在中性偏酸性的土壤和岩石花园中；氧化锌和氧化钛是最受欢迎的防晒成分；金属氧化物混合成聚合物，生产饮料包装、食品包装和化妆品。 |
| 物 理 | 雪绒花根状茎呈圆柱形；除了波长为0.18微米的紫外线，大部分光都会被雪绒花的白色绒毛层反射掉，雪绒花的绒毛吸收紫外线，防止它穿透植物的"皮肤"；雪绒花的绒毛是透明的，但它看起来是白色的，就像北极熊，这是因为绒毛之间的空隙散射光的缘故；臭氧吸收太阳紫外线辐射；悬浮在空气中的水粒子会阻挡太阳；越高的地方空气越稀薄，更多的阳光可以穿透大气层；紫外线会破坏微生物；鸟类、蜜蜂、爬行动物可以看到紫外线，紫外线可以改善它们的视力；紫外线能帮助昆虫导航。 |
| 工程学 | 大自然有时不是抵御而是顺应自然力量。 |
| 经济学 | 奥地利面值2分欧元的硬币图案设计是雪绒花；防止紫外线和使用紫外线都是数十亿美元的商机；世界各地都在保护雪绒花，现在它被种植以用于为化妆品提供原料。 |
| 历 史 | 《雪绒花》这首歌出自1959年美国百老汇的歌舞剧《音乐之声》，今天许多奥地利人认为这是一首古老的民歌。 |
| 地 理 | 阿尔卑斯山、喀尔巴阡山、侏罗山和比利牛斯山都在欧洲；臭氧层位于距地面30～50千米高的平流层。 |
| 生活方式 | 我们已经意识到要保护自己免受太阳过度照射，但是我们却忽视了我们也需要把自己暴露在太阳下。 |
| 社会学 | 德国军队组成的阿尔卑斯部队使用雪绒花作为制服的徽章，但在第二次世界大战结束时，雪绒花已经成为德国人民抵抗纳粹主义的象征。 |
| 心理学 | 人们视雪绒花这种植物为勇气和高贵纯洁的象征，因为它在严酷的气候下仍能生长，而且它拥有纯白色的外表。 |
| 系统论 | 雪绒花象征即使在恶劣的条件下也能生活，它散发着美丽并激励着人们；雪绒花作为生物医学产品的价值已被认可，而且它也是化妆品的一个关键成分。 |

# 教师与家长指南

## 情感智慧
### Emotional Intelligence

**雪绒花**

雪绒花在整个对话中都是有自知之明和自信的。对于奥地利赋予它的荣誉以及在百老汇音乐剧中起重要作用,她感到很荣幸。她的话语风格简明扼要。她不自夸,也不说太多自己的细节。雪绒花把重点从自己身上转移到环境上,表现出对生态系统的理解。雪绒花准备分享她如何保护自己免受紫外线过度照射。她提供了一些细节,同时保持语言简洁,展示了很好的沟通能力。一旦雪绒花揭示了她保护自己抵御阳光的秘密,龙胆草只能向她解释了藏在根里的能量。

**龙胆草**

龙胆草在对话的开始就表示了他对雪绒花的钦佩和共鸣。龙胆草的任务是:了解雪绒花如何保护自己,并以极大的信心和坚持直接提出问题。当雪绒花把话题转向环境时,龙胆草用一句简单的请求把话题拉了回来:告诉我你的秘密。当雪绒花给出说明后,龙胆草迅速跟进使问题更加清晰。当雪绒花表示他对武术的解释不正确时,龙胆草不仅没有生气,反而为雪绒花巧妙的解决方案表示祝贺。龙胆草愿意回报雪绒花的分享,解释说,他的主要优势在根上,它含有一种能用于治疗消化疾病的很好的药。

## 艺术
### The Arts

找到《雪绒花》这首歌,找几个朋友和你一起在唱诗班唱这首歌。你可以清唱,就是说没有音乐伴奏地唱。这是一首短小而浪漫的歌曲。然而,不能只是唱这首歌,还要讨论在第二次世界大战背景下这首歌的幕后故事。你甚至可以花点时间看看《音乐之声》这部电影。

# TEACHER AND PARENT GUIDE

## 思维拓展
### Systems: Making the Connections

在最困难的环境下也有生命。雪绒花接受了潜在的威胁生命的生存条件，通过进化以一种杰出的适应能力把这些条件变成了一个赞美她的美丽的机会。她生活在海拔3000米的地方，那里常年有风，冬天天气寒冷，夏天空气稀薄干燥、氧气有限、阳光猛烈。雪绒花证明了在极端的条件下也会产生极致的创新。雪绒花的茎和装饰花瓣的成千上万的绒毛分散掉了紫外线，紫外线的影响在其到达植物的"皮肤"之前已经消散。一些潜在的伤害（紫外线）已被转化为中立的力量，这种力可能有温和的细菌控制效果。此外，雪绒花给近30种昆虫提供营养，但只有几种昆虫给她授粉作为回报。如此巨大的慷慨在自然界里是很少见的。然而，这些少数授粉昆虫依赖雪绒花来增加生活所需要的氨基酸，这证明了共生和进化。

雪绒花起源于亚洲，后定居在欧洲各大山脉，激励年轻的登山者攀爬岩石去寻找雪绒花，在19世纪这种行为成为一种表达爱的流行方式，一些人勇于登峰并最终摘取雪绒花送给他们的情侣。雪绒花这个纯爱和奉献的符号后来成为了抵抗纳粹忠于国家的象征，这是在德国吞并奥地利之后。这种花至今仍然继续激励着人们。雪绒花让我们深刻地理解，在一个复杂的环境下，技术、自然和社会行为是如何结合的。这种花富含象征意义，代表承诺、毅力和美德。美既在艺术中，也在科学中。

## 动手能力
### Capacity to Implement

龙胆草很容易获得。去买一些根部完整的龙胆草。确保它的花没有暴露于恶劣的化学杀虫剂和杀真菌剂中。然后拿一块根啃一啃。不要感到惊讶：它很苦！找出使用这个健康产品的最好方式。记住，它刺激消化，是促进消化的药物的主要成分。不用它来做饭，而是酿造一种健康的饮料，将它尽可能地与其他成分融合。确保你在家长的监督下进行尝试，在交给别人尝试前先检查一下。

# 教师与家长指南

故事灵感来自

## 让-波尔·维涅龙
## Jean-Pol Vigneron

让-波尔·维涅龙（1950—2013）曾在那慕尔大学和列日大学学习物理学，并探索超越旧的科学学科定义的新科学前沿。他先研究半导体，后来进入美国IBM进行研究生学习。他的兴趣从计算机技术发展到光学和不同种透明材料的研究。在他与安德鲁·帕克（本童书第53个故事《不用画的色彩》就是受他的启发）的合作中，他发现了结构性色彩，这引导他开始对雪绒花进行研究。他是那慕尔圣母和平学院的教授，他教导学生：科学需要严谨和训练，只有当我们探索梦想和欢迎不确定性时，科学中的新领域才会被发现。

### 更多资讯

http://www.researchgate.net/profile/Jean_Pol_Vigneron

http://onlinelibrary.wiley.com/doi/10.1111/j.1095-8339.1993.tb01900.x/abstract

**图书在版编目（CIP）数据**

阳光下的雪绒花：汉英对照 /（比）冈特·鲍利著；
（哥伦）凯瑟琳娜·巴赫绘；高芳，李原原译. —— 上海：
学林出版社，2016.6
（冈特生态童书. 第三辑）
ISBN 978-7-5486-1066-3

Ⅰ. ①阳… Ⅱ. ①冈… ②凯… ③高… ④李… Ⅲ.
①生态环境－环境保护－儿童读物－汉、英 Ⅳ.
① X171.1-49

中国版本图书馆 CIP 数据核字 (2016) 第 126068 号

---

Ⓒ 2015 Gunter Pauli
著作权合同登记号 图字 09-2016-309 号

**冈特生态童书**
**阳光下的雪绒花**

| | |
|---|---|
| 作　　者—— | 冈特·鲍利 |
| 译　　者—— | 高　芳　李原原 |
| 策　　划—— | 匡志强 |
| 责任编辑—— | 匡志强　蔡雩奇 |
| 装帧设计—— | 魏　来 |
| 出　　版—— | 上海世纪出版股份有限公司 学林出版社 |
| | 地　址：上海钦州南路81号　电话／传真：021-64515005 |
| | 网址：www.xuelinpress.com |
| 发　　行—— | 上海世纪出版股份有限公司发行中心 |
| | （上海福建中路193号　网址：www.ewen.co） |
| 印　　刷—— | 上海丽佳制版印刷有限公司 |
| 开　　本—— | 710×1020　1/16 |
| 印　　张—— | 2 |
| 字　　数—— | 5万 |
| 版　　次—— | 2016年6月第1版 |
| | 2016年6月第1次印刷 |
| 书　　号—— | ISBN 978-7-5486-1066-3/G·401 |
| 定　　价—— | 10.00元 |

（如发生印刷、装订质量问题，读者可向工厂调换）

# 胃里的塑料
## Plastics in my Tummy

**Gunter Pauli**

冈特·鲍利 著
凯瑟琳娜·巴赫 绘
隋淑光 译

学林出版社
www.xuelinpress.com

## 丛书编委会

主　任：贾　峰

副主任：何家振　郑立明

委　员：牛玲娟　李原原　李曙东　吴建民　彭　勇
　　　　冯　缨　靳增江

## 丛书出版委员会

主　任：段学俭

副主任：匡志强　张　蓉

成　员：叶　刚　李晓梅　魏　来　徐雅清　田振军
　　　　蔡雯奇　程　洋

特别感谢以下热心人士对译稿润色工作的支持：

姜竹青　韩　笑　贾　芳　刘　晓　张黎立　刘之杰
高　青　周依奇　彭　江　于函玉　于　哲　单　威
姚爱静　刘　洋　高　艳　孙笑非　郑莉霞　周　蕊

# 目录

| | |
|---|---|
| 胃里的塑料 | 4 |
| 你知道吗？ | 22 |
| 想一想 | 26 |
| 自己动手！ | 27 |
| 学科知识 | 28 |
| 情感智慧 | 29 |
| 艺术 | 29 |
| 思维拓展 | 30 |
| 动手能力 | 30 |
| 故事灵感来自 | 31 |

# Contents

| | |
|---|---|
| Plastics in my Tummy | 4 |
| Did you know? | 22 |
| Think about it | 26 |
| Do it yourself! | 27 |
| Academic Knowledge | 28 |
| Emotional Intelligence | 29 |
| The Arts | 29 |
| Systems: Making the Connections | 30 |
| Capacity to Implement | 30 |
| This fable is inspired by | 31 |

一条小梭鱼正在为几天来的肚子疼而抱怨,他的爸爸束手无策,只能在海底游来游去,试图找出造成他儿子难受的原因。这时他注意到了一只正在转圈的海龟。

梭鱼爸爸问道:"打扰一下,你为什么要转圈呢?"

"哦,幸亏碰到你了,你能帮我弄掉这根缠住我的渔线吗?"海龟说道。

A young barracuda is complaining that his tummy has been very sore for quite a few days now. His dad is desperate to help him and looks around the seabed for any clues to what could have caused his son's distress. He spots a turtle swimming in circles.

"Excuse me, why are you swimming in circles?" asks the dad.

"Oh, thank goodness you came along," says the turtle. "Can you please help me get rid of this fishing line I'm stuck in?"

你为什么要转圈呢?

Why are you swimming in circles?

让我看看能不能帮你摆脱它

Let me see if I can set you free

## 梭鱼

爸爸回答道:"什么渔线?我什么都没看到。"

"嗯,人们把渔线做成透明的,这样鱼就看不到它了。它很结实,一旦你被它缠住了,就无法摆脱。"

"让我靠近点看看。"梭鱼爸爸回答道,"是的,我看到了,这条线割伤了你的脖子和腿。伤得很严重!让我看看能不能帮你摆脱它。"

"What fishing line? I can't see anything," replies the barracuda.

"Well, people make fishing line transparent so fish cannot see it. They also make it so strong that once you get caught in it, there's no escape."

"Let me take a closer look," says the barracuda. "Yes, I see the line wound around your neck and your legs. That must hurt! Let me see if I can set you free."

"谢谢你。"海龟说,"这些线感觉比钢丝还结实,被它缠住后,我既不能游泳又不能捕食。"

"这真是生死攸关!我尽全力来帮你。"

"你觉得你能咬断它吗?"海龟问道。

"Thank you," says the turtle. "These lines feel stronger than steel, and tangled up like this, I can't swim or feed."

"This is a matter of life and death then! Let me do the best I can."

"Do you think you will be able to bite through all this?" asks the turtle.

被缠住后,我既不能游泳又不能捕食

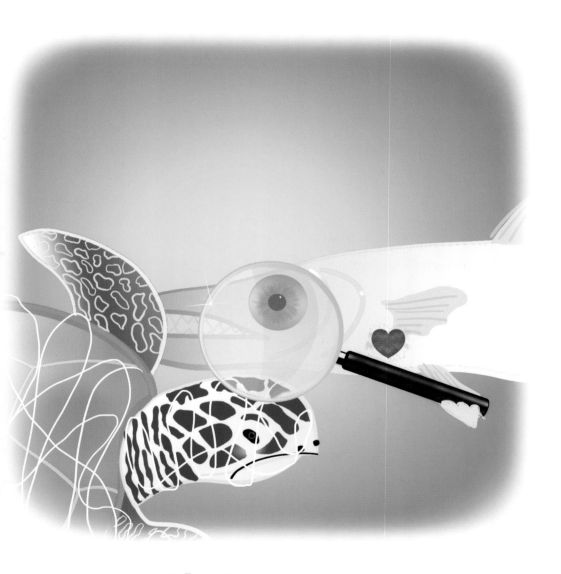

Tangled up like this, I can't swim or feed

幸运的是，我有非常锋利的牙齿

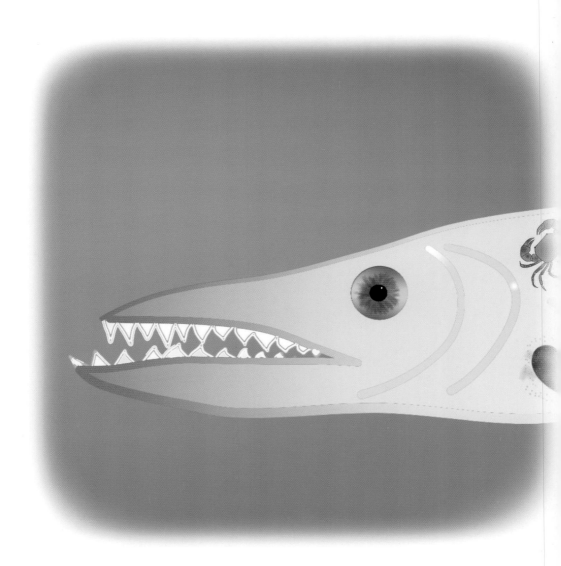

Fortunately, I have very sharp teeth

梭鱼爸爸说:"幸运的是,我有非常锋利的牙齿。"然后他迅速咬断了渔线。海龟自由了,浮出海面大口呼吸。

"太感谢了!"海龟说道,"请告诉我,渔线咬起来感觉怎样?"

"它不像任何其他东西,倒像是人们倾倒进海里的尼龙和塑料。"梭鱼爸爸回答道。

"Fortunately, I have very sharp teeth," says the barracuda, quickly snapping the lines, setting the turtle free to swim to the surface for air.

"I'm so grateful, thank you," says the turtle. "But tell me, how does it taste?"

"It doesn't taste like anything at all, just like the nylon and plastics that people dump in the sea," replies the barracuda.

"我不理解人们为什么不把海里的渔线、塑料垃圾、破渔网收集起来,我听说这些能用来生产服装,扩大就业!"海龟评论道。

"生产服装?我想人们更喜欢种棉花来做服装原料……可种棉花需要大量的水和农药。"

"I don't understand why people don't collect all their fishing lines, plastic garbage, and broken nets from the sea. I've been told it can be used to make clothing and generate jobs!" remarks the turtle.

"Clothing? I thought people are sticking to growing cotton for that … cotton that needs a lot of water and chemicals to grow."

这些能用来生产服装，扩大就业！

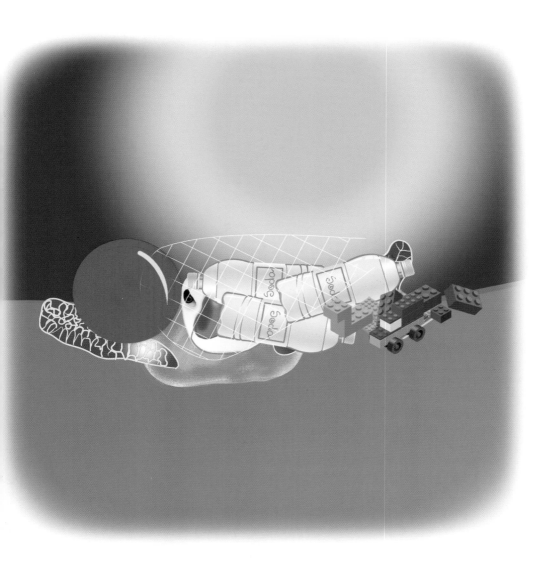

It can make clothing and generate jobs!

你消化不良吗?

Are you not digesting your food well?

"嗯，一些人已经意识到要节约用水，并且很快就可以用其他材料，比如海藻、海带来生产纺织品了。"

"好吧，陆地上的人面临的问题让他们自己去应对吧，我们要想办法解决海里的问题，比如说我儿子经常肚子疼。"

"你看上去也不太好，很臃肿。你是不是消化不良？"海龟说道。

"Well, some people have realised that they need to save water and now textiles will soon be made of other materials, like algae and seaweed."

"Oh well, let them worry about the problems they have on land. We have to find ways to deal with our problems here in the sea, like my son's frequent tummy aches!"

"You don't look that well either," remarks the turtle. "You are very bloated. Are you not digesting your food well?"

"我的胃胀得厉害，好像不能正常地吃食物和排泄了。"梭鱼爸爸叹息道。
"你是不是吃了漂在海里的塑料片？"
"塑料？你是说那些在阳光下亮闪闪的、看上去很诱人的东西是塑料？天啊，我要告诉孩子们离它们远点。"

"My stomach feels blocked. It seems nothing that goes in, comes out again," sighs the barracuda.
"Have you been eating those pieces of plastic that float around the ocean?"
"Plastic? You mean those flickering things that look so appetising in the sunlight are bits of plastic? Oh dear, I will tell my children to stay away from it."

你是不是吃了那些塑料片？

Have you been eating those pieces of plastic?

为什么不用可降解的生物塑料？

Why can't they use degradable bioplastics?

"你知道吗？"
海龟说，"现在是时候完全停止塑料袋的生产了，那些高吸水性材料会堵塞我们的肠道。人们为什么不用可降解的生物塑料代替塑料袋来装垃圾呢？"
"我担心的是，在人们变得明智之前，我的家人和我恐怕还要遭受几年消化不良的痛苦。"梭鱼爸爸叹息道。

"Didn't you know?" asks the turtle. "It's about time people stopped making plastics bags altogether. And those super absorbent materials they block our intestines too. Why can't they use degradable bioplastics to collect their waste, instead of plastic bags?"
"I'm afraid my family and I may still suffer from indigestion for years before people become that smart," sighs the barracuda.

"嗯，别绝望，朋友。已经有人在用蓟生产塑料了。"

"蓟？我过去以为那是一种杂草！我知道有些真正有创造力的人非常关心环境和我们的生存，他们取得的成绩非常惊人。"

"就像你今天救了我一样惊人！"

……这仅仅是开始！……

"Well, don't despair, my friend. There's already someone making plastics from thistles."

"Thistles? I thought that was a weed! I know there are some truly creative people who deeply care for the environment and our survival, and what they have achieved is truly amazing."

"As amazing as you saving my life today!"

... AND IT HAS ONLY JUST BEGUN!...

……这仅仅是开始！……

... AND IT HAS ONLY JUST BEGUN! ...

# Did You Know?

## 你知道吗？

For every three tonnes of fish in the oceans, it has been estimated that there is one tonne of plastics. These accumulate into gyres of trash that disintegrate very slowly.

据估计，海里每三吨鱼就有一吨塑料，这些塑料汇聚成许多条垃圾带，降解得非常缓慢。

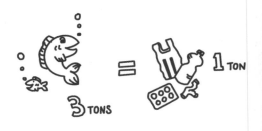

After decades of pollution, some 100 to 150 million tonnes of waste have accumulated as floating plastic islands, often made up of small plastic particles suspended in the sea, called "plastic soup". Every year between 5 and 12 million tonnes are added.

经过数十年的污染，共有大约1亿~1.5亿吨的垃圾汇集成了塑料浮岛，并且规模以每年500万~1200万吨的速度在扩大。塑料浮岛通常由小塑料颗粒组成，漂在海面上，人们称其为"塑料汤"。

The size of these huge plastic islands is estimated to have grown to 15 million km2. It is expected to double in size in a decade.

据估计，巨型塑料浮岛的面积已经达到1500万平方公里，预计在十年中将会增大一倍。

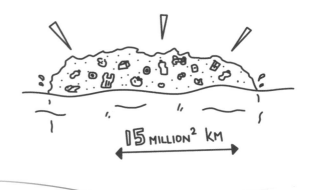

The concentration of small plastic particles in this gyre is up to seven times larger than the concentration of zooplankton.

垃圾带中塑料颗粒的浓度已经超过浮游生物的浓度七倍以上。

Long-lasting plastics end up in the stomachs of marine mammals, fish, and birds. One third of albatross chicks die due to plastics ingested when fed by their parents. The surviving marine life suffers from hormone disruptions due to the additives.

不易降解的塑料的最终归宿是进入海洋哺乳动物、鱼和鸟的胃里。有三分之一的信天翁幼鸟因为父母喂食时混入了塑料而死亡。那些幸存的海洋生物则正在遭受由塑料引起的激素紊乱。

When washed down the drain, microbeads from cosmetics and toothpastes pass through sewage treatment plants unfiltered and cause particle water pollution everywhere, from the Great Lakes of North America and elsewhere to the oceans of the world.

牙膏和化妆品里的微小颗粒被冲进下水道，如果不经过污水处理厂处理，就会到处造成"颗粒水污染"，从北美等地的大湖到全世界的海洋都不能幸免。

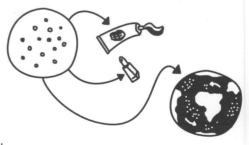

Thistles, which grow prolifically, especially in fallow fields, have previously been considered a weed. As renewable resources, the flower, stem, and root can be converted into raw materials for biodegradable chemicals.

曾经被视作杂草的蓟在闲置的土地上生长得尤其繁茂。作为一种可再生资源，它的花、茎和根能被转化成生产生物降解药品的原材料。

The Netherlands was the first country in the world to plan doing away with microbeads in cosmetics and Italy was the first country to ban non-biodegradable plastic bags. Milan is the city that recovers most biomass in the world by using bioplastics to collect waste.

荷兰是世界上第一个计划禁止在化妆品中添加微颗粒的国家，意大利是第一个禁止使用非生物降解塑料袋的国家，米兰是世界上通过使用生物塑料袋收集垃圾，回收生物量最多的城市。

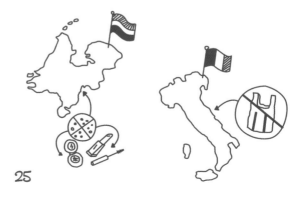

# Think About It
# 想一想

Would you like to eat fish that have been feeding on plastics?

你愿意吃以塑料为食的鱼吗?

Do you think that recycling plastics is enough to solve the problem of plastic-waste gyres in the sea?

你觉得通过回收塑料能解决海洋中的塑料废物洋流问题吗?

蓟能用来生产在阳光、水和土壤中降解的塑料,这样的植物是不是很令人惊叹?

Do you find it amazing that weeds, like thistles, are used to make plastics that degrade in the sun, in water, and in the soil?

If you or any of your family members are sick, are you likely to be worrying about other people's problems?

如果你或你的家庭成员生病了,你会因此担心其他人吗?

# Do It Yourself!
# 自己动手！

Have a look at everything you have in your home, and that you use on a daily basis, that is made of plastic or contains plastics. Be on the lookout for especially the small plastic particles, known as the microbeads (mainly found in your mother's cosmetics and in toothpaste). Make a list of these items and figure out how these plastics get from your toothpaste or sunscreen into the sea. Now make a drawing that offers a detailed flow chart showing how these plastics pass from your mouth or face all the way into rivers and the sea.

看一下家里由塑料构成或者含有塑料的每一样东西，特别是那些日常用品。继续寻找被称作"微球"的小塑料颗粒（主要存在于你妈妈的化妆品和牙膏里），列一张清单，并搞清这些塑料是如何从防晒油和牙膏里进入海洋的。现在画一张详细的流程图，标明这些塑料是如何从你的嘴里或脸上进入河流和海洋里的。

# TEACHER AND PARENT GUIDE

## 学科知识
### Academic Knowledge

| | |
|---|---|
| 生物学 | 梭鱼会把闪闪发光的物体误当成猎物；塑料降解为小颗粒后被小的水生生物摄入，并在体内聚集，塑料通过这种方式进入了食物链；海龟需要浮出水面呼吸氧气，当它被钓线或渔网困住时就会死去；褐海藻是最好的天然纤维生产者。 |
| 化学 | 塑料浸泡在水里不到一年就可以降解，释放出双酚A、多氯联苯(PCBs)等有毒化学物质；弹性体是橡胶中的一种添加剂。 |
| 物理 | 塑料光降解的最终目标是降解到分子水平，但是对大多数塑料来说，仅能分解成小颗粒，仍保持完整的聚合物状态；涡流是一种环形或螺旋形的运动形式；单丝具有较低的光密度，因此不容易被看到；超强吸水聚合物能吸收相当于自身重量500倍、自身体积30～60倍的水。 |
| 工程学 | 渔线由合成纤维制成，即使是由单丝制成的细渔线，也比钢丝坚韧10倍；在热、氧和光的作用下，塑料会随着时间的推移而降解。 |
| 经济学 | 塑料产品的半衰期长达数十年甚至数百年，而其用途只有一天或几周；数十亿计的免费塑料袋仅仅用过一次就被扔掉了；回收高性能的尼龙并将它转化为高性能的服装生产材料是一种共同的社会责任；一次性用品提高了食品的卫生程度但导致了过度包装。 |
| 伦理学 | 塑料只能被风化而不能完全降解，当类似产品的使用寿命结束后，我们怎么可以对其引发的问题放任不管？我们希望由其他人来解决我们自己造成的问题吗？我们生产和使用塑料，在污染地球的同时，最终也会使自己受害，为什么我们不生产耐用的产品来取代一次性用品呢？ |
| 历史 | 1988年，人类首次发现了海洋中的塑料废物洋流；一次性尿布是在瑞典政府的资助下，于1955年发明的。 |
| 地理 | 北太平洋环流、南太平洋环流、北大西洋环流、南大西洋环流、印度洋环流；产品包装艺术是从亚洲文化（尤其是日本文化）中发展出来的。 |
| 数学 | 每平方公里海洋中分布着8平方米的塑料颗粒，其粒径为5毫米×5毫米×1毫米。 |
| 生活方式 | 现代社会已经成了一次性用品社会；尿布的舒适度取决于超吸收性能的化学材料对尿的吸收能力，这种材料不能降解，只会在水中被侵蚀。 |
| 社会学 | 社会上把一次性用品视作提高生活质量的需求；包装产品和打开包装的乐趣。 |
| 心理学 | 由于认知上的错误，人们既没有回收塑料垃圾，也没有想办法避免或补救这一问题，需要引入经济激励机制，或者采取征税这样的抑制措施来提高塑料回收效率。 |
| 系统论 | 未经充分考虑的、不成功的产品设计会降低生活质量，并从根本上影响生活。 |

# 教师与家长指南

## 情感智慧
## Emotional Intelligence

**梭 鱼**

梭鱼一家正在遭受病痛,但他们并不知道原因。尽管这样,梭鱼爸爸还是很关心别人,当他发现海龟行为异常时表现出了同情。但是他并不同情人类。梭鱼爸爸不知道渔线是什么,但是一旦意识到海龟面临生命危险,他立即准备用牙齿解救她。梭鱼爸爸不相信人类会欢迎那些真正有利的发明,因此他仍然想继续关注海洋所面临的问题,而不是去考虑陆地上人们所面临的抉择。当海龟提到塑料的替代品,如用蓟生产的生物塑料时,梭鱼爸爸发现这超出了他的理解。

**海 龟**

海龟被渔线缠住了,处于生死存亡的紧急关头,梭鱼爸爸伸出援手解救了她,对此她非常感激。在指挥梭鱼爸爸救她时,海龟表现得既平静又有自控力。被解救出来后,她表达了谢意,然后询问塑料的味道,并且对人们不采用已知的方法来解决塑料污染问题感到不理解。海龟非常清楚人类所面临的挑战和抉择。她关注梭鱼一家的健康情况,并指出问题是因为误食塑料引起的。尽管梭鱼对她提到的新发明不抱信心,但海龟出于对救命之恩的感激,还是提供了更深入的信息,分享了自己的智慧。

## 艺术
## The Arts

塑料通常被认为是废品,但可以回收再利用。试试能不能收集25片废塑料,并用来制作某种形式的艺术品。比如说用从塑料提袋上拆下的塑料条编成小垫子;从塑料瓶上切下圆片,来做一件时尚的衣服。邀请你的家人和朋友一起收集废塑料,然后探讨可以把它们做成什么形式的艺术品。只要实用就可以了,不一定非得华丽。

# TEACHER AND PARENT GUIDE

## 思维拓展
### Systems: Making the Connections

给我们带来便利的塑料产品已经成为现代生活的一部分。在食品包装方面，它们以低廉的成本以及改善了卫生程度的优势，正在逐步替代金属产品。一次性塑料购物袋也已经取代了纺织品购物袋这样的耐用品。但是，当塑料产品的优势越来越明显时，我们对它的危害却视而不见。如此多的不可降解的塑料产品（其半衰期即使没有几百年也有几十年）遍布于我们的生活中，正在渐渐汇集成一个沉重的负担。人们努力推动塑料回收，并纠正填埋和焚烧的过激做法。然而我们的环境中仍有大量塑料，遍布街道、山坡、峡谷、河流、海洋。尽管我们早就被警告过塑料不易降解，但只有在得到令人信服的统计学数字后，在看到海龟、鱼和鸟死亡的令人不安的照片后，这个问题的严重性才凸显出来。有毒的化学品正在进入食物链，威胁着一切生命。塑料曾经被视为现代化的标志，现在却成为一个令人不安的现实问题。人们必须重新设计塑料，使其既保留功能，又不会对土地和水生生物以及我们赖以生存的食物链造成威胁。目前新的产品已经出现了，它所带来的意义已经超越了用一个产品代替另一个产品，而是提示人们要选择更合适的经济发展模式。将蓟作为一种可再生资源来生产生物塑料或者其他复合物，这不仅宣告了无污染产品设计的出现，而且还把容易获得的资源转变为经济增长和扩大就业的机遇。

## 动手能力
### Capacity to Implement

你在家里和学校使用过多少种生物塑料产品？检查一下当地商店使用的塑料袋和食品包装物是否是用可再生材料做成的。然后研究一下这些材料在阳光下、水中和土壤中进行生物降解的可行性数据。列出可以用生物塑料取代的塑料清单。也去车库看看汽车所用的润滑油，确认一下是不是只是用石油生产的。一旦你列出了所有可以用生物塑料替代的产品清单，看看哪个供应商保证他们提供的塑料产品和润滑油是用可生物降解的可再生资源生产的。现在你可以制订一个促进可持续发展的日程。

# 教师与家长指南

## 故事灵感来自
## 卡狄亚·巴斯蒂奥利
### Catia Bastioli

卡狄亚·巴斯蒂奥利是一位科学家出身的意大利企业家。在佩鲁贾大学获得化学学位后，她作为一个材料学家，在蒙特爱迪生化工集团开始了研究生涯，致力于从可再生资源中寻找可生物降解的塑料。当公司所遵循的传统商业模式失败，业务解体后，她把她的研究部门转变为一个独立单位，然后转型为一个叫诺瓦蒙特的新企业。她继续致力于用一系列可生物降解和合成的生物塑料（被称为"Mater–Bi"）来发展生物化学工业，这是一种不会与食物生产产生竞争的资源。

卡狄亚·巴斯蒂奥利及其团队在生物塑料研制以及基于本地的原材料（包括农业废弃物和杂草）规划当地经济发展方面取得了进展。她的努力促进了工业的进一步转型，把无效的工业生产单位转变成有效的生产单位。卡狄亚·巴斯蒂奥利拥有超过一百项与生物聚合物有关的专利，证明人们通过重新设计塑料购物袋，可以实现创新，从而为海洋和陆地的问题提供了新的解决思路。2007 年，卡狄亚·巴斯蒂奥利荣获"欧洲发明者"称号。

### 更多资讯

http://archive.theplastiki.com/wp-content/uploads/2010/04/Plastiki-_5gyres_A3_HR_RGB.jpg

http://www.matrica.it/article.asp?id=26&ver=en#.VhlGWc6m2r8

http://thesocietypages.org/socimages/2011/01/22/the-economic-injustice-of-plastic/

图书在版编目（CIP）数据

胃里的塑料：汉英对照／（比）冈特·鲍利著；
（哥伦）凯瑟琳娜·巴赫绘；隋淑光译．－－上海：学林
出版社，2016.6
　（冈特生态童书．第三辑）
　ISBN 978-7-5486-1048-9

Ⅰ．①胃… Ⅱ．①冈… ②凯… ③隋… Ⅲ．①生态环
境－环境保护－儿童读物－汉、英 Ⅳ．① X171.1-49

中国版本图书馆 CIP 数据核字 (2016) 第 125824 号

————————————————————————

ⓒ 2015 Gunter Pauli
著作权合同登记号　图字 09-2016-309 号

**冈特生态童书**
**胃里的塑料**

| 作　　者—— | 冈特·鲍利 |
| 译　　者—— | 隋淑光 |
| 策　　划—— | 匡志强 |
| 责任编辑—— | 匡志强　蔡雯奇 |
| 装帧设计—— | 魏　来 |
| 出　　版—— | 上海世纪出版股份有限公司 学林出版社 |
| | 地　址：上海钦州南路 81 号　电　话／传　真：021-64515005 |
| | 网址：www.xuelinpress.com |
| 发　　行—— | 上海世纪出版股份有限公司发行中心 |
| | （上海福建中路 193 号 网址：www.ewen.co） |
| 印　　刷—— | 上海丽佳制版印刷有限公司 |
| 开　　本—— | 710×1020　1/16 |
| 印　　张—— | 2 |
| 字　　数—— | 5 万 |
| 版　　次—— | 2016 年 6 月第 1 版 |
| | 2016 年 6 月第 1 次印刷 |
| 书　　号—— | ISBN 978-7-5486-1048-9/G·383 |
| 定　　价—— | 10.00 元 |

（如发生印刷、装订质量问题，读者可向工厂调换）

# Water 89

# 矿井水

## Mine Water

**Gunter Pauli**

冈特·鲍利 著

凯瑟琳娜·巴赫 绘

郭光普 译

www.xuelinpress.com

## 丛书编委会

主　任：贾　峰
副主任：何家振　郑立明
委　员：牛玲娟　李原原　李曙东　吴建民　彭　勇
　　　　冯　缨　靳增江

## 丛书出版委员会

主　任：段学俭
副主任：匡志强　张　蓉
成　员：叶　刚　李晓梅　魏　来　徐雅清　田振军
　　　　蔡雪奇　程　洋

特别感谢以下热心人士对译稿润色工作的支持：

姜竹青　韩　笑　贾　芳　刘　晓　张黎立　刘之杰
高　青　周依奇　彭　江　于函玉　于　哲　单　威
姚爱静　刘　洋　高　艳　孙笑非　郑莉霞　周　蕊

# 目录

| | |
|---|---|
| 矿井水 | 4 |
| 你知道吗？ | 22 |
| 想一想 | 26 |
| 自己动手！ | 27 |
| 学科知识 | 28 |
| 情感智慧 | 29 |
| 艺术 | 29 |
| 思维拓展 | 30 |
| 动手能力 | 30 |
| 故事灵感来自 | 31 |

# Contents

| | |
|---|---|
| Mine Water | 4 |
| Did you know? | 22 |
| Think about it | 26 |
| Do it yourself! | 27 |
| Academic Knowledge | 28 |
| Emotional Intelligence | 29 |
| The Arts | 29 |
| Systems: Making the Connections | 30 |
| Capacity to Implement | 30 |
| This fable is inspired by | 31 |

一个干冷的下午，一群鹌鹑正在一片金黄的油菜花田里享用晚餐。

"我们在你这里收获了很多，为什么你开的花比以往更亮丽呢？"一只鹌鹑问道。

"这要感谢矿井水。"油菜花答道。

It is a crisp spring afternoon. Some quail are enjoying their dinner in a field of beautiful yellow rapeseed flowers.

"We are getting a great harvest from you. Why are you blossoming brighter than ever before?" asks one of the quail.

"It's thanks to the mine water," responds the rapeseed flower.

比以往更亮丽

Blossoming brighter than ever before

# 4000米深

## As deep as four thousand metres

"矿井水？我知道雨水与河水，还有从大坝和地下来的水，但矿井水是什么？对我来说可真新鲜！"

"你知道这里的矿井有多深吗？"

"当然，钻机已经往地下打了几十年了，有些都打到4000米深了！"

"Mine water? I know about rainwater and river water. And there's also water from dams and aquifers, but mine water? That's new to me."

"Do you know how deep the mines are around here?"

"Yes, the shafts were dug over decades and some may be as deep as four thousand metres."

"什么?都快打到地狱了!"油菜花惊讶地说。

"噢!即使温度可能上升到50℃,那里也没有地狱。但是你能想象要乘多久电梯才能到那里吗?"鹌鹑问道。

"What? That's getting close to hell!" exclaims the flower.

"Well, even though the temperature could rise to fifty degrees Celsius, of course there is no real hell there. But can you imagine how long it must take to get down there in a mine elevator?" asks the quail.

想象要多久才能到那里

Imagine how long it must take to get there

卡车在地球的肚子里开来开去?

Trucks driving around in the belly of the Earth?

"我想肯定很长！而且在这么深的地下工作肯定每天都像在洗桑拿！"

"我同意，那真是一种挑战！当人们取出矿石时，机器把空气冷却了。在地下工作也不再像以前那样艰难了。还有卡车和火车在地球的肚子里运输物品。"

"卡车在地球的肚子里开来开去？那排出的废气怎么办呢？"油菜花问道。

"Quite long, I suppose. But working so deep underground must be like being in a sauna all day."

"It's quite a challenge, I agree. While people are taking out the ore, machines cool the air though. Working underground is not as hard as it used to be either. There are even trucks and a train for transport inside the belly of the Earth."

"Trucks driving around in the belly of the Earth? What happens to all the exhaust fumes?" asks the flower.

"为了清洁空气，矿井里有世界上最精致的空调系统，就安装在我们脚下几公里的地方。"

"唯一的问题是当矿井关闭的时候，水会流进去。时间长了，所有东西都会坍塌。"

"不过，矿井里肯定有泵把水抽出去吧？"鹌鹑说，"你想想，这可是地球内部的纯净水呀！"

"To freshen the air, mines have some of the most elaborate air conditioning systems in the world, located only a few kilometres under our feet."

"The only problem is that when a mine is closed down, water flows into it. Over time everything could collapse."

"But surely mines have to pump the water out?" remarks the quail. "Imagine, pure water from the inner sanctum of the earth."

世界上最精致的空调系统

Most elaborate air conditioning systems

# 如何正确处理废物

How to properly manage the waste

"噢，是的，从地球深处来的水不仅纯净，还很温暖。它能让我的根保持温暖，所以冬天我也能很容易生长在矿井附近的田地里，而这些地没人想来。"

"人们为什么不愿意生活在这里呢？"

"因为以前矿井周围都是人们扔掉的垃圾。那时，人们还不知道如何正确处理这些从地下挖出的废物。"油菜花解释说。

"Oh yes, the water that comes from deep, deep down is not only pure, but also warm. It helps to keep my roots warm so I can easily grow throughout the winter in the fields around the mines – on land that no one wants."

"Why don't people want to live on that land?"

"Because before, a lot of rubbish was just left lying around the mines. Back then, people didn't know how to properly manage the waste that comes out of the earth," explains the flower.

"现在，我们可以用废弃的岩石和沙子造纸，还可以用抽上来的水灌溉农田。但是，我不知道这些水是不是能喝。水里是不是含了矿井里的一些化学物质？"

"这些水除了岩石和金子，从没和其他东西接触过。水里甚至可能会有几个小金片呢！"

"难怪你们有这么金灿灿的颜色！"鹌鹑大喊道。

"These days, we make paper from the rocks and sand that's left behind and we know that we could use the water for irrigation. But, I wonder if this water is safe for us to drink. Isn't it contaminated with chemicals from the mine?"

"This water has never been in contact with anything except rocks and gold. It could even have a few tiny flakes of gold in it!"

"No wonder you rapeseeds have such a bright golden colour!" exclaims the quail.

这么金灿灿的颜色!

Such a bright golden colour!

用来做生物燃料或蜡烛

Used as biofuel and to make candles

"我知道我们鹌鹑是为数不多的可以安全地吃油菜籽的动物。我还知道人类和其他动物不会吃长在这里的庄稼。所以,有时候我在想我们是否应该冒这个险……"
"不用担心,对你来说这是非常安全的!"
"你知道吗?从生长在矿井附近的农作物中提取的油可以用来做生物燃料和蜡烛。"油菜花问道。

"I know that us quail are some of the only animals that can safely eat rapeseed. And I know that people and animals should not eat crops that are grown here. So, I sometimes wonder if we should take the risk …"
"Don't worry, it's perfectly safe for you!"
"And did you know that oil from the crops grown near the mines can be used as biofuel and to make candles?" asks the flower.

"这太棒了!"小鹌鹑喊道。

"我在想你的油是不是可以用来做唇膏,让女人们把自己打扮得更漂亮。"

……这仅仅是开始!……

"That's fantastic!" exclaims the quail. "Now I wonder if your oil can be used in the lipstick that women could use to make themselves even more beautiful."

... AND IT HAS ONLY JUST BEGUN!...

……这仅仅是开始！……

...AND IT HAS ONLY JUST BEGUN!...

# Did You Know?
## 你知道吗?

The deepest mine in the world is 3,9 km deep and is located to the southwest of Johannesburg, South Africa. Of the 10 deepest mines in the world, 8 are found in the same region. The other two are in Ontario, Canada.

世界上最深的矿井在南非约翰内斯堡的西南部，有3900米深。世界上最深的10个矿井中的8个都在这个地区，另外两个在加拿大的安大略省。

The deeper the mine, the hotter it gets down below and also the more costly it becomes to pump fresh air into the shafts. This fact drives the mining industry towards automation.

越往矿井深处就越热，而且要把新鲜空气打到矿井里的费用就越高。这一事实推动了采矿工业的自动化。

The first steam-powered engine to perform mechanical work was the Newcomen engine, designed by Thomas Newcomen in 1712. Hundreds of these engines were built in the 18th century and were used to pump water out of mines.

第一个进行机械工作的蒸汽动力发动机是1712年由托马斯·纽卡门设计的纽卡门发动机。18世纪人们制造了数百台这样的发动机把水从矿井里抽出来。

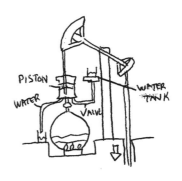

Rapeseed, also called canola (after one of the varieties), is the third largest source of vegetable oil. In China and Southern Africa, the rapeseed plant is eaten as a vegetable. The oil used to be used as a bio-lubricant for steam engines.

油菜籽也被叫做芥菜籽，是植物油的第三大来源。在中国和南非，油菜还可以作为蔬菜食用。菜籽油曾经被用作蒸汽发动机的润滑油。

The rapeseed plant gets its name from the Latin name for 'turnip', i.e. rapa. For marketing purposes a new name was created: canola, which is a contraction of Canada and oil low acid.

油菜这个词来源于拉丁名字"turnip"。出于商业目的,人们发明了一个新名字:canola,这个词是"加拿大低酸油"的缩略。

One tonne of rapeseed yields 400 litres of oil. It takes the oil 150 days to mature.

1吨油菜籽可以榨出400升油。油菜籽需要生长150天才能成熟。

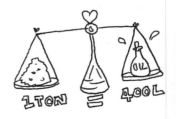

Nearly 70% of all accessible fresh water is used in agriculture – that is more than triple that used for industrial purposes (23%). Only 8% is used for municipal use.

人们把近 70% 的可用淡水都用在了农业上——相当于工业用水（23%）的三倍多，只有 8% 用于生活。

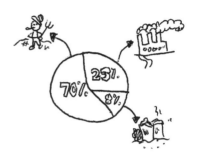

Quail are hardy, disease-resistant birds. Hens lay 270-300 eggs per year, for three to four years. The quail egg has been described as a mineral cocktail, containing many of the essential minerals people require.

小鹌鹑是勤奋又抗病的鸟类。母鹌鹑每年可以产 270 ~ 300 个蛋，而且可以产三四年。鹌鹑蛋被誉为矿物质鸡尾酒，含有许多人体必需的矿质元素。

# Think About It
# 想一想

Would you like taking an elevator that drops you four kilometres down a mineshaft?

你愿意乘坐电梯到达 4 千米深的矿井吗?

How would it be not to know if it is morning, noon or night, or whether it is raining or hot outside?

如果分不清现在是早晨、中午还是晚上，或者外面是下雨还是炎热，你会有什么感觉?

你会吃种在原来是矿井的土地上、现在又用矿井里的水灌溉的食物吗?

Would you be prepared to eat food grown on land that was previously mined and is now irrigated with water from a mine?

When you irrigate plants on contaminated land with pure water from a mine, could it be possible to purify the soil over time, distributing the pollutants in very small amounts?

如果长时间地用矿井里的纯净水浇灌种在污染的土地上的植物，是否可能把污染物分解成很小的颗粒，进而净化土壤?

# Do It Yourself!

# 自己动手!

The colour yellow symbolises energy, positivity and hope. Do a test to see if yellow does indeed have a positive, energising effect on you, your friends and your family. Collect pictures of yellow flowers such as sunflowers or rapeseed flowers as well as pictures of blue and purple flowers, like the gentian. Show them to others and ask them how looking at the different coloured flowers makes them feel. Compare your feedback with what your friends got when they did the same.

黄色代表了能量、积极和希望。做一个测试，看看黄色是否真的对你、你的朋友和你的家人有积极的、激励的效果。搜集黄色花朵的照片，比如向日葵或油菜花，同时也搜集蓝色和紫色花朵的照片，比如龙胆。让其他人看看这些照片，询问他们看到不同照片时的感觉。比较一下你的感受和朋友们的反馈。

# TEACHER AND PARENT GUIDE

## 学科知识
### Academic Knowledge

| | |
|---|---|
| 生物学 | 油菜籽可以产出大量花蜜；鹌鹑具有强大的免疫系统，不易患病，所以不需要抗生素来治疗；小鹌鹑刚孵出后吃虫子，5~6周以后就只吃植物了；人工修建的湿地在一定程度上可以修复酸性废水的污染。 |
| 化 学 | 油菜籽含有50%的芥子酸；种植油菜需要高浓度的氮肥；矿井里的废水是由大量的天然硫矿，尤其是黄铁矿的氧化造成的；当氧化反应发生时，氢离子被释放出来，pH值就降低了；用石灰、硅酸钙或者碳酸盐中和的方式恢复pH值。 |
| 物 理 | 托马斯·纽卡门设计的空气发动机。 |
| 工程学 | 采矿工程是一门综合学科，包括地质学、测绘学、选矿学、冶金学以及岩土工程学；芬兰桑拿浴和土耳其蒸气浴在结构上的区别。 |
| 经济学 | 利用现有的设备在现有的矿中获得更多的资源可以减少资本投入；投资一个采矿业最重要的不可知因素是停止开采后的花费，即使该地区恢复到原来的生态状态的费用；碎石头（废弃物）可以用来制造纸张；金属元素是可以100%循环使用的，而塑料的商业使用是难恢复的。 |
| 伦理学 | 采矿是危险的职业，所以在古埃及和古罗马时期都是由罪犯和战俘来做的；国际公约中已经禁止使用强迫劳动。 |
| 历 史 | 詹姆斯·瓦特设计的（蒸汽）发动机是托马斯·纽卡门设计的（空气）发动机的改进版，只要用一半的燃料就能产生相同的动力；大约4000年前埃及的沉井就达到了地下90米深，罗马的矿井则达到了200米深；在炸药发明之前，矿工用火力碎石；桑拿产生于拜占庭王国，并被斯拉夫商人带到了芬兰。 |
| 地 理 | 格陵兰和纽芬兰有最古老的桑拿；南非的约翰内斯堡是围绕金矿而建的，也是世界上少数几个不是建立在森林或大河附近的主要工业城市之一。 |
| 数 学 | 纽卡门发动机的工作频率为每分钟12次，每次能够泵出10加仑水；鹌鹑每天需要喂食20~25克，而鸡需要120~130克，而且鹌鹑在5~8周后就能达到上市体重（140~180克），之后将生产200~300只鹌鹑蛋。 |
| 生活方式 | 一个健康的生活方式包括卫生和保健：从外部（桑拿）和内部（喝水里的金片）保持身体清洁。 |
| 社会学 | 意大利诗人但丁写了一部《神曲》，他对地狱毛骨悚然的描写已经成为西方思想中根深蒂固的地狱形象，并且赋予了米开朗琪罗、弥尔顿和艾略特等人创作的灵感；"桑拿"在芬兰语中是指浴室。 |
| 心理学 | 黄色是使人振奋的颜色，可以提升人们的乐观、自信、自尊和创造力，尤其是绽放着油菜花和向日葵的黄色田野。 |
| 系统论 | 现代文明需要矿山、矿石和金属来驱动，然而这经常会对环境造成过度危害；由于采矿的社会成本也很高，所以现存的经营模式需要重新设计。 |

# 教师与家长指南

## 情感智慧
### Emotional Intelligence

**小鹌鹑**

小鹌鹑知道油菜花是他们的食物来源之一，他赞美油菜花，并且还问她们为什么比以前长得更好看更鲜艳。小鹌鹑对水很好奇，他问了很多问题以期能发现更多。小鹌鹑和油菜花的对话就是信息的交换。小鹌鹑愿意回答所有问题并无偿地分享他的知识，包括一些实用的方法，比如用石头造纸。然而，小鹌鹑担心水的质量，并勇担责任，建议水不应该用于生产供人类食用的粮食。小鹌鹑了解到用菜籽油做蜡烛和生物燃料是安全的，但他想知道菜籽油是否也可以用来做唇膏。

**油菜花**

油菜花很谦虚。当她收到赞美时马上把功劳归于纯的矿井水。油菜花是无知的，但她很爱学习，所以她用自己的问题来回应小鹌鹑的问题。油菜花喜欢利用矿井水，但她很惊讶矿井竟然这么深，人们不得不在那么闷热的条件下工作。小鹌鹑解释了空气是如何保持新鲜的。油菜花立刻形成了自己的观点，并认识到关闭矿井是个问题，因为里面的水还得抽出来。油菜花重新综合了已知的所有信息，并思考如何利用矿井水度过寒冷的冬天。油菜花的金黄色代表了乐观和自尊。即使鹌鹑提出了警告，油菜花依然保持乐观，并注意到了使消极变成积极的机会。

## 艺术
### The Arts

你能想象黑暗中的生活是什么样子吗？你现在在地下4000米的地方，很闷热，水的温度已经上升到了威胁你的生存的程度。听起来像是恐怖小说的开头！你能不能以你深入到地下数千米开头，写一篇惊悚故事？

# TEACHER AND PARENT GUIDE

## 思维拓展
### Systems: Making the Connections

开矿和做外科手术的逻辑是一样的，要多为病人的健康考虑，尽量不留下不良后果。地下采矿比露天开采导致的危害要少得多，但产生的酸性矿井水却是严重的环境威胁。采矿工程能够按照最严格的健康、安全和环境标准进行。然而，黄铁矿的出现推动了化学采矿的应用，不断产生的污染会毁坏水路、土地，对地球和人类的健康造成了很大的风险。同时，裂隙水的出现威胁到了矿井和地面的安全（地下水会淹没矿井）。这种水不一定受到了污染，可能很纯净，也可以饮用，但从矿井安全的角度看，必须把水抽出来。这不仅是一种很大的能源耗费，尤其是从几千米的矿井下抽到地面；也使矿主承担了很大的责任，要为矿井关闭后的水处理负责。在干旱的地区突然降下了大量的水，既带来了机会，也带来了挑战。矿井周围的地区遭受着污染物导致的严重污染，如放置碎石的尾矿坝形成的粉尘。另一个主要污染源是依赖有毒化学反应的处理技术。结果，本来可以用来灌溉土地的营养水现在却被污染了，只能用来灌溉生产蜡烛用油和生物燃料的农作物，而不适合被人类食用。把水和土地结合起来生产能提供燃料的农作物，有很多好处。第一，燃料立刻可以使用；第二，油可以从种子中提取（每吨可以生产400升），而废弃的、不能喂动物的油渣饼现在经处理后可以生产沼气。如此将进一步增加能源产出。灌溉和燃料生产的连续过程可以慢慢减少土地的污染并使土地恢复到原本的状态，化负面效果为正面效果：生产燃料、清洁土地、增加就业，这将弥补采矿自动化造成的失业，而自动化对在困难条件下开采越来越深的岩矿是很有必要的。

## 动手能力
### Capacity to Implement

你见过鹌鹑蛋吗？去食品店或商场买一些这种小小的蛋，给人们看看并请他们告诉你鹌鹑蛋和鸡蛋的区别。除了大小不一样，你的朋友和家人还能告诉你这两种蛋和这两种鸟有什么不同吗？记下你学到的内容，然后做出明智的决定：你要吃哪个蛋？为什么？

故事灵感来自

## 吉尔·马库斯
### Gill Marcus

吉尔·马库斯的祖辈是移民到南非的立陶宛人。由于她的父母是反种族隔离的激进分子，她在1969年便和兄弟姐妹跟着父母流亡海外。1976年她获得了南非大学工业心理学的商学士学位。加入非洲国民大会以后，她开始在它设在伦敦的办公室工作。她于1994年当选为议员，并在1996–1999年间任财政部副部长，1996–2009年间被任命为储备银行副行长。她还担任了西部矿业公司的总裁，后来成为金田有限公司的董事会成员，并敦促管理层考虑在实践中进行改革。

更多资讯

http://physicstasks.eu/1289/mine-shaft-elevator

图书在版编目（CIP）数据

矿井水：汉英对照／（比）冈特·鲍利著；（哥伦）凯瑟琳娜·巴赫绘；郭光普译． —— 上海：学林出版社，2016.6
（冈特生态童书．第三辑）
ISBN 978-7-5486-1045-8

Ⅰ．①矿… Ⅱ．①冈… ②凯… ③郭… Ⅲ．①生态环境－环境保护－儿童读物－汉、英 Ⅳ．① X171.1-49

中国版本图书馆 CIP 数据核字（2016）第 125647 号

------------------------------------------------

ⓒ 2015 Gunter Pauli
著作权合同登记号 图字 09-2016-309 号

## 冈特生态童书

### 矿井水

| | |
|---|---|
| 作　　者—— | 冈特·鲍利 |
| 译　　者—— | 郭光普 |
| 策　　划—— | 匡志强 |
| 责任编辑—— | 匡志强　蔡雩奇 |
| 装帧设计—— | 魏　来 |
| 出　　版—— | 上海世纪出版股份有限公司 学林出版社 |
| | 地　址：上海钦州南路81号　电　话／传真：021-64515005 |
| | 网　址：www.xuelinpress.com |
| 发　　行—— | 上海世纪出版股份有限公司发行中心 |
| | （上海福建中路193号 网址：www.ewen.co） |
| 印　　刷—— | 上海丽佳制版印刷有限公司 |
| 开　　本—— | 710×1020　1/16 |
| 印　　张—— | 2 |
| 字　　数—— | 5万 |
| 版　　次—— | 2016年6月第1版 |
| | 2016年6月第1次印刷 |
| 书　　号—— | ISBN 978-7-5486-1045-8/G·380 |
| 定　　价—— | 10.00元 |

（如发生印刷、装订质量问题，读者可向工厂调换）

# 冈特生态童书（73-108）

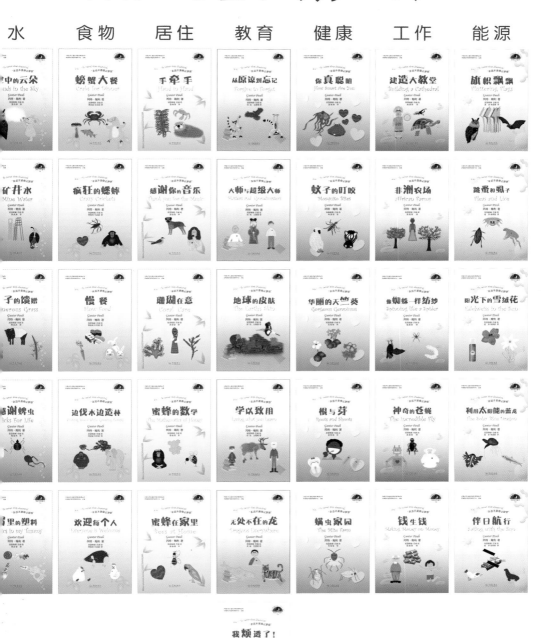

记得要和身边的小朋友分享环保知识哦！
八喜冰淇淋祝你成为环保小使者！

联合国开发计划署支持

## 永远不要停止梦想

你拥有的梦想越多,你能重温的梦想就越多,实现梦想就越容易……永远不要停止梦想!

冈特·鲍利是世界零排放研究创新基金会(ZERI)的创始人,该基金会的理念是:"不要奢望让地球为我们创造更多资源,而要更善于利用地球现有的资源。"《冈特生态童书》正是这一理念的结晶。书中的一系列童话,看似充满奇思怪想,但却揭示了现实生活中的许多道理。孩子们通过这些童话学习科学知识,开发情商,提高艺术表达能力、动手能力和对复杂系统的理解能力。他们将成为真正的地球保护者。孩子们的内心充满着梦想和希望,千万不要让这些梦想和希望落空。未来一定会更加美好,因为孩子们的优秀总是超乎父母的想象。

绿色印刷产品

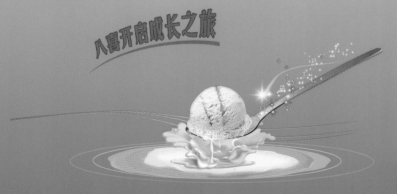

上架建议:教育

ISBN 978-7-5486-1045-8

定价:10.00元
易文网:www.ewen.co

中国少年儿童生态意识教育丛书
环境保护部宣传教育中心　主编

冈特生态童书
103

ZERI

"To never stop dreaming"
"永远不要停止梦想"

# 感谢蜱虫
# Ticks For Life

**Gunter Pauli**
冈特·鲍利 著

凯瑟琳娜·巴赫 绘
隋淑光 译

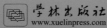

学林出版社
www.xuelinpress.com

# 亲爱的小朋友们

小朋友们，你们知不知道竹子其实是一种特殊的草？知不知道海里有一种蓝龙，它还能利用太阳能呢？冈特·鲍利先生为你们创作了许多有趣的童话故事，带着你们去奇妙的生态王国寻宝，相信在这段神奇的旅程中，你们一定会大开眼界。

如果读完故事还意犹未尽，"你知道吗？"会告诉你们更多的科学知识，"自己动手"会指导你们开辟自己的一片天地。最后，邀请你们的爸爸妈妈和老师们一起加入奇妙之旅吧！"教师和家长指南"为他们准备了各种背景知识，当然，遇到困难时，还是要请他们向工具书或者专家求助哦！

扫一扫，就能听冈特小故事。
更多惊喜，等你发现哦！

Water
103

# 感谢蜱虫
# Ticks For Life

Gunter Pauli

冈特·鲍利 著
凯瑟琳娜·巴赫 绘
隋淑光 译

学林出版社
www.xuelinpress.com

## 丛书编委会

主　任：贾　峰
副主任：何家振　郑立明
委　员：牛玲娟　李原原　李曙东　吴建民　彭　勇
　　　　冯　缨　靳增江

## 丛书出版委员会

主　任：段学俭
副主任：匡志强　张　蓉
成　员：叶　刚　李晓梅　魏　来　徐雅清　田振军
　　　　蔡雩奇　程　洋

特别感谢以下热心人士对译稿润色工作的支持：

姜竹青　韩　笑　贾　芳　刘　晓　张黎立　刘之杰
高　青　周依奇　彭　江　于函玉　于　哲　单　威
姚爱静　刘　洋　高　艳　孙笑非　郑莉霞　周　蕊

# 目录

| | |
|---|---|
| 感谢蜱虫 | 4 |
| 你知道吗？ | 22 |
| 想一想 | 26 |
| 自己动手！ | 27 |
| 学科知识 | 28 |
| 情感智慧 | 29 |
| 艺术 | 29 |
| 思维拓展 | 30 |
| 动手能力 | 30 |
| 故事灵感来自 | 31 |

# Contents

| | |
|---|---|
| Ticks For Life | 4 |
| Did you know? | 22 |
| Think about it | 26 |
| Do it yourself! | 27 |
| Academic Knowledge | 28 |
| Emotional Intelligence | 29 |
| The Arts | 29 |
| Systems: Making the Connections | 30 |
| Capacity to Implement | 30 |
| This fable is inspired by | 31 |

在非洲中部热带草原上的一个小池塘里，一条鱼正在游泳。一只孤伶伶的蝌蚪也在游来游去，他快要变成青蛙了。

𝓐 fish is swimming in a small pool of water in the middle of the African savanna. A lonely tadpole swims around. He is about to turn into a frog.

一只蝌蚪在游来游去……

A tadpole swims around ...

我是怎么来到这里的?

How did I ever get here?

"你注意到附近没有河吗?"蝌蚪问道。

"我为什么要关心那个?我在这里过得很快乐。"鱼回答道。

"嗯,我很奇怪自己是怎么来到这里的,还有,你是怎么来到这里的?这附近没有我们的族群。"

"Did you know that there is no river anywhere close by?" asks the tadpole.

"Why should I care? I am quite happy here," responds the fish.

"Well, I wonder, how did I ever get here? And, how did you get here? There is no family around."

"肯定是我的爸爸妈妈送我来的,可能我是个淘气的孩子。"
"你会飞吗?"
"我是鱼,又不是鸟。你能飞吗?"

"I am sure my parents sent me off. Maybe I was a very naughty kid."
"Can you fly?"
"I am a fish, not a bird. Can you fly?"

我是鱼，又不是鸟

I am a fish, not a bird

……被鹳鸟吃掉了!

... eaten by a stork!

"不，根本不行，我只是一只蝌蚪。我在找我的爸爸妈妈和兄弟姐妹，但是没有他们的任何痕迹，甚至连骨架都没有。"

"他们很可能被鹳鸟吃掉了，我知道鹳鸟喜欢吃青蛙。"

"No, not at all, I am just a tadpole. But I have been looking around for my mom and dad, brothers and sisters. There is no sign of them, nothing, not even a skeleton."

"They were most likely eaten by a stork! I know they love to eat frogs."

"他们也喜欢吃鱼。那么，如果我们的爸爸妈妈不在附近，是谁带我们飞到这儿的呢？"

"哦，你认为自己是坐喷气式飞机来的？真是异想天开。"

"And they love to eat fish as well. So if our parents are not around, who flew us here?"

"Oh, now you think that you arrived on a jet plane? You are such a dreamer."

是谁带我们飞到这儿的?

Who flew us here?

在飞往欧洲的途中……

On his way to Europe...

"也许不是飞机，可能是一只鹳鸟在某处觅食，那时你我也许还是一只卵，被嵌进了他的脚趾缝里。在他飞往欧洲的途中，他在这个池塘里驻足休息，我们就被留下来了。"

"很有趣的想法，但是这个池塘是谁挖的呢？"

"Maybe not a plane, but perhaps a stork was having a meal somewhere. It could be that you and I, as eggs, were stuck to his feet. On his way to Europe, he made a pit stop here at this pool and we got off."

"Interesting thought. But who dug this pool?"

"你看到过大象、犀牛和水牛洗尘土澡吗?"
"噗,当他们试图通过洗尘土澡来刮掉那些蜱虫的时候,尘土飞扬,你什么都看不到。"

"Have you ever seen an elephant, a rhino or a buffalo take a dust bath?"
"Pffff, you cannot see anything when he tries to scratch his butt and back clean to get rid of those ticks."

# 洗尘土澡

# A dust bath

# 微小的、黑乎乎的生物

Tiny, black creatures

"你是说那些微小的、黑乎乎的、样子丑陋、有坚硬头部的吸血生物吗?"

"我们很幸运,我们生活在水里,而那些害虫不能在这里生存。"

"那为什么大象和水牛不能通过洗澡来摆脱这些讨厌的小动物呢?"

"因为一年中的大部分时间里,这个池塘里没有水。"

"You mean those tiny, black, ugly-looking, hard-headed, bloodsucking creatures?"

"We are so lucky. We live in water where these pests don't survive."

"So why don't the elephants and buffaloes take a nice soaking bath, and drown the little menaces?"

"Because most of the year there is no water in this pool."

"但是那意味着池塘会干涸!我们也会死的!"

"那就是生命循环。大象洗尘土澡时,用身体压出一个布满灰尘的干坑,然后雨水会填满它,再后来鹳鸟会带来新的生命体。那就是这个故事告诉我们的。"

"也就是说我们之所以能享受生命的快乐,应该感谢蜱虫?"

……这仅仅是开始……

"But that means this pool will dry out! And we will then die!"

"That is the cycle of life. The elephants take a dry bath, create a solid dust pool, the rain will fill it, and the stork will bring new life. That is what this fable tells us."

"And this means we have the ticks to thank for this moment of joy?"

... AND IT HAS ONLY JUST BEGUN!...

……这仅仅是开始！……

… AND IT HAS ONLY JUST BEGUN! …

# Did You Know?

## 你知道吗？

Ticks need blood to survive. Ticks unable to find a host will die.

蜱虫依靠吸血来生存，如果找不到寄主就会死亡。

Ticks have four pairs of legs. Larval ticks have only three pairs of legs; the fourth pair only grows after it has had its first meal of blood.

蜱虫有四对足。幼虫阶段只有三对，第四对要在第一次吸血后才开始发育。

The first pair of legs can smell, sense vibration, notice increased moisture and feel the body heat of potential hosts.

第一对足的功能有嗅闻、感觉振动、分辨湿度变化,以及感受潜在寄主的体温。

Ticks love a warm and humid climate, as they need moisture in the air to undergo a metamorphosis from a larva to a nymph and then to a full-grown tick. Low temperatures inhibit their growth.

蜱虫喜欢温暖潮湿的气候,因为它在从卵蜕变成幼虫以及成虫时,需要吸收空气中的水汽。低温会抑制它的发育。

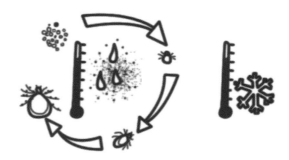

除了剧毒的DDT以外，尚不知道有其他化学药品能杀死蜱虫。珍珠鸡是最高效的蜱虫"终结者"，几只珍珠鸡就可以把一公顷土地上的蜱虫消灭干净。

No known chemicals except the very toxic DDT kill ticks. Guinea fowl are the most successful tick eliminators. Just a few guinea fowl will clear a hectare of all ticks.

蜱虫寄生在大象身上，安居在皮肤的褶皱里，或者通过伤口进入血液。大象洗尘土澡是为了摆脱蜱虫。

Ticks infest elephants and nestle into the wrinkles of the skin or enter through a cut to get to the blood. Elephants shower with dust and sand to dislodge ticks.

牛椋鸟能帮助犀牛清理蜱虫，但是它偏爱那些吸饱了血的胖胖的蜱虫，会忽略那些同样有威胁的瘦家伙。

The little oxpecker cleans the rhino by picking off ticks, but the bird prefers thick ticks full of blood, ignoring the little ones that are just as much of a nuisance.

鹳鸟从南非西开普省飞到欧洲要花26天时间，返程时要花49天。它利用空气热气流升空，升到一定高度后，滑翔到目的地。

The stork flies from the Western Cape (South Africa) to Europe in 26 days and returns in 49 days. Storks use the lift of air thermals to gain height and glide to their destination.

# Think About It
## 想一想

Would you ever consider a tick as a friend?

你会把蜱虫看作朋友吗?

How do you think the fish and the frog ended up in this isolated pool of water?

你怎么看待鱼和青蛙在孤伶伶的池塘里结束生命这件事?

水牛、犀牛和大象的身体重到洗个尘土澡就能挖出一个池塘吗?

Are buffaloes, rhinos and elephants heavy enough to dig a pool by taking a dust bath?

Do you like the idea that life emerges again every year again from a dust pool that fills with water during the rainy season, but turns into a dust pool again in the summer?

在雨季蓄满雨水、夏季干涸的池塘里,生命年复一年地出现,你喜欢这种设想吗?

# Do It Yourself!
## 自己动手！

It is springtime. Go and look for a pool of water and search for tadpoles. Take a net with you and catch a few. Keep them in clean water. Feed them lettuce and leftovers. If the tadpole is still very young, cook the lettuce first, as that makes it easier to eat. It can take 6 to 12 weeks before a tadpole turns into a frog and you will be able to observe all the stages in the development of the tadpole. If you want you can catch a small fish and watch it grow too. Find one that is not a carnivore, so that the tadpole and the fish can happily live alongside each other. Once they grow big, release them back into nature and enjoy the memories!

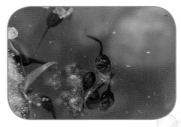

春天来了。到池塘里找一找蝌蚪，用网捞几只，把它们养在净水里，用生菜和剩饭剩菜喂它们。如果蝌蚪一直不发育，就把生菜煮熟再喂，以方便它们进食。蝌蚪要花6到12周时间才能变成青蛙，你可以观察到整个发育过程。如果你想捉一条小鱼来观察它的发育，试着捉一条非肉食性的鱼，这样它就能与蝌蚪和平共处了。一旦它们长大了，将它们放回到自然界中，享受喂养它们的美好回忆吧。

# TEACHER AND PARENT GUIDE

## 学科知识
### Academic Knowledge

| | |
|---|---|
| 生物学 | 我们所认为的有害物种在生态系统中也有作用；寄生虫和病菌承担"生态系统工程师"的角色；寄生植物和寄生动物在养分循环中的作用及其对生物多样性的影响；寄生生物对于物种间遗传物质的转移起作用，并引起了生物进化；水是影响生物生存的重要因素。 |
| 化　学 | 鸟类必须快速地定居、摄食和消化，在长期的进化中它们失去了牙齿，颚部也退化了；蜱虫通过产生抗凝血肽来防止血液堵塞，这对医药工业很有启发。 |
| 物　理 | 鹳鸟通过拍动翅膀升空并飞到目的地，它们所消耗的能量是那些利用热气流升空并飞行的鸟类的23倍多；在准备迁徙时，为方便飞行，鹳鸟会把生殖器官的重量降低到准备生殖时的1%；经过干旱季节后土壤变得致密，雨水很难渗入。 |
| 工程学 | 利用水的压力来清洗物体表面的方式被称作喷流，利用砂的压力清洗物体表面的方式被称作喷砂。 |
| 经济学 | 商业具有季节性，在一年时间内消费行为会有变化。 |
| 伦理学 | 根据好和坏来快速区分物种的做法，并没有真正理解每一个现存物种所承担的角色和责任，我们需要花时间来理解每一个现存物种在每一种生态背景下的价值。 |
| 历　史 | 鹳鸟从开始迁徙到返回的时间和人类从女性排卵到婴儿诞生的时间（9个月）一致，因此在维多利亚时期有"一只鹳鸟带来一个婴儿"的故事，而在古希腊神话中则有鹳鸟偷婴儿的传说。 |
| 地　理 | 鹳鸟在迁徙时要避开大面积的水域，因此从南非到欧洲的路线要么是飞越黎巴嫩（东线），要么是飞越摩洛哥（西线），因为大面积的水域上空的空气中没有可供它们自由上升的热气流。 |
| 数　学 | 鹳鸟飞往欧洲和返回时所花的时间不同，前者需要26天，后者需要49天，这是由风、气流和食物三个因素决定的。 |
| 生活方式 | 在伊索寓言中首次提到了"狐狸和鹳鸟"的故事，然后安徒生进行了再创作；父母常用这些故事来向小孩解释婴儿是如何来到世上的，而不必深入讲生育的细节。 |
| 社会学 | 在一些亚洲文化中，时间被认为是往复的，失去的还可以回来；而西方文化则认为时间是线性的，流逝的就永远流逝了。 |
| 心理学 | 每个人都问过"我是从哪里来的"这个问题，并向父母、长辈和同龄人寻求答案，从这一问题出发可以进一步质疑生命以及我们的父母是从哪里来的；寻根使我们明确生活的目的并且有稳定感。 |
| 系统论 | 生命网中有很多纵横交错的联系，它们看起来似乎是随机的、偶然的，但是一旦深入研究就会对这种联系和连接产生更好的理解，这对维持自然界中生存和死亡的循环是非常有必要的。 |

# 教师与家长指南

## 情感智慧
### Emotional Intelligence

**鱼**

鱼非常不愿意和蝌蚪进行深入交流,对于蝌蚪提出的更深层次的问题,她快速简单地作答,语气冷淡甚至有些讽刺的意味,似乎是想终止这个话题。但是蝌蚪接连不断的提问最终吸引了她的注意力,她开始和蝌蚪讨论起她们生活环境的起源问题,即这个孤伶伶的小池塘是她们仅有的栖息地。蝌蚪的话触动了她的心弦,使她开始认真对待这场谈话。于是鱼开始分享自己对生命周期的见解,她对自身起源和未来的深入理解使蝌蚪感到惊讶。在这个故事将要结束时,鱼的认识越来越接近问题的核心,并开始进行哲学上的思考。

**蝌蚪**

蝌蚪好奇自己是怎么来到这里的,并开始寻求答案。但她除了和共处于这块栖息地的鱼探讨外别无选择,因此她没有考虑鱼的感受,就觉得有必要讨论这个问题。鱼有些粗暴的回答没有使她泄气,她坚持追问她们是从哪里来的。蝌蚪认可鱼对她的生存困境的判断,她准备深入思考一些问题,这促使她和鱼进行了第一次接触。她认识到一个可怕的事实,即她所鄙视的一些物种却和她的生存密不可分。然而,现实情况使她焦虑,她意识到应该对蜱虫表示感激,接受尽管她希望能活着,但她的生命最终会结束这一现实。

## 艺术
### The Arts

你看到过特写镜头下的蜱虫吗?准备一台显微镜,再想办法捉一只蜱虫,小心地把它浸在水里。如果能用酒精浸泡更好,这样它就不可能咬到你并吸你的血了。请一位科学家拍一张蜱虫头部的照片,你会觉得似乎看到了外星人。在A4纸上画一画这张照片,用不同的颜色突出头部或者四对足上最不可思议的部位。你现在看到的是一个让人产生不适并且能传播疾病的生物体,但它同时又是生物多样性最有效的传播者之一。

# TEACHER AND PARENT GUIDE

## 思维拓展
### Systems: Making the Connections

　　生命体充满了惊喜。从蜱虫的故事能联想到未知的生命起源和进化。我们都知道生态系统复杂多样，但是仍然对于微小的寄生生物竟和地球上的大型哺乳动物连接在一起而感到震惊。这个故事能引导我们理解候鸟的特殊作用。蜱虫和非洲的五种大型动物（水牛、犀牛、大象、狮子和豹）之间的关联，一条鱼、一只青蛙、一只鹳鸟和一只蜱虫的关联，是对生命之网的真实揭示。我们想要弄清生命是如何起源和进化的，尽管我们知道自己将永远无法完全掌握现实。这激励我们从哲学上思考我们是如何出现在地球上的，以及我们为什么会生存在现有的条件下。这个故事告诉我们，生命是有限的，而生命的循环和延续是无限的。微小的寄生生物对其他生物的影响强化了这样的观念——我们必须尊重所有形式的生命体，即使是那些微小的、我们还没有理解的生命体。生物多样性越丰富，我们的生活就会越乐观，就会充溢着我们还没意识到的机会。

## 动手能力
### Capacity to Implement

　　你注意到了身边的生命之网了吗？站在至少10个人组成的圈里，用手捏紧一个线团的线头，每当某个人能说出一个和前面的人所说的词语关联的词，就把线团传递给他。可以从你想到的任何一个词语（也许是与你家有关的词）开始这个游戏。每个人要逼自己至少想出50种关联，想的时候要打开思路，不要局限于纯科学，可以扩展到神话、文化象征，甚至纯想象，只要你能说服其他人认同这种想象即可。

# 教师与家长指南

## 故事灵感来自

## 彼得·雷蒙多
### Peter Raimondo

自从 2003 年起,彼得·雷蒙多一直作为指导老师在位于南非赫卢赫卢韦国家禁猎区的野外领导能力学校工作。他曾经在开普敦大学就读,主要学习哲学、人类学和环境地理科学,但是他更喜欢在户外工作,向人们提供获得直观知识和认识自然规律的机会。他组织了穿越南非荒野的徒步活动,让参加者在没有舒适的帐篷、预加工食品以及通讯设施的条件下体验生活。

### 更多资讯

www.esajournals.org/doi/abs/10.1890/110016

www.wildernesstrails.org.za/about-us

图书在版编目（CIP）数据

感谢蜱虫 ：汉英对照 /（比）冈特·鲍利著 ；
（哥伦）凯瑟琳娜·巴赫绘 ；隋淑光译. —— 上海 ：学林
出版社，2016.6
（冈特生态童书. 第三辑）
ISBN 978-7-5486-1047-2

Ⅰ. ①感… Ⅱ. ①冈… ②凯… ③隋… Ⅲ. ①生态环
境－环境保护－儿童读物－汉、英 Ⅳ. ① X171.1-49

中国版本图书馆 CIP 数据核字 (2016) 第 125821 号
----------------------------------------------------

ⓒ 2015 Gunter Pauli
著作权合同登记号 图字 09-2016-309 号

## 冈特生态童书
### 感谢蜱虫

| | |
|---|---|
| 作　　者—— | 冈特·鲍利 |
| 译　　者—— | 隋淑光 |
| 策　　划—— | 匡志强 |
| 责任编辑—— | 匡志强　蔡雯奇 |
| 装帧设计—— | 魏　来 |
| 出　　版—— | 上海世纪出版股份有限公司 学林出版社 |
| | 地　址：上海钦州南路81号　电话／传真：021-64515005 |
| | 网址：www.xuelinpress.com |
| 发　　行—— | 上海世纪出版股份有限公司发行中心 |
| | （上海福建中路193号 网址：www.ewen.co） |
| 印　　刷—— | 上海丽佳制版印刷有限公司 |
| 开　　本—— | 710×1020　1/16 |
| 印　　张—— | 2 |
| 字　　数—— | 5万 |
| 版　　次—— | 2016年6月第1版 |
| | 2016年6月第1次印刷 |
| 书　　号—— | ISBN 978-7-5486-1047-2/G·382 |
| 定　　价—— | 10.00元 |

（如发生印刷、装订质量问题，读者可向工厂调换）

# Water 86

# 空中的云朵

## Clouds in the Sky

**Gunter Pauli**

冈特·鲍利 著
凯瑟琳娜·巴赫 绘
郭光普 译

www.xuelinpress.com

## 丛书编委会

主　任：贾　峰

副主任：何家振　郑立明

委　员：牛玲娟　李原原　李曙东　吴建民　彭　勇
　　　　冯　缨　靳增江

## 丛书出版委员会

主　任：段学俭

副主任：匡志强　张　蓉

成　员：叶　刚　李晓梅　魏　来　徐雅清　田振军
　　　　蔡雩奇　程　洋

特别感谢以下热心人士对译稿润色工作的支持：

姜竹青　韩　笑　贾　芳　刘　晓　张黎立　刘之杰
高　青　周依奇　彭　江　于函玉　于　哲　单　威
姚爱静　刘　洋　高　艳　孙笑非　郑莉霞　周　蕊

# 目录

| | |
|---|---|
| 空中的云朵 | 4 |
| 你知道吗? | 22 |
| 想一想 | 26 |
| 自己动手! | 27 |
| 学科知识 | 28 |
| 情感智慧 | 29 |
| 艺术 | 29 |
| 思维拓展 | 30 |
| 动手能力 | 30 |
| 故事灵感来自 | 31 |

# Contents

| | |
|---|---|
| Clouds in the Sky | 4 |
| Did you know? | 22 |
| Think about it | 26 |
| Do it yourself! | 27 |
| Academic Knowledge | 28 |
| Emotional Intelligence | 29 |
| The Arts | 29 |
| Systems: Making the Connections | 30 |
| Capacity to Implement | 30 |
| This fable is inspired by | 31 |

小老鼠和猫头鹰

又见面了。猫头鹰喜欢吃老鼠,但不吃这一只,因为他喜欢和这只小老鼠聊天。小老鼠虽然不喜欢科学和数学,却问了很多聪明的问题。这些问题让猫头鹰思考。

"嗨!你好!"猫头鹰说,"你这几天好吗?你为什么看着天空发呆?"

The rat and the owl meet again. The owl likes to eat rats, but not this one, as he likes to talk to him. Although the rat does not like science and mathematics he asks a lot of smart questions. Questions that make the owl think.

"Oh, hello," says the owl. "How are you and why are you looking at the sky?"

你为什么看着天空发呆?

Why are you looking at the sky?

由亿万个很小很小的水滴形成

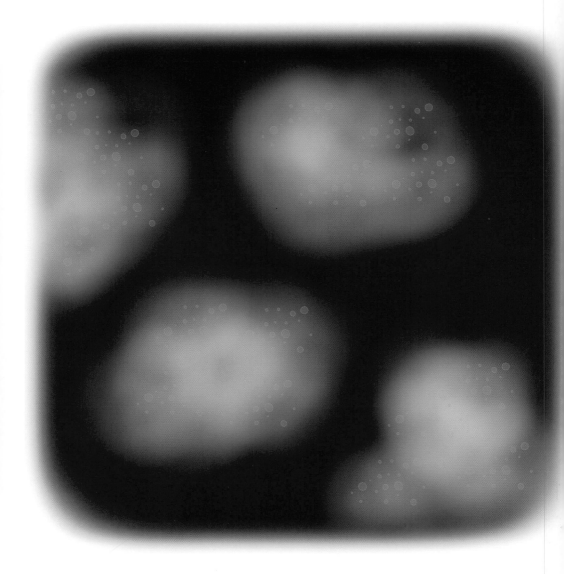

Made up of millions of tiny water drops

"我对云着迷了。"小老鼠回答道,"云使重力看上去好像失效了。"

"云是由天空中亿万个很小很小的水滴形成的。"

"我们知道苹果不会飞,它们成熟后就会掉到地上,但我还不知道水居然会飞!"

"I'm fascinated by the clouds," responds the rat. Clouds show that the law of gravity does not really work."

"Clouds are made of millions of tiny water drops up there in the sky."

"We know apples don't fly, they just drop to the ground when ripe, but I didn't know water could fly!"

"水当然不会飞,但是云只要比周围的空气温度高就会漂浮在空中。"

"但你刚才说云是由水做的,水可比空气重呀!"

"但你看,那些水滴太小了,它们遇到的空气阻力很大,减慢了它们的下降速度。"

"Water doesn't fly, but clouds will float as long as they are warmer than the air around them," Owl explains.

"But you just said clouds are made of water. That's heavier than air."

"But you see, those droplets are so small that they experience a lot of air resistance and that slows their fall."

比周围的空气温度高

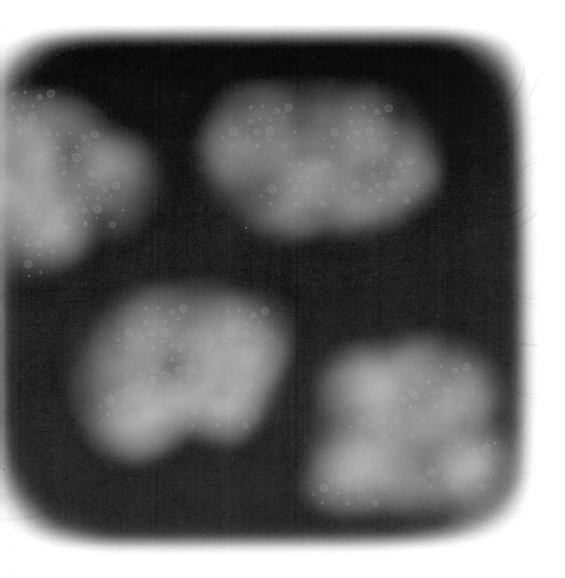

Warmer than the air around them

因为还有风

Because there is also wind

"那你怎么解释云漂浮在空中,很少掉下来呢?"

"因为还有风呀。"猫头鹰说道。

"天上到处都有蒸发的水汽,为什么从这里到那里只有一片云,而不是很多云到处漂浮呢?"

"So how do you explain clouds floating in the sky and seldom dropping to the ground?"

"Because there is also the wind," says Owl.

"And when there's evaporated water everywhere in the sky, why does only one cloud form here or there, and not lots of clouds everywhere?"

"因为小水滴需要粘在小灰尘上才能形成云。"

"我明白了。"小老鼠说,"所以云是水和灰尘形成的。那你告诉我,为什么有些云是白色的,还有些是灰色的呢?"

"因为当太阳光照透云层时,云看起来就是白色的。但是当云层太厚时,穿透的太阳光太少了,云看起来就是灰色的了。"

"Because water droplets need to stick to tiny dust particles to form a cloud."

"I see," says Rat. "So a cloud is water and dust. Now tell me why are some clouds white and others grey?"

"Because when light shines through clouds, they look white. But when clouds get too thick, less light shines through and they look grey."

云是由水和灰尘形成的

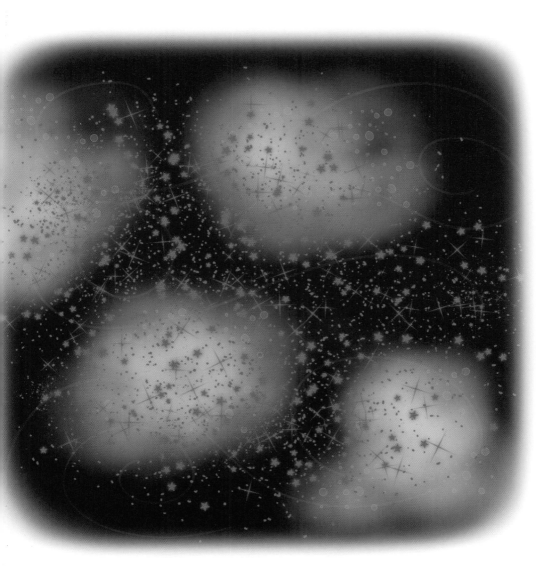

A cloud is water and dust

# 水是一个简单的分子

Water is a simple molecule

"哎呀!猫头鹰先生,你对云可真了解呀!"

"好吧,其实我真的只知道一点。另一方面,人们觉得自己对水很了解,但我却觉得他们一点儿也不懂水。"

"有人告诉我,水是一个简单的分子,是由2个氢原子和1个氧原子组成的。"

"My word, Mr Owl. You do know a lot about the clouds!"

"Well, I really only know a little. People, on the other hand, think they know a lot about water but I don't think that it is true."

"I've been told water is a simple molecule, made up of two hydrogen atoms and one oxygen atom."

"噢，是的，我们完全了解什么是分子，但我们需要进一步了解的是很多分子聚在一起是如何相互作用的。只看分子，就像只看一根头发却不看发型，而发型却能反映一个人的性格。"

"太对了。很多水分子紧紧挤压在一起时会像石头一样坚硬，我们把它称为冰。而当水以液体形式在河流里流动时，我们就可以喝了。但是水还可以以气体形式存在，我们称之为蒸汽。"

"Oh yes, of course, we know all about molecules, but we need to know more about how a collection of many of these molecules interact. Only looking at the molecule is like looking at only one hair on your head without seeing the hairstyle that reflects your character."

"Too true. Lots of water molecules packed closely together can be as solid as a rock – we call that ice. And water flows through rivers as a liquid – that we drink. But water can also be in the form of a gas – we call that vapour."

很多分子聚在一起是如何相互作用的

How a collection of molecules interact

这你都知道，太让我吃惊了

It amazes me that you know this

"这你都知道，太让我吃惊了。"猫头鹰说。

"但云到底是什么呢？是由很多小冰粒、小水珠和水蒸气组成的吗？"

"嗯，要是磁铁的两个正极靠近会发生什么事呢？"

"这个我知道！它们会互相排斥；两个负极也会相互排斥。"

"It amazes me that you know this," says the owl.

"But what is a cloud really? Are clouds made of lots of tiny balls of ice, drops of water, and also vapour?"

"Well, what happens when two positive poles of a magnet come close to each other?"

"I know! They repel each other. And negatives repel each other as well."

"太对了!但云是靠什么把这些带负电的水分子聚集在一起的呢?"

小老鼠沉默了一会儿,看着猫头鹰说道:"我觉得我得再观察云一段时间才能搞清楚这个问题。"

"好极了!就让美丽的天空激励你去进行新的科学探索吧!"

……这仅仅是开始!……

"Exactly, but what makes the clouds with all these negative water molecules hold themselves together?"

After a moment of silence, the rat looks at the owl and says, "I think I will have to look at the clouds a little bit longer to figure that one out."

"Great. Let the beauty of the sky inspire you to make new scientific discoveries!"

... AND IT HAS ONLY JUST BEGUN!...

……这仅仅是开始!……

…AND IT HAS ONLY JUST BEGUN!…

# Did You Know? 你知道吗？

About 80% of your body weight is water. However, as water molecules are the lightest molecules in our body, approximately 99% of our body's molecules are water.

我们身体重量的80%都是水。然而，因为水分子是我们身体中最轻的分子，所以其实我们身体中99%的分子都是水分子。

A water molecule ($H_2O$) has two hydrogen molecules and one oxygen molecule. However, when water molecules cluster together it also behaves like $H_3O_2$.

一个水分子（$H_2O$）有两个氢原子和一个氧原子。但是，当水分子聚集在一起的时候，也能形成 $H_3O_2$。

Fog is a type of cloud. It is created when warm, moist air flows over colder soil. If the air is saturated, the moisture condenses and forms fog that could reduce visibility to near zero.

雾是云的一种形式。它是由温暖潮湿的空气流过温度较低的地面形成的。如果空气湿度饱和了，湿气就会凝结并形成雾，并使能见度几乎降为零。

There are 10 types of clouds and some can move at more than 100 km/hr.

云有10种，其中一些云的移动速度能达到100千米/小时。

When air is heated by the sun, it rises and slowly cools down, and when it reaches saturation point it condenses, forming a cloud. The forces that create a cloud are pressure and temperature. Clouds are held together as a result of charged molecules.

当空气被太阳加热时,就会上升并慢慢冷却,当达到饱和点时就会凝结形成云。形成云的原因是压力和温度。云聚在一起是由于带电分子的作用。

While two negative or positive charges repel each other, two negative charges surrounded by positive charge attract each other.

两个负电荷或正电荷互相排斥,而两个负电荷被正电荷包围时就会互相吸引。

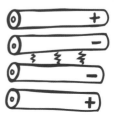

Scientists argue that clouds, water bridges, and insects running on water can be explained through an understanding of the "fourth phase" of water.

科学家认为云、水桥和昆虫在水上行走可以用对水的"第四相"的理解来进行解释。

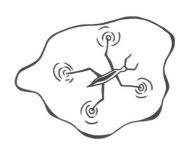

Water is one of the simplest molecules and also the most abundant molecules on Earth. Yet water is the least understood of all substances.

水是最简单的分子之一,也是地球上含量最多的分子;但它也是所有物质中最不被理解的。

# Think About It 想一想

**W**ater expands as it gets warmer and contracts as it gets colder, but if it gets colder than 4 °C, it starts expanding again. Is this the rule or the exception?

水在变得温暖时就会膨胀，而在变冷时就会收缩，但是当温度降到4℃以下时又开始膨胀。这是规律还是例外呢？

**T**he owl knows a lot and still claims he knows very little. Is he modest?

猫头鹰懂得很多却仍然说自己知道得很少，他是在谦虚吗？

如果云可以制造闪电，那么压力和温度差是不是也能制造闪电呢？

**I**f clouds produce lightning, can lightning be produced through pressure and temperature differentials only?

**D**o you think new scientific discoveries can be made by looking at the beauty in nature?

你认为通过观察自然界的美可以获得新的科学发现吗？

# Do It Yourself!

# 自己动手!

Let us make a bridge with water.

Fill two cups with water. Put thecups on isolated padding.

Make the water inside the cups connect. Then send a small electric current through the cups. Slowly move the cups apart. What will be left behind is a bridge of floating water.

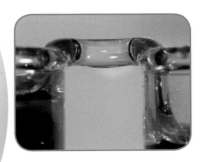

Water flows from the positively charged cup to the negatively charged cup.

You can make the water bridge up to 2 cm long. If the water gets hot, the bridge will collapse.

http://phys.org/news/2007-09-bridge-exposed-high-voltage.html

https://www.youtube.com/watch?v=1iJrRaUc8Q8

让我们用水造一座桥吧。在两个杯子中装满水，放到隔离的衬垫上。让杯子里的水连接起来；再通一个低压电流，慢慢把杯子移开，就会发现杯子间有一条漂浮的水桥。水将从正极连接的杯子流向负极连接的杯子。你可以把水桥做到2厘米长。如果水变热了，水桥就会断掉。

请参考网址：http://phys.org/news/2007-09-bridge-exposed-highvoltage.html https://www.youtube.com/watch?v=1iJrRaUc8Q8

# TEACHER AND PARENT GUIDE

## 学科知识
### Academic Knowledge

| | |
|---|---|
| 生物学 | 鸟如何利用暖气流而不是自己的能量飞得更高；撒哈拉沙漠的尘土如何来到亚马逊三角洲并成为表层土壤；生命离不开水，水是所有生命形式的主要成分。 |
| 化　学 | $H_2O$和$H_3O_2$的区别；氢原子（H）和氧原子（O）。 |
| 物　理 | 云的作用；重力规则；悬浮在空中的水滴产生的摩擦力；灰尘颗粒和小水滴之间的吸引力；比较一个分子的行为和一堆分子的行为；热空气的膨胀；空气中水的饱和；云的色彩有白色和灰色；云的10种形态；原子和分子的不同之处；磁性物质同极相斥，异极相吸；比较压力和温度的结合与压力、温度和电荷的结合；水滴的大小决定其物理特性；风是怎样形成的；打雷和闪电；水的禁区——上文提到的水的"第四相"对人类健康和地球健康的核心作用；水的第四相的发现超越了物理学的范畴，进入了化学和生物学的范畴；水和光的相互作用。 |
| 工程学 | 天气和气候工程学能提高天气预测能力。 |
| 经济学 | 云计算是储存和处理数据和信息的新形式。 |
| 伦理学 | 要接受科学上新的视角，这将不断激励我们彻底改变我们周围的世界。 |
| 历　史 | 威廉姆·阿姆斯特朗在1850年第一次展示了水桥（也被称作水绳）。 |
| 地　理 | 利用气象学决定农事。 |
| 数　学 | 当更多的参数互相作用时，复杂性会增加；光和声的速度。 |
| 生活方式 | 俗话说："干打雷不下雨。"就是说：只有承诺却没有实现。 |
| 社会学 | 在希腊神话中，仙女和气象之神生活在云里；基督教中的天使形象生活在云中；云被用作美国土著人（包括霍皮族）的宗教符号；在德国神话中托尔是雷神，而在希腊神话中涅斐勒是云神；禅宗认为云和虚空有关；儒家学说认为通过不道德手段所获得的财富和地位只是浮云。 |
| 心理学 | 发型可以表达个性；艺术和科学的协同作用：科学家已经发现音乐，尤其是莫扎特的音乐能够促进智力发展，如数学思考。 |
| 系统论 | 自然系统中的相互作用不能被简单化，就像云的形成不能仅仅用温度和压力来解释，还需要考虑灰尘颗粒和水滴的摩擦力，此外还有磁力。 |

# 教师与家长指南

## 情感智慧
### Emotional Intelligence

**猫头鹰**　　猫头鹰变得更谦虚了。开始的时候，猫头鹰认为自己是个聪明人，他让小老鼠活了下来。现在猫头鹰认识到小老鼠的存在并准备和他交流。他兴趣盎然地用一个问题开始了这次交谈。猫头鹰无私地和小老鼠分享了自己的知识，甚至很高兴回答小老鼠的所有问题。当小老鼠抛开细节看到整个问题时，猫头鹰很是吃惊。于是他冒险走进了一个未知的领域，把云和磁联系起来，还引入了一些虽然已经研究了一个多世纪但对于大多数科学家还很新的概念。然后猫头鹰表示了对小老鼠的支持并鼓励他自由思考，去拥抱自然和艺术的美丽。

**小老鼠**　　小老鼠表现得比较成熟，并且在猫头鹰面前很少挑衅。他没有挑战猫头鹰，而是从猫头鹰那里获得了知识，而且学得很快。小老鼠对猫头鹰的聪明才智表现出了敬佩和赞赏。当猫头鹰让小老鼠进一步思考的时候，小老鼠愉快地接受了挑战；当猫头鹰对小老鼠丰富的知识表示祝贺时，小老鼠并没有为此表示感谢，而是继续提出了更困难的问题。但他很快便用完了自己的知识。然而，他依然保持着自信，他抓住了移动的云朵给他的灵感，继续学习更多的知识。

## 艺术
### The Arts

在一大张白纸上画出至少6种不同形状的云。你不用知道这些云的科学名称。现在在你相信会先落下雨的云朵下面画上水滴。再选一张你认为一定会产生闪电的，画上闪电。最后选一张，画上彩虹。

# TEACHER AND PARENT GUIDE

## 思维拓展
### Systems: Making the Connections

　　世界是复杂的。为了更好地理解组成整个现实生活的很多小的部分，科学已经为我们把现实简化了。我们拥有的所有知识比所有独立部分的总和还要巨大。仅仅描述氢原子和氧原子的特点无法解释水分子的行为。很多水分子结合在一起时，在不同的压力和温度环境下会有不同的表现。然而，这还仅仅是影响物质行为的两个变量，我们知道还有其他因素（如磁力和电荷）也影响着所有物体，包括水和云，正如我们在打雷和闪电时看到的。但是，由于我们不知道如何测量云中的电荷，所以我们忽略了一些事实，尽管我们知道它们的存在。虽然一些现象在明确定义的假说中得到了观察，但是我们知道现实世界要复杂得多。根据万有引力定律，自然的力量能使苹果从树上掉下来。但我们是否理解使苹果首先从树上长出来的所有力量是什么？我们将现实过度简单化以更好地理解自然现象，去理解水为什么在冰冻时膨胀，但同时我们也要面对另外一些无法解释的现象。还有很多现象我们无法理解，比如为什么很多水汽蒸发到空中各处却只形成了一朵云？为什么水桥在水温上升时会断掉？为什么蜥蜴、蜘蛛和水黾能在水上行走？

## 动手能力
### Capacity to Implement

　　对于那些用我们学过的简单科学知识无法解释的现象，我们必须仔细考虑现象的复杂性和它们之间的相互联系。这就是为什么我们要理解我们每天都能观察到的云。当我们大胆地对生活的复杂性和所有关系进行深入理解时，必须认识到今天的科学还不能解释一切。我们将有新的发现并将创立基于更多高深的假说上的各种新理论。这需要强烈的求知欲、创造力和灵感：在幻想、想象和现实之间探索。有时艺术家能够帮助我们进行探索并超越已经建立起来的科学的边界。他们就像社会的天线，能感觉到无法解释的事物，并提供找到答案的途径，一条不同寻常的途径。然后就由科学家进行深入研究，提出理论和证据，对我们周围的现实给出了新的理解。

故事灵感来自

# 杰拉尔德·波拉克
## Dr Gerald Pollack

　　杰拉尔德·波拉克博士本来在纽约大学电气工程专业学习，后来他转入宾夕法尼亚大学并获得了生物医学工程学位。他大胆地开始了有关水、细胞和健康的新的科学研究。现在他是华盛顿大学西雅图分校波拉克实验室的负责人。他研究运动、细胞生物学以及生物表面和水溶性的相互作用。他的两部著作《肌肉和分子》及《胶体和生命的动力》都得到了很高的评价。他发现了水的第四相，这为新设备的创新和发明提供了巨大的机会。

更多资讯

　　http://faculty.washington.edu/ghp/cv/

　　https://www.youtube.com/watch?v=i-T7tCMUDXU

　　http://www.weatherwizkids.com/weather-clouds.htm

## 图书在版编目（CIP）数据

空中的云朵 ：汉英对照 ／（比）冈特·鲍利著 ；
（哥伦）凯瑟琳娜·巴赫绘 ；郭光普译 . －－ 上海 ：学林
出版社 ，2016.6
 （冈特生态童书．第三辑）
 ISBN 978-7-5486-1044-1

Ⅰ．①空… Ⅱ．①冈… ②凯… ③郭… Ⅲ．①生态环
境－环境保护－儿童读物－汉、英 Ⅳ．① X171.1-49

中国版本图书馆 CIP 数据核字（2016）第 125658 号

---

ⓒ 2015 Gunter Pauli
著作权合同登记号 图字 09-2016-309 号

**冈特生态童书**
空中的云朵

| | | |
|---|---|---|
| 作　　　者—— | 冈特·鲍利 | |
| 译　　　者—— | 郭光普 | |
| 策　　　划—— | 匡志强 | |
| 责任编辑—— | 蔡雩奇 | |
| 装帧设计—— | 魏　来 | |
| 出　　　版—— | 上海世纪出版股份有限公司 学林出版社 | |
| | 地　址：上海钦州南路 81 号　电话／传真：021-64515005 | |
| | 网址：www.xuelinpress.com | |
| 发　　　行—— | 上海世纪出版股份有限公司发行中心 | |
| | （上海福建中路 193 号 网址：www.ewen.co） | |
| 印　　　刷—— | 上海丽佳制版印刷有限公司 | |
| 开　　　本—— | 710×1020　1/16 | |
| 印　　　张—— | 2 | |
| 字　　　数—— | 5 万 | |
| 版　　　次—— | 2016 年 6 月第 1 版 | |
| | 2016 年 6 月第 1 次印刷 | |
| 书　　　号—— | ISBN 978-7-5486-1044-1/G·379 | |
| 定　　　价—— | 10.00 元 | |

（如发生印刷、装订质量问题，读者可向工厂调换）

# Water
## 90

# 竹子的馈赠
## Generous Grass

**Gunter Pauli**

冈特·鲍利 著
凯瑟琳娜·巴赫 绘
隋淑光 译

学林出版社
www.xuelinpress.com

## 丛书编委会

主　任：贾　峰
副主任：何家振　郑立明
委　员：牛玲娟　李原原　李曙东　吴建民　彭　勇
　　　　冯　缨　靳增江

## 丛书出版委员会

主　任：段学俭
副主任：匡志强　张　蓉
成　员：叶　刚　李晓梅　魏　来　徐雅清　田振军
　　　　蔡雯奇　程　洋

特别感谢以下热心人士对译稿润色工作的支持：

姜竹青　韩　笑　贾　芳　刘　晓　张黎立　刘之杰
高　青　周依奇　彭　江　于函玉　于　哲　单　威
姚爱静　刘　洋　高　艳　孙笑非　郑莉霞　周　蕊

# 目录

| | |
|---|---|
| 竹子的馈赠 | 4 |
| 你知道吗？ | 22 |
| 想一想 | 26 |
| 自己动手！ | 27 |
| 学科知识 | 28 |
| 情感智慧 | 29 |
| 艺术 | 29 |
| 思维拓展 | 30 |
| 动手能力 | 30 |
| 故事灵感来自 | 31 |

# Contents

| | |
|---|---|
| Generous Grass | 4 |
| Did you know? | 22 |
| Think about it | 26 |
| Do it yourself! | 27 |
| Academic Knowledge | 28 |
| Emotional Intelligence | 29 |
| The Arts | 29 |
| Systems: Making the Connections | 30 |
| Capacity to Implement | 30 |
| This fable is inspired by | 31 |

在树木稀疏的小山坡上,一只貘正看着一只松鼠在树木间跳来跳去。

"你还记得这些山上曾经长满了竹子吗?"貘问道。

"我没看到过,但是我祖母曾经说过。她还说那时候这里更凉爽,每年的降雨量更大。"松鼠回答道。

On a sparsely wooded hillside, a tapir is watching a squirrel leap from tree to tree.

"Remember when these mountains were covered in bamboo?" asks the tapir.

"I never saw it myself, but my grandma told me about it. She also said that it used to be much cooler during the day and that we had more rain all year round," replies the squirrel.

山上长满了竹子

Mountains were covered in bamboo

桉树种得越来越多

Eucalyptus tree became so popular

"嗯,自从桉树种得越来越多,土地就变得越来越干燥了。"

"为什么人们要种这种疯狂吸水的树,并且放任它生长呢?这对土地一点好处也没有。"

"你知道,农民们听说这种外来的树长得比其他树更快。"

"Well, since the eucalyptus tree became so popular, the soil is much drier."

"How can anyone plant a tree that guzzles so much water? And then allow it to keep on growing? It's not good for the soil."

"You see, farmers were told that this alien tree would grow faster than any other tree."

"甚至比竹子长得还快?种子公司肯定有一位出色的推销员!"

"是的。你知道种上竹子后,多长时间才能收获一根25米长的竹材?"

"我听说只要三年就可以。"松鼠回答道。

"那么再次收获需要多长时间呢?"

"Even faster than bamboo? The seed company must have had a excellent salesman!"

"Right. And do you know how long after planting it you will be able to harvest a twenty-five-metre tall bamboo?"

"I hear you can do so in just three years," the squirrel replies.

"And when can you harvest it again?"

三年后就可以收获竹材

Harvest bamboo in just three years

# 海里康花

Heliconias

"第二年就可以。"

"没错,更重要的是,竹子可以连续收获70年以上。在这个时间段内,它们产生的纤维素是速生树种的60倍。另外,它们还寄生了最美丽的花——海里康花!"

"是啊,那些海里可花真是美丽无比。"

"The following year."

"Exactly. And what's more, is that bamboo plants can be harvested for up to seventy years. During that time, they produce sixty times more cellulose than the fastest growing tree in the world. Plus, they play host to the most beautiful flowers: heliconias!"

"Ah yes, those helicopters are gorgeous."

"不是海里可花,是海里康花!听着松鼠,我们要更深入地了解竹子。"

"你说得对。树木被砍伐以后还要重新种植,而且它们不断地消耗水,而竹子却能保护水土不流失,除了关爱以外,不需要其他任何东西。"

"你知道谁住在高地上吗?"松鼠问道。

"Not helicopters, heliconias! But listen here, Squirrel, we need to find out more about bamboo."

"You're right. Trees need to be replanted after harvesting and they need water all the time, while bamboo can hold water and needs nothing else, except some loving care."

"Do you know who lives in the highlands?" asks the squirrel.

除了关爱以外，不需要其他任何东西

Needs nothing else, except some loving care

# 种豆子的农民

The bean farmers

"种豆子的农民。"

"对，所以这种25米长的草顶端的3米对他们很有用，可以搭豆秧支架。"

"草？"貘吃惊地问道。

"没错，竹子是一种草，所以它被收割后还可以再长出来，就像你的草坪一样。"

"The bean farmers."

"Exactly, so the first three metres of the twenty-five-metre-long grass can go to the bean farmers, who can use them as posts."

"Grass?" asks the tapir, surprised.

"Yes, grass. Bamboo is a grass. You cut it and it grows again, just like your lawn."

"再切下3米可以做竹竿,种豆子的农民用他们的毛驴把竹竿运下山,用来盖房子。"

"盖那种能随着地壳运动的节奏起舞的神奇竹屋?我喜欢这种房子。"

"剩下的东西可能有19米长,它们是很好的造纸材料。"

"Poles can then be cut from the next three metres of bamboo and the bean farmers can bring them down the mountain on their donkeys. These can be used to build houses."

"Those incredible houses that can move – dancing to the rhythms of the Earth? I love those."

"And whatever is left over, and that can be as much as nineteen metres, is good for making paper."

会跳舞的房子

Moving houses

# 竹林里更凉爽

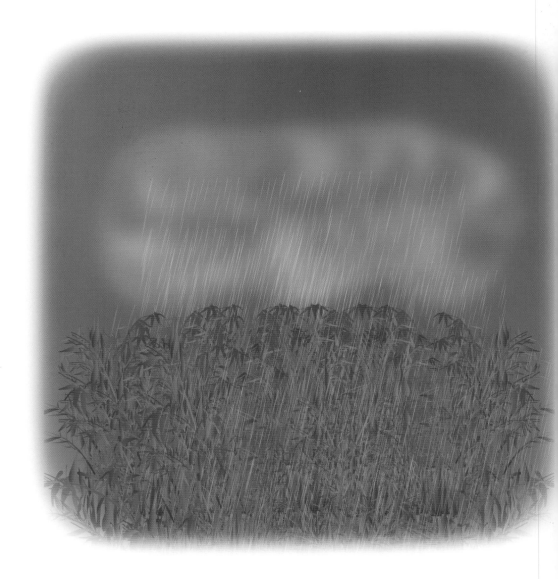

Bamboo forests are much cooler

"造纸？太神奇了！竹子可以用来建造既舒适又成本低廉的房子，还能创造就业机会，让农民安居乐业。"

"它甚至能让这座山感到快乐。"貘笑着说。

"为什么？"松鼠很好奇。

"因为竹子能在下雨后蓄存雨水，还因为竹林里更凉爽，云朵乐意在这里洒下雨滴。"

"Paper to write on? That is amazing! Bamboo can be used to build good, inexpensive houses, and it creates work and keeps the farmers happy."

"Even the mountains will be happy," says the tapir, smiling.

"Why?" the squirrel wants to know.

"Because bamboo holds water after it has rained. And because bamboo forests are much cooler, clouds are happy to let their raindrops fall there."

"这样我们就有更多的雨水了!土地也能蓄存更多的水!你们貘也会因此更快乐吧。"
"竹子不仅能蓄存水,还能滋养土地。土地是最重要的生命资源。有了富含水分的土壤,就意味着有足够多营养丰富的食物!"
……这仅仅是开始!……

"So, we get more rain and the soil retains more water! Now that should make you tapirs happy as well."
"Not only does bamboo hold water, it also nourishes the soil. And soil is the most important source of life. With good soil and plenty of water, there will be an abundance of nutritious food!"
... AND IT HAS ONLY JUST BEGUN!...

……这仅仅是开始!……

...AND IT HAS ONLY JUST BEGUN!...

# Did You Know?

## 你知道吗？

Bamboo buildings are quite earthquake resistant, because they move (dance) along with the rhythms of the earth.

竹建筑物有很好的防震性能，因为它们能随着地壳运动的节奏晃动（跳舞）。

Bamboo is a grass; when cut, it grows again. Unfortunately, in Colombia, bamboo has been classified as a tree, so a special permit is required in order to cut it down.

竹子是一种草，砍掉后能再生。不幸的是，它们在哥伦比亚被视作树，只有获得特殊许可后才能砍伐。

*B*amboo grows to maturity in only three years, and after being cut, grows to the same height again within a year. It keeps on growing for 70 years. It is the most efficient producer of cellulose in nature.

竹子只要三年就能成材，砍伐后在一年内又能长到同样高度。它能持续生长70年，是自然界中最高效的纤维素生产者。

*W*hen bamboo grows in small, localised clumps, it holds water and regenerates topsoil. It grows in symbiosis with heliconias and arbolocos (also called 'crazy trees').

竹子是一簇簇丛生的，它能蓄存水分，再生表层土壤。它与海里康花及蒙坛菊（也叫"疯狂树"）共生。

When cultivated, eucalyptus grows with a fast harvesting and planting rotation cycle. Consequently, the soil is quickly depleted of nutrients. Biodiversity is then rapidly reduced in that area.

桉树被种下后，就进入了快速收获和重新种植的循环。结果，土壤养分会被迅速耗尽，生物多样性也会急剧下降。

Bamboo is an excellent source of fibre for making paper, especially for water-absorbent paper like toilet paper and tissue paper (e.g., Kleenex).

竹子能提供优质的造纸纤维，特别适合用来生产像卫生纸和纸巾这样的吸水性纸。

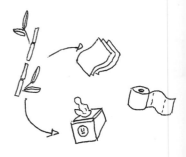

Tapirs resemble pigs and have trunk-like snouts. The tapir is an herbivore and can weigh between 150 and 300 kg. Tapirs are closely related to horses and rhinos.

貘外表像猪，鼻子像大象鼻子，它是一种草食性动物，体重在 150 公斤到 300 公斤之间。貘是马和犀牛的近亲。

The tapir uses his extended, prehensile snout as a snorkel when swimming. The tapir is an extraordinary seed disperser; therefore local extinction of tapirs leads to a loss of biodiversity.

貘的鼻子能伸缩、缠卷，在游泳时可伸出水面换气。它是一个勤劳的种子散播者，因此灭绝后会引起当地生物多样性的丧失。

# Think About It
# 想一想

**I**f a structure is flexible enough, it can move along with the force of an earthquake and absorb some of its impact. Can you do that?

如果一种结构足够有韧性，就能在地震时随着力的作用晃动，并化解掉一些冲击力。你能试验一下吗？

**W**ould you like to have a tapir as a pet, or would you prefer a pig?

如果可以养一只宠物，你喜欢养一只貘还是一头猪？

通过种植桉树来获得造纸纤维的做法是否有意义？

**D**oes it make sense to plant eucalyptus trees as a source of fibre to make paper?

**B**amboo not only helps hold water and regenerate soil, it also provides building materials in the form of poles. It can also be used to make paper, while increasing biodiversity. Would you say that bamboo is a generous grass?

竹子不仅能蓄存水分、再生表土，还可以提供建筑材料、造纸原料，以及提高生物多样性。你认为它是一种有价值的草吗？

# Do It Yourself!
# 自己动手！

Let's find out how easy it is to make paper or cardboard. Take a newspaper and remove all the inserts printed on glossy paper, as it contains too much coloured ink and plastic. Soak the newsprint in a little water for a day or two. Stir it until the paper had completely dissolved in the water. Now spread a thin layer of this pulp onto a cloth and let it dry. You will now have a thick sheet of paper or a thin sheet of cardboard.

让我们来看看造纸或硬纸板有多么容易。取一张报纸，去掉上面的彩色墨水和塑料膜，用少量水浸泡一两天，搅拌直到完全化开，然后把纸浆倒在一块布上铺成薄层，晾干后，你就能得到一张厚纸或薄纸板了。

# TEACHER AND PARENT GUIDE

## 学科知识
### Academic Knowledge

| | |
|---|---|
| 生物学 | 貘是食草动物，它的摄食和排泄对植物生长有重要的促进作用；貘有很好的潜水能力，也吃水生植物；当受到美洲虎或美洲狮攻击时，貘会逃进密林中的狭窄通道里躲避，低垂的树枝会不断刮擦攻击者的脸，使它们很难捉到貘；桉树大约有700个品种，都是澳大利亚本地物种；人们发现了约200种海里康属的外来热带植物，它们耐强降雨但不耐干旱；海里康属植物是蜂鸟的重要食物来源，蜂鸟和一些种类的蝙蝠几乎是它们专一的授粉者。 |
| 化学 | 桉树分泌一种能杀菌的萜类化合物，弥漫在树的周围，以此来对抗竞争；纤维素是一种聚合物。 |
| 物理 | 竹子茂密的枝叶能使周围环境温度降低大约10摄氏度，从而营造出清新凉爽的空间，这是熊猫、老虎和貘喜欢的栖息地；桉树能吸收地表下10米处的黄金，并通过枝叶排放出来。 |
| 工程学 | 貘依靠伸展的鼻子和长长的上唇抓住树枝、剥离树叶，并且不弄伤树就可摘到水果；人们把纤维素转变成纤维，再把纤维转变成人造丝。 |
| 经济学 | 桉树是一种速生树种，用于生产木材、清洁油、天然杀虫剂、染料和药物（特别是治疗关节痛疼的药物）；蜜蜂采集桉树花粉酿蜜；桉树耐火，种植桉树可以降低火灾风险；桉树不耐霜冻，只能种在无霜区。 |
| 伦理学 | 外来物种虽然能提高经济收益，但也会和本地生物竞争水，并危及生物多样性，这一问题值得思考。 |
| 历史 | 在南美洲发现了始新世时期（5100万年前）的桉树化石，在新西兰发现了中新世时期（500万至2500万年前）的桉树化石；桉树成为澳大利亚当地的优势树种与大约5万年以前，原住民祖先的到来有关；1770年，詹姆斯·库克船长首次把桉树从澳大利亚带到了欧洲。 |
| 地理 | 貘栖息在南美洲、中美洲和东南亚（马来西亚和印度尼西亚）的丛林里。 |
| 数学 | 竹子中含有60%的纤维素；经过三年的生长，此后70年，每公顷竹林（根据每根竹子重60公斤，每公顷竹林包含500根竹子来计算）每年可获得至少30吨的生物量，这一数字远远超过了桉树。 |
| 生活方式 | 在汉语、韩语和日语中，"貘"这个词是相同的，它在中国神话中被认为是一种能吃梦的神兽；竹子被哥伦比亚人和印度尼西亚人称作"植物钢材"。 |
| 社会学 | "eucalyptus"（桉树）是一个合成词，"eu"的意思是"好"，"calyptos"在希腊语中的意思是"覆盖"。 |
| 心理学 | 推销员利用心理学知识来说服人们购买某个产品，包括赢得信任和使顾客的决定合理化，面对不断的拒绝，推销员不能气馁；《推销员之死》这部剧，说的是一个好心的推销员因为看不到工作的意义以及无力实现美国梦而崩溃的故事。 |
| 系统论 | 桉树和其授粉者——蜂鸟、蝙蝠和负鼠是共生关系，这些动物不受桉树分泌的酚类化合物的影响。 |

# 教师与家长指南

## 情感智慧
### Emotional Intelligence

貘

貘很怀旧，喜欢回忆森林过去的样子。他观察敏锐，作出的批评言之有据，难怪他觉得受到了欺骗。当谈论种植竹子时，他很克制，但是当谈到海里康花时，他变得情绪化。他热衷于通过发掘竹子的更多用途，来提高它的经济价值和接受度。当他描绘竹子超越其他树种的大量商业用途时，他很享受松鼠的补充说明。他感兴趣的不仅是种竹子能造福人类这一点，而是由此展开的更宏大的场景：竹子对这座山以及对构成其栖息地的生态系统至关重要。貘具备自我认知能力，知道自己的需求来自他与这块栖息地中其他生物和有机体的生态关系。

松鼠

松鼠从祖母那里知道了这块栖息地过去的样子。他很难相信会有人认为引进外来树种是个好主意，除非这个人被巧舌如簧的推销员洗脑了。松鼠很有学问但很谦虚，当貘问他问题时，他很有耐心，也对貘表现出了足够的尊重。他思维活跃，对于竹子能造福人类这个话题，他热心于寻找答案，并为貘提供了思路。但他做得很隐蔽，并不是想炫耀自己的智慧。通过回答貘的问题，他也获得了一些新的认识。松鼠知道貘不仅是在思考这座山和种豆农民的处境，也是在寻求更好的栖息地。他支持貘去达成这个愿望。

## 艺术
### The Arts

海里康花是一种美丽的热带花，几乎只由蜂鸟授粉。这种花有各种各样奇特的形状。找一些海里康花的图片，挑选三种你认为最稀奇古怪的形状画下来，然后考虑一下这种形状是否有利于传粉。

# TEACHER AND PARENT GUIDE

## 思维拓展
### Systems: Making the Connections

貘在其栖息地的生态系统中扮演了最重要的角色，这意味着如果它的生存受到威胁，整个生态系统就会面临崩溃的危险。作为草食性动物，貘以各种水果和浆果为食，广泛散播其种子，因此它是热带森林生物多样性的维护者。它是一种温顺的动物，在进化中发展出了通过逃进密林来摆脱食肉动物这样的防御策略，以避免冲突。通过貘，我们可以了解到生态系统对关键物种的依赖性，以及生态系统内部是如何解决冲突的。

在一个生态系统中，人们要在种植本土植物和为了经济利益种植外来植物之间做出选择。推销员会尽力宣扬他所销售产品的优势，并诱导人们忽略他人的理性观点。当错误决定所带来的不利后果开始彰显时，往往已经太迟了。正如这则童话所示，我们要先集合一切现有的信息和智慧，要依靠我们今天的知识以及前人的知识，来发现基于本地现有材料以及所需材料的一系列机遇。这能增加就业机会，确保物种有更好的生存机会，提高生物多样性，增加水的蓄积，提高美丽的共生植物的欣赏价值。我们做决定时不能只聚焦于某个产品，而是需要关注到整个生态系统，深入理解某个建议的优势和不足，以保有更大的选择空间。

## 动手能力
### Capacity to Implement

推销员必须基于逻辑、事实和数据来说服别人。让我们来扮演推销员吧。分成两组，一方支持销售桉树种子，另一方支持销售竹子种子，然后双方清晰、简洁地发表意见来论证自己的观点，尽量不要涉及太多技术细节。确保每个推销员都有机会陈述他的观点。然后讨论一个有没有可能推销员总是对的，而另一个总是错的……我们是否可以得出一个很有意义的结论：在某个地方正确的观点在另一个地方可能是不正确的？

故事灵感来自

## 路易斯·米格尔·阿尔瓦雷斯·梅希亚
## Luis Miguel Alvarez Mejía

路易斯·米格尔·阿尔瓦雷斯·梅希亚毕业于哥伦比亚国立罗萨里奥大学，获得遗传学硕士学位，并在卡尔达斯大学获得农业工程学士学位。他是研究竹林及其生态系统物种再生方面的专家，是卡尔达斯大学植物园的负责人，对生物多样性的动态、水和本地物种推动经济增长方面有深入理解。他相信即使森林被破坏了，依靠热带生物的力量也可以恢复原貌。

更多资讯

http://tapirs.org/tapirs/

http://www.guaduabamboo.com/forum/comparing-biomass-of-beema-with-guadua-bamboo

http://scienti.colciencias.gov.co:8081/cvlac/visualizador/generarCurriculoCv.do?cod_rh=0000427918

图书在版编目（CIP）数据

竹子的馈赠：汉英对照 /（比）冈特·鲍利著；
（哥伦）凯瑟琳娜·巴赫绘；隋淑光译 . — 上海：学林
出版社，2016.6
（冈特生态童书 . 第三辑）
ISBN 978-7-5486-1046-5

Ⅰ．①竹… Ⅱ．①冈… ②凯… ③隋… Ⅲ．①生态环
境－环境保护－儿童读物－汉、英 Ⅳ．① X171.1-49

中国版本图书馆 CIP 数据核字（2016）第 125820 号

———————————————————————————

ⓒ 2015 Gunter Pauli
著作权合同登记号 图字 09-2016-309 号

冈特生态童书
竹子的馈赠

| 作　　者―― | 冈特·鲍利 |
| 译　　者―― | 隋淑光 |
| 策　　划―― | 匡志强 |
| 责任编辑―― | 蔡雩奇 |
| 装帧设计―― | 魏　来 |
| 出　　版―― | 上海世纪出版股份有限公司学林出版社 |
|  | 地　址：上海钦州南路 81 号　　电话／传真：021-64515005 |
|  | 网　址：www.xuelinpress.com |
| 发　　行―― | 上海世纪出版股份有限公司发行中心 |
|  | （上海福建中路 193 号 网址：www.ewen.co） |
| 印　　刷―― | 上海丽佳制版印刷有限公司 |
| 开　　本―― | 710×1020　1/16 |
| 印　　张―― | 2 |
| 字　　数―― | 5 万 |
| 版　　次―― | 2016 年 6 月第 1 版 |
|  | 2016 年 6 月第 1 次印刷 |
| 书　　号―― | ISBN 978-7-5486-1046-5/G·381 |
| 定　　价―― | 10.00 元 |

（如发生印刷、装订质量问题，读者可向工厂调换）

# Health 107
# 螨虫家园
## The Mite Farm

**Gunter Pauli**

冈特·鲍利 著
凯瑟琳娜·巴赫 绘
何家振 译

学林出版社
www.xuelinpress.com

## 丛书编委会

主　任：贾　峰
副主任：何家振　郑立明
委　员：牛玲娟　李原原　李曙东　吴建民　彭　勇
　　　　冯　缨　靳增江

## 丛书出版委员会

主　任：段学俭
副主任：匡志强　张　蓉
成　员：叶　刚　李晓梅　魏　来　徐雅清　田振军
　　　　蔡雪奇　程　洋

特别感谢以下热心人士对译稿润色工作的支持：

姜竹青　韩　笑　贾　芳　刘　晓　张黎立　刘之杰
高　青　周依奇　彭　江　于函玉　于　哲　单　威
姚爱静　刘　洋　高　艳　孙笑非　郑莉霞　周　蕊

# 目录

| | |
|---|---|
| 螨虫家园 | 4 |
| 你知道吗？ | 22 |
| 想一想 | 26 |
| 自己动手！ | 27 |
| 学科知识 | 28 |
| 情感智慧 | 29 |
| 艺术 | 29 |
| 思维拓展 | 30 |
| 动手能力 | 30 |
| 故事灵感来自 | 31 |

# Contents

| | |
|---|---|
| The Mite Farm | 4 |
| Did you know? | 22 |
| Think about it | 26 |
| Do it yourself! | 27 |
| Academic Knowledge | 28 |
| Emotional Intelligence | 29 |
| The Arts | 29 |
| Systems: Making the Connections | 30 |
| Capacity to Implement | 30 |
| This fable is inspired by | 31 |

螨虫一家站在一座巨大的崭新的大楼前面,他们正要去寻找一个好住处。

"我们进去看看地板上是否有地毯。"螨虫爸爸说。

"你知道,我不喜欢这些毛皮地毯。"螨虫妈妈说,"走在上面感觉很硬。"

A family of mites are standing in front of a grand, new building, looking for a good place to live.

"Let's go inside to see if it has carpets on the floors," says the dad.

"I don't like these furry floor tiles, you know," the mom replies. "They're hard to walk on."

我不喜欢这些毛皮地毯

I don't like these furry floor tiles

现在让我们看看窗户吧

Now let's have a look at the windows

"但是,这种地毯的好处也很多呢。我们一家人将拥有所需的全部食物。"

"是的,有很多人住在这里,他们的皮肤不停地脱落死细胞,为我们提供了食物。"

"现在让我们看看窗户吧。"螨虫爸爸说。

"But think of all the advantages. We will have all the food we need for our big family."

"That's true. There are a lot of people living here, and their skins shed dead cells all the time, providing us with food."

"Now let's have a look at the windows," says Dad Mite.

"你觉得他们会为了呼吸新鲜空气打开窗户吗?"
"不会的,在这些现代化的大楼里,他们从来不开窗户,他们为了节能不计代价,甚至不惜牺牲健康。"
"我们太幸运了,人们关上所有的窗户,强迫所有人在相同的空气中呼吸,天天如此。"螨虫爸爸说。

"Do you think they'd want to open a window for fresh air?"
"Oh no, in these modern buildings, they never open windows. They want to save energy at all costs, even at the cost of their own health."
"We are so lucky that people seal all their windows and force everyone to breathe the same air, day in and day out," remarks Mother Mite.

……从来不开窗户……

... never open the windows ...

……地毯是躲藏的好地方……

... carpets are good places to hide ...

"我不知道他们是怎么活下来的。但是我看到他们在空调系统中装上强大的过滤装置，滤掉他们不想要的东西。"

"这些地毯是我们躲藏的好地方。你知道，如果他们看见我们，就会设法杀死我们。"

"I don't know how they survive. But I see that they use strong filters in the air conditioning system to get rid of things they don't like."

"These carpets are good places for us to hide then. You know, if they see us, they will try to kill us."

"别担心,人们从来不看他们脚下有什么,特别是不看地毯里有什么。"

"人们允许我们住在这里,还允许我们把粪便留在这里,这真令人惊讶。他们把地毯弄得这么脏,有些人在这里根本无法呼吸。"

"Don't worry, people never look at what's below their feet, especially not at what's in the carpet."

"It's amazing how people would allow us to live here and leave our droppings behind. They make the carpets so dirty and make it hard for some people to breathe in here."

人们从来不看他们脚下有什么

People never look at what's below their feet

# 我们最坏的敌人——太阳

Our worst enemy - the sun

"住在这儿,我们世代都会受到保护,因为人们在窗户上贴上了防紫外线膜,以保护地毯和艺术品不会被太阳晒褪色。"

"我们真幸运!那就是说,他们保护我们不受最坏的敌人——太阳的照射,而且这里还有丰富的美食。"

"既然我们不能清除自己的粪便,那就得等着一年被迫洗一次化学浴了。"螨虫爸爸警告说,"那玩意儿毒性很大,就连人类接触它也会生病。"

"Well, we'd be protected for generations to come, because people put ultra-violet film on the windows to protect their carpets and artwork from being faded by the sun."

"Lucky us again! That means they protect us from our worst enemy – the sun – while we enjoy the abundant food available here."

"As we cannot clean up after ourselves, we have to be prepared to have a chemical bath once a year," warns Dad Mite. "And that stuff is so toxic that even people can get sick from it."

"等到他们弄来那些化学杀毒剂时,我们会变得更强大、更有抗药性。"螨虫妈妈说道。

"是的,暴露在这种化学喷雾剂中几年之后,我们就适应了,到时候它对我们根本不管用。我们懂得生存之道,绝不是好对付的!"

"By the time they come with that terrible onslaught of chemicals, we'd have become much stronger and more resistant," says Mother Mite.

"That's true. After being exposed to those chemical sprays for years, we've become used to it and it hardly bothers us anymore. We know how to survive and have become a force to be reckoned with!"

更强大、更有抗药性

Stronger and more resistance

我们的聚居地就越多

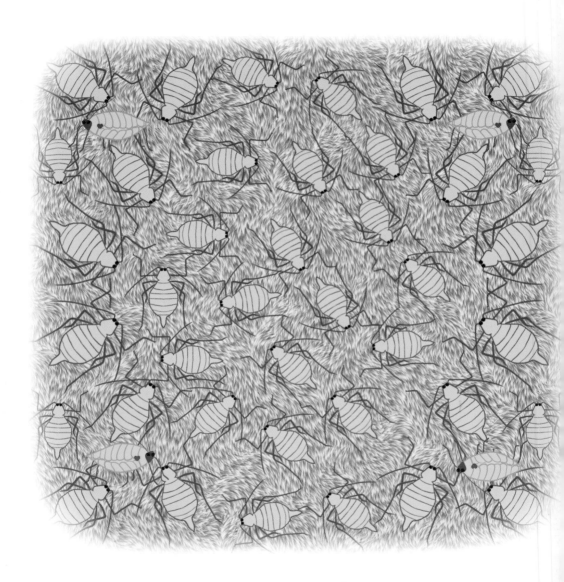

The more colonies we can establish

"是的，不管人们怎么做，比如，用可再生纤维做地毯，并且在本地制作以减少因交通运输产生的污染排放，我们都能轻易地生存，甚至更加兴旺！"

"地上铺的地毯越多，我们的聚居地就越多。"

"Yes, despite what people do, like making carpets from recycled fibres and doing so locally, to reduce emissions by not unnecessary transporting things around the world, we can easily survive – and even thrive!"

"The more carpets that are being laid, the more colonies we can establish."

"我知道。总部设在亚特兰大的全世界最大地毯制造商之一,不是曾经说过'可持续发展路途漫漫,我们只是迈出了很小的第一步'吗?"

"他说得对。不管他们用什么制造地毯,在这场游戏中的每一步,我们都注定会赢!"

……这仅仅是开始!……

"I know. Wasn't it one of the great carpet makers in the world, the one from Atlanta, who used to say that sustainability is only the small beginning of a long process?"

"And he was right. Whatever they make their carpets from, we'll certainly win at every step of this game!"

... AND IT HAS ONLY JUST BEGUN!...

……这仅仅是开始!……

... AND IT HAS ONLY JUST BEGUN! ...

# Did You Know?
# 你知道吗？

There are 48 000 different species of mites on Earth. Many eat plants, some tunnel through leaves, others feed on animals such as birds and deer, and some infest other insects like bees.

地球上有48 000种螨虫。很多螨虫吃植物，有些啄食叶子，另外一些以鸟、鹿等动物为食，还有些寄生于蜜蜂等其他昆虫身上。

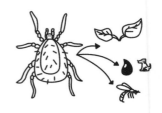

Dust mites live in houses, especially in warm places like beds. They feed on the dead cells of human skin. Other types feed on pollen and pet dander.

尘螨生长在房间里，特别喜欢待在温暖的地方，比如床上。它们以人皮肤的死细胞为食。另一些种类的螨虫以花粉和宠物毛屑为食。

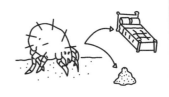

Dust mites are only half a millimetre in size. Even though mites do not bite, sting or burrow, their droppings is the cause of allergies, asthma, and eczema. The longer we are exposed to mites, the more sensitive we become.

尘螨只有半毫米大。尽管螨虫不咬、不叮、不打洞，但它们的粪便是过敏、哮喘和湿疹的病原。人接触螨虫的时间越长，就会变得越敏感。

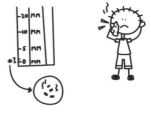

Mites are members of the spider family. Nearly 100 000 mites can live in one square metre of carpet. A person loses about 1 g of dead skin every day.

螨虫是蜘蛛家族的成员。1平方米地毯能住10万只螨虫。一个人一天会脱掉大约1克死皮。

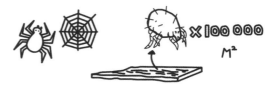

Dust mites do not like cold and dry weather. Temperatures under 20 °C and humidity below 50% lead to a decrease in the mite population. Their greatest enemy is sunlight.

尘螨不喜欢冷而干的天气。温度低于20℃，湿度低于50%时，螨虫数量会减少。它们最大的敌人是阳光。

The male dust mite lives for 10-19 days, while a female will live for up to 70 days. This gives her enough time to produce 100 eggs, and 2 000 droppings or faecal particles (bits of waste expelled from her body).

雄性尘螨能活10～19日，而雌性尘螨最多能活70天。这使它有足够的时间产100个卵，排2000粒粪便或颗粒物（从它身体里排出的小碎片）。

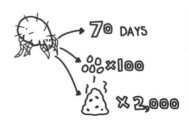

Lanolin (wool wax or wool grease) repels dust mites, giving us good reason to change from synthetic to natural fibres.

羊毛脂（羊毛蜡或羊毛干油）排斥尘螨，使我们有很好的理由用天然纤维替换合成纤维。

In the USA alone, 5 million tonnes of used clothing and 2.4 million tonnes of used carpets go to landfills every year. A 200 m² house creates 4 tonnes of demolition waste.

仅仅在美国，每年就有500万吨旧衣服和240万吨旧地毯被送进垃圾填埋场。一个200平方米的房子会产生4吨拆迁垃圾。

# Think About It
# 想一想

How does it feel to realise that there are 100 000 tiny creatures sleeping with you in your bed?

当你意识到有10万个小生物与你一起在床上睡觉,你感觉如何?

你是愿意用化学药品杀死螨虫,还是愿意住在经过设计使螨虫无法生存的屋子里?

Would you prefer to kill the mites with chemicals or would prefer to live in a house that is designed in such a way that mites cannot survive there?

When someone in the room sneezes, can you imagine that the sneeze was caused by thousands of tiny mite droppings?

如果有人在房间里打喷嚏,你能想象打喷嚏是由数千颗螨虫粪便微粒导致的吗?

太阳只是提供阳光和温暖,使植物生长吗?还是它也会造成损害呢?

Does the sun only give light and warmth and make plants grow, or does it do any damage as well?

# Do It Yourself!
# 自己动手！

It is time to have a look at the carpets in your house and at your school. Where are they placed? Are those rooms vacuum cleaned every day? And are they ventilated well? Are there many windows that can be opened and can the sun shine into the room and onto the floor? Or is there ultraviolet (UV) film on the windows to keep the sun from fading the carpets or the artwork on the walls? Once you have checked the room, take some dust scrapings from the carpet. Make sure you take dust from the deepest layer, where the nylon or woollen fibres are attached to the backing material. Place the dust under a microscope and watch the mites enjoy their meal.

看看你家里和学校里的地毯吧。地毯被放在哪里呢？那些房间每天都用真空吸尘器打扫吗？那里通风良好吗？有很多可以打开的窗户吗？阳光可以照到屋里、照到地板上吗？窗户上贴了防紫外线膜以防止地毯和艺术品褪色吗？等你完成了检查，取一些地毯上掉下的尘屑。确保在地毯的最深层（就是在尼龙或者羊毛织物紧贴背面材料的地方）取尘屑。把尘屑放在显微镜下，观察螨虫享受大餐的情景。

# TEACHER AND PARENT GUIDE

## 学科知识
### Academic Knowledge

| | |
|---|---|
| 生物学 | 螨虫的生命周期从虫卵开始，随后变成了有六条腿的幼虫、蛹，最后成为八条腿的成虫；螨虫与真菌之间存在共生关系：螨虫吃的多数食物是已经被真菌部分毁坏的东西；羊毛脂是产毛动物的皮脂腺产生的；羊毛脂有8000到20 000种酯类。 |
| 化学 | 生丝没有有害细菌和真菌；表层土在固碳和减缓气候变化中非常重要；从可再生资源中扩展蛋白质资源，以制造出多种多样的功能聚合物。 |
| 物理 | 紫外线是一种消毒剂，因为它能摧毁核酸以杀死微生物或者降低它们的活性；紫外线是导致褪色的主要原因，因此办公楼的玻璃都会加上保护膜以保护地毯和艺术品不褪色，但是，使用保护膜会导致细菌、真菌和螨虫的增长。 |
| 工程学 | 在地毯制造中，用黄麻、剑麻、椰壳纤维和羊毛等天然纤维代替合成材料。 |
| 经济学 | 人们对紫外线保护膜制造产业和成本更高的空气过滤系统的需求不断上升；紫外线会使汽车油漆失去光泽，因此金属被加进车漆以保持光泽。 |
| 伦理学 | 不应只聚焦于单个问题的解决（如褪色），只重视表面效果，而不考虑可能给健康带来的副作用以及由此产生的额外成本。 |
| 历史 | 1964年，尘螨被确认为主要的过敏源；古往今来，人们一直在寻找最明亮和最稳定的颜色，这使得胭脂红染料（从一种在墨西哥发现的甲虫身上提取的染料）在16世纪非常流行。 |
| 地理 | 亚特兰大是美国佐治亚州的首府。 |
| 数学 | 人们往往不考虑计算副作用（即外部化）所带来的资本成本和运转费用，如果把目前被社会承担的这些额外成本也列入原始模型一并考虑，投资回报将会减少。 |
| 生活方式 | 一些人盲目相信化学药剂是解决虫害问题的唯一方案。 |
| 社会学 | 人类对自然系统的无知以及对看不见的事物的漠视。 |
| 心理学 | 只有面对数据（如一个枕头中有10万只螨虫）和看到螨虫的照片，人们才会改变行为。 |
| 系统论 | 优先考虑健康而非外表的必要性，以及如何处理好二者之间的关系。 |

# 教师与家长指南

## 情感智慧
## Emotional Intelligence

螨虫

螨虫爱冒险，他们乐于尝试寻找新的居住地，对自己好恶态度鲜明。螨虫擅长构建论据，讨论利弊得失。他们喜欢调查研究，不只看到了事物的表面，他们乐于亲自验证事实并获取第一手信息。他们有自知之明，知道体型非常小在某种程度上是他们自我保护的利器：如果没有人能够看到他们，就没有人能伤害他们。螨虫非常清楚他们对环境的影响。他们掌握了一系列事件的逻辑：人们不希望他们的地毯褪色，这给螨虫带来意外的保护，使他们免受太阳（螨虫最糟糕的敌人）照射。螨虫们做了最坏的准备——忍受化学剂喷洒，但是他们知道作为一个家庭和一个物种，他们会继续生存下去，因为他们对化学剂已经具有耐药性，这使他们在这场斗争中占了上风。

## 艺术
## The Arts

颜色对我们非常重要。紫外线会影响颜色，使其褪色。我们鉴定一下哪种颜色能在太阳每天都照射的情况下保持鲜亮和稳定。谁是创造不褪色的稳定颜色的大师？他使用了什么材料？找到更多可以制成颜料的东西，以及那些被持续用了成百上千年的颜料。现在与你的朋友一起组织一次绘画活动吧。

# TEACHER AND PARENT GUIDE

## 思维拓展
### Systems: Making the Connections

建筑设计的重点是功能。由于制热制冷是主要的能源成本，建筑是隔热设计的，空气在建筑内循环使用，并通过过滤清除颗粒物和微生物。空调系统是用来保持恒温的，避免房间太冷或太热。吸音的地毯被用来改善对噪音和回音的控制，改善建筑物的整体氛围。特别是学校喜欢采用满铺地毯或拼接地毯来降低噪声。这些方法虽然有利于节能和降噪，却给居住者带来了患病态建筑综合症的风险。以照明营造良好氛围也已成为总体设施设计的一部分。而照明带来了额外的能源成本。建筑设计正朝着更加整体化的方向发展，将各种因素都考虑进来，包括健康。建造成本一直是考虑的重点，接下来是运转成本，最后是美学价值。很不幸，健康很少被优先考虑。促进进驻人员生产率的提高，往往是后期才想到的事，而不是建筑设计阶段的重点。最近的观察显示，很多从工程角度做出的决策，没有把居住者的健康考虑进来，例如，封闭窗户阻止新鲜空气流入。这一设计提高了能源效率。然而如果考虑到居住者的呼吸系统健康，这并非最优方案。此外，在窗户玻璃上贴上紫外线保护膜虽然有利于提高能源效率和保护房间内部设施，但是有很严重的健康隐患。这表明设计者缺乏对各种不同设计相互之间关联性的洞察力，而且把短期资本支出和年运行成本看得比居住者的长远健康更重要。人们不优先考虑健康，对决策如何影响健康也一无所知。我们需要一种更好的方案来同时达成多重目标，而无需在能源成本和健康之间寻找平衡。

## 动手能力
### Capacity to Implement

做一个成本分析。是使用地毯成本低，还是使用无螨虫问题的其他地面铺装方式低呢？地毯需要每日吸尘，不能被紫外线照射，以免掉色。试着给生活质量估价：对你而言，低哮喘风险或低皮肤过敏风险的价值是多少？你应该有能力为你的选择、你对健康与外表的估价和你的结论而辩护，这是非常重要的。如果不同小组得出了不同的结论，与你的估价比较一下。然后，与你的父母和学校校长分享你的建议。

# 教师与家长指南

## 故事灵感来自

## 雷·安德森
## Ray Anderson

雷·安德森（1934—2011）毕业于佐治亚理工学院，获工业工程学位。他在一家美国地毯龙头企业工作时，获得了地毯制造方面的第一手经验。1973年，雷创办了英特飞——一家专门制造地毯的公司。总部设在佐治亚州的英特飞公司从小做起，发展为地毯制造行业的领头羊之一，在4个国家设立工厂，生产的地毯销往100多个国家。雷提出了"零排放愿景"，承诺到2020年，该公司将通过产品和流程的再设计，消除对环境的所有负面影响，同时增加可再生材料和可再生能源的使用。雷著有《一个激进的实业家的商业经验》一书。

**更多资讯**

http://www.raycandersonfoundation.org

http://www.inc.com/magazine/20061101/green50_industrialist.com

### 图书在版编目（CIP）数据

螨虫家园：汉英对照 /（比）冈特·鲍利著；
（哥伦）凯瑟琳娜·巴赫绘；何家振译. -- 上海：学林
出版社，2016.6
（冈特生态童书. 第三辑）
ISBN 978-7-5486-1063-2

Ⅰ. ①螨… Ⅱ. ①冈… ②凯… ③何… Ⅲ. ①生态环
境－环境保护－儿童读物－汉、英 Ⅳ. ① X171.1-49

中国版本图书馆 CIP 数据核字 (2016) 第 125787 号

--------------------------------------------------

### 冈特生态童书
### 螨虫家园

| | |
|---|---|
| 作　　者—— | 冈特·鲍利 |
| 译　　者—— | 何家振 |
| 策　　划—— | 匡志强 |
| 责任编辑—— | 匡志强　蔡雪奇 |
| 装帧设计—— | 魏　来 |
| 出　　版—— | 上海世纪出版股份有限公司 学林出版社 |
| | 地　址：上海钦州南路81号　电话／传真：021-64515005 |
| | 网　址：www.xuelinpress.com |
| 发　　行—— | 上海世纪出版股份有限公司发行中心 |
| | （上海福建中路193号　网址：www.ewen.co） |
| 印　　刷—— | 上海丽佳制版印刷有限公司 |
| 开　　本—— | 710×1020　1/16 |
| 印　　张—— | 2 |
| 字　　数—— | 5万 |
| 版　　次—— | 2016年6月第1版 |
| | 2016年6月第1次印刷 |
| 书　　号—— | ISBN 978-7-5486-1063-2/G·398 |
| 定　　价—— | 10.00元 |

（如发生印刷、装订质量问题，读者可向工厂调换）

# Health 96

# 华丽的天竺葵
## Gorgeous Geraniums

**Gunter Pauli**

冈特·鲍利 著
凯瑟琳娜·巴赫 绘
何家振 译

学林出版社
www.xuelinpress.com

## 丛书编委会

主　任：贾　峰

副主任：何家振　郑立明

委　员：牛玲娟　李原原　李曙东　吴建民　彭　勇
　　　　冯　缨　靳增江

## 丛书出版委员会

主　任：段学俭

副主任：匡志强　张　蓉

成　员：叶　刚　李晓梅　魏　来　徐雅清　田振军
　　　　蔡雩奇　程　洋

特别感谢以下热心人士对译稿润色工作的支持：

姜竹青　韩　笑　贾　芳　刘　晓　张黎立　刘之杰
高　青　周依奇　彭　江　于函玉　于　哲　单　威
姚爱静　刘　洋　高　艳　孙笑非　郑莉霞　周　蕊

# 目录

| | |
|---|---|
| 华丽的天竺葵 | 4 |
| 你知道吗？ | 22 |
| 想一想 | 26 |
| 自己动手！ | 27 |
| 学科知识 | 28 |
| 情感智慧 | 29 |
| 艺术 | 29 |
| 思维拓展 | 30 |
| 动手能力 | 30 |
| 故事灵感来自 | 31 |

# Contents

| | |
|---|---|
| Gorgeous Geraniums | 4 |
| Did you know? | 22 |
| Think about it | 26 |
| Do it yourself! | 27 |
| Academic Knowledge | 28 |
| Emotional Intelligence | 29 |
| The Arts | 29 |
| Systems: Making the Connections | 30 |
| Capacity to Implement | 30 |
| This fable is inspired by | 31 |

在南非桌山脚下, 两株天竺葵正在享受着阳光。"你知道,三百多年前我们的祖先第一次被带到欧洲。"白色天竺葵说道。

粉色天竺葵回答:"是的,从那以后我们广受人们喜爱。一些天竺葵秋天枯萎了,但春天又开始发芽。人们就是喜欢我们复生的能力!"

Two geranium plants are enjoying the sun at the foot of Table Mountain in South Africa.
"You know, our ancestors were sent to Europe for the first time more than three hundred years ago," says the white geranium.
"Yes, and we've become very popular since," answers the pink geranium. "Some of us die back in autumn and then reappear in spring. People just love us for our ability to spring back to life!"

两株天竺葵正在享受着南非的阳光

Two geranium plants in sunny South Africa

桉树吸干了我们的水

Eucalyptus trees suck up all our water

"说得对，我们中的一些人确实很怕冷。我们需要充足的阳光，不能有太多树荫。待在这些高山硬叶灌木林中就挺好，这里几乎没有树木，恰好适合我们生存。"

"几乎没有树木？你没看见附近种了那么多桉树吗？它们把我们的水全吸干了！"

"You're right, some of us do suffer from the cold weather. We need a lot of sun and not too much shade, so we do well living in the fynbos, where there are hardly any trees. That suits us just fine."

"Hardly any trees? Haven't you seen how many eucalyptus trees have been planted around here? They suck up all our water?"

"嗯,是的,我看见了。人类是多么没远见啊!荷兰人把我们带到荷兰,然后英国人又把澳大利亚的桉树带到这里!"

"荷兰人把我们带到欧洲是因为我们的花很美,而且种类丰富。"

"对,不过我们还得有其他让人喜欢的理由!"

"Oh yes, I have. How short-sighted of people to do that! The Dutch took us to Holland, but then the Brits brought Australian trees here!"

"The Dutch took us to Europe because we have such a wide variety of beautiful flowers."

"That's right, but surely we should be liked for other reasons as well!"

我们还得有其他让人喜欢的理由！

We should be liked for other reasons as well!

# 呼吸问题

## Problems with their breathing

"是的！非洲人，如科伊桑人、科萨人，他们在这里居住了很久很久，久到没人记得他们是什么时候来的，他们从来不只把我们当花瓶。他们知道我们的价值。"

"开普省的很多人，特别是年轻人，都患有鼻塞或者呼吸问题，你知道吗？"

"So true! African people, like the Khoisan and the Xhosa, who have lived here for as long as anyone can remember, never look at us for our beauty only. They know that we are good for them."

"Did you know that many people, especially young people, here in the Cape have blocked noses or problems with their breathing?"

"鼻塞？他们不知道我们南非天竺葵能够治疗感冒吗？我们还能帮助治疗支气管炎甚至扁桃体炎呢。"

"什么是扁桃体炎？

"嗯，扁桃体是人喉咙后边的两个淋巴腺。它们是人们抵御呼吸系统疾病的第一道防线，需要保持完好才能保护人类的健康。如果受到感染，人就会得扁桃体炎。"

"Blocked noses? Don't they know that us Southern African geraniums can help cure colds? We are also good for bronchitis too and can even cure tonsillitis."

"What is tonsillitis?"

"Well, tonsils are two lymph glands people have at the back of their throats. As they are the first line of defence against disease, they need to be in great shape to protect a person's health. It is when these become infected, that a person gets tonsillitis."

能帮助治疗感冒和扁桃体炎

Can help cure colds and tonsillitis

用我们的叶子泡茶喝，能够减轻症状

Make tea from our leaves, it will offer relief

"这是不是说扁桃体在防止人生病的时候,它们自己也生病了?"

"恐怕是这样。如果孩子患了扁桃体炎,在急着送医院之前,家长可以用我们天竺葵的叶子泡茶给孩子喝,能减缓症状。这是最好的方法,因为我们的叶子是天然的药物,而且还免费呢!"

"Does that mean that when they prevent a person from getting sick, these tonsils get sick themselves?"

"I am afraid so. But there is something parents can do before rushing a child with tonsillitis to the doctor. If they make tea from our leaves, it will offer relief. It's the best, as it is natural and free!"

"我听说我们细小的油腺一年能产两到三次油,这种油能杀死使人类生病的细菌。"

"是的,还不止这些呢。你知道吗?我们的几滴油就能使人的头发变得美丽、闪亮,而且我们的油对干性皮肤也很好。"

"真神奇!我听说,人类还把我们用在食物中。我们能让食物变得更美味可口。"

"I've heard that we produce oil from our tiny oil glands two or three times a year and that this oil can destroy the microbes that make people sick."

"Yes, and that's not all. Did you know that a few drops of our oil give people beautiful, shiny hair? And that it also helps for dry skin?"

"Incredible! And they can use us in food too, I hear. We provide fragrance and flavour."

还把我们用在食物中

Can use us for food too

# 独特的非洲风味

Unique African taste

"就这么简单。只需要在把我们的叶子放进罐子里,加上橄榄油,放到阳光下晒两星期,然后加到沙拉里。我们提供了独特的非洲风味,可以与最好的地中海美食媲美。"

"完全正确。所以,如果只看重我们美丽的外表,岂不是很可悲吗?"

"And it is so easy. You simply put our leaves in a jar with olive oil, leave it in the sun for two weeks, and then add it to a salad. We provide a unique African taste that can compete with the best Mediterranean cuisine."

"Exactly. So, isn't it a pity that we are usually judged only on our looks?"

"就是！我们被某些人称为老鹳草，而另一些人称我们天竺葵，太可悲了，不同的叫法引起了很多混乱。更糟糕的是，在大洋的彼岸，我们被称为鹤嘴！"

"对我来说怎么称呼并不重要，只要我能以最好的方式为大家服务就好。我得为世界做点什么。"

……这仅仅是开始！……

"Absolutely. And it's also a pity that we are called geraniums by some and pelargoniums by others, which causes some confusion. And to top it all, on the other side of the ocean, we are called crane's-bills!"

"It doesn't matter to me what I'm called, as long as I can be used in the best way for all I have to offer the world."

… AND IT HAS ONLY JUST BEGUN!...

……这仅仅是开始!……

...AND IT HAS ONLY JUST BEGUN!...

# Did You Know?

## 你知道吗？

天竺葵被用来治疗支气管炎（因为它有抗菌作用），还可以被用作免疫系统的增强剂。它是祖鲁人、科萨人、科伊桑人使用的一种天然药物。

Pelargonium plants are used to treat bronchitis (as it has an antibiotic effect) and as an immune system stimulant. It is a natural remedy used by Zulu, Xhosa, and Khoisan people.

天竺葵根的提取物具有抗菌属性（预防和治疗细菌感染）和抗病毒属性（预防和治疗病毒感染），而且是一种化痰剂（帮助清除呼吸系统的粘液）。

Pelargonium root extract has anti-bacterial properties (preventing and curing bacterial infections) and anti-viral properties (preventing and curing viral infections), and is an expectorant (aiding in clearing mucus from the airways).

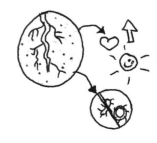

天竺葵这个名字起源于古希腊语单词"pelargós",意思是"鹳",因为其果实看起来像鹳的嘴。天竺葵的原产地是非洲南部和西南部,在其280个亚种中,90%是在这里发现的。剩下的10%是在澳大利亚发现的。

The name Pelargonium is derived from the Greek word pelargós, meaning 'stork', as its seedpod looks like a stork's bill. The Pelargonium is native to southern Africa and south-western Africa where 90% of the 280 varieties are found. The other 10% are found in Australia.

天竺葵因为其美丽和芳香而被种植。天竺葵的叶子和花可以食用,可被用于制作蛋糕、果冻和茶,其花瓣还可以用作天然颜料。

Pelargonium species are grown for their beauty and fragrance. The leaves and flowers are edible and are used in cakes, jellies, and teas. The petals can also be used as a natural pigment.

Geraniums disperse their seeds by catapulting them up to 10 m away. This is made possible by the shape of the seedpod, which resembles a crane's long bill. Other Pelargonium species have feathered seedpods that allow seeds to float on a breeze for up to 100 m.

老鹳草通过弹射传播种子，弹射距离可达10米。老鹳草的这种能力来源于其果实的形状，它的果实形状像鹤的长嘴。另外一些种类的天竺葵羽状的荚果能在微风中漂浮100米。

The original inhabitants of the Kalahari Desert, Namaqualand, and the Karoo were the foraging San people and the pastoral Khoi people. They also inhabited the southern and western Cape coasts. They made use of the Pelargonium for its nutritional value and health benefits.

喀拉哈里沙漠、纳马夸兰、卡鲁的原始居民是以采集为生的桑族人和以放牧为生的科伊人。他们也居住在南开普省和西开普省的海岸地区。他们利用天竺葵的营养价值和医疗作用为自己服务。

在班图语中，有非常精确的词汇为不同种类的天竺葵属植物命名。

In the Xhosa language there are very precise words such as *umtetebu*, *umuncwane*, *ibhosis*, and *iyeza lesikhalih* for naming different pelargonium species.

天竺葵属植物被顶端长有油腺的细绒毛覆盖，这些油腺发出很强的气味。天竺葵的气味有很多种，如橙子、玫瑰、杏仁、苹果、椰子、柠檬、肉豆蔻、薄荷、草莓，甚至还有一种被叫做"火味"。

Pelargonium plants are covered in fine hairs that have oil glands at the end, producing strong scents. Scented geraniums range from orange to rose, almond, apple, coconut, lemon, nutmeg, peppermint, and strawberry scents, and there is even one called "fire".

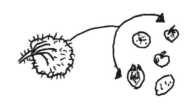

# Think About It
# 想一想

If you get sick, would you prefer to use antibiotics from a pharmaceutical company or would you first try natural products that have been used successfully for millennia?

如果你生病了，你愿意用药厂制造的抗生素，还是先尝试一下千百年来已被证明有效的天然药物？

你认为为什么人们会给同一种植物起不同的名字，如老鹳草、天竺葵和鹤嘴？

Why do you think people give the same plants different names, as has happened with the geranium, Pelargonium, and crane's-bill?

Do you like people based on their beauty, personality, friendship, or workmanship?

你喜欢一个人，是因为他（她）外表好看，还是因为他（她）的品格、手艺或者你们之间的友谊？

你愿意别人仅以外表评判你吗？

Would you like to be judged by your looks only?

Pelargonium, geranium, and crane's-bill flowers come in a variety of different colours. Nearly all these plants grow in one small region of the planet, a unique floral kingdom of the Western Cape Province in South Africa called fynbos. See how many different geranium colours you can identify by doing an Internet search for images of these plants.

天竺葵属植物的花有各种各样的颜色。几乎所有天竺葵都生长在地球上一个很小的区域，这是南非西开普省内一个独特的植物王国，叫作南非高山硬叶灌木林。上网搜索一下天竺葵的图片，看看天竺葵的花有多少种颜色。

# TEACHER AND PARENT GUIDE

## 学科知识
### Academic Knowledge

| | |
|---|---|
| 生物学 | 天竺葵属植物的果实是有五瓣双悬果的分裂果；天竺葵属植物被有油腺的细绒毛覆盖，这些绒毛主要用于防卫，因为香味可以阻止食草动物；桉树需要大量的水。 |
| 化 学 | 天竺葵属植物的花瓣含有使君子氨酸——一种能够通过干扰其神经递质使甲壳虫麻痹的化学物质；花葵苷（天竺葵宁）是一种取自猩红色天竺葵属植物的颜料。 |
| 物 理 | 带"羽毛"的天竺葵属植物的种子成熟时会从荚中喷射出来，在太阳照射下变干；荚的各部分的连结组织以不同的速度收缩，导致荚如同受到弹簧作用一样突然崩裂，将种子弹射出去。 |
| 工程学 | 为了达到最远的水平喷射距离，应该以45度角度发射。 |
| 经济学 | 天竺葵属植物的经济价值远远超过仅仅作为花卉的观赏价值；天竺葵生长在干燥的环境里，只需要很少的水维护。 |
| 伦理学 | 喜欢某人是因为他的长相或者他是谁以及他能为我们的幸福作什么贡献；世界各地的天竺葵属植物一般都是从南非引进的，但是南非人从天竺葵身上得到的经济利益远低于其他地区的人；人们对外来物种对当地生物多样性的破坏作用缺乏了解，而且很多人在知道其不利影响后，也不会做任何事消除不良后果。 |
| 历 史 | 天竺葵属植物在16世纪90年代被引进荷兰。 |
| 地 理 | 纳马夸兰高山硬叶灌木林是南非西开普地区一个独特的生物圈。 |
| 数 学 | 伽利略算出当一个抛射物体垂直运动时，重力拉动物体以每秒9.8米的加速度落向地面，而水平运动则保持不变，从而建立了精确的数学曲线，称为抛物线。 |
| 生活方式 | 欧洲城市在举行最美花卉比赛时，使用最多的就是天竺葵；空气污染主要是由柴油发动机产生的烟气和颗粒物造成的。 |
| 社会学 | 在远古时期，人们用几种不同颜色的天竺葵来测试色盲；科伊桑语变得越来越濒危，目前只有25万人说这种语言。 |
| 心理学 | 天竺葵属植物整个夏季都开花，不需要园丁们付出太大的努力，就能为人们提供美景和快乐；天竺葵的美丽对人们的情绪健康有积极效果，并且有利于增进亲密关系；天竺葵的花香能促进社交互动。 |
| 系统论 | 植物不仅是我们共享的生态系统的一部分，也是我们物质、精神和情绪健康的重要组成部分。 |

# 教师与家长指南

## 情感智慧
## Emotional Intelligence

天竺葵

天竺葵表现了对自身历史的了解。他们知道自己倍受欢迎的原因是不需要太多照料，耐干旱，还能提供美丽的花朵。他们也知道自身的不足：怕冷，依赖阳光。天竺葵承认荷兰人在将他们从南非运往欧洲并进行交易的过程中所起的作用。他们也承认英国人引进外来树种（如桉树）对生物多样性具有消极影响。天竺葵质疑他们的价值只限于"颜值"，提醒大家，南非地区传统的居民非常珍视天竺葵的多重价值。现在，天竺葵已经准备好让人类重新发现他们独特的优势。他们认为名称混乱并不要紧，真正重要的是因为他们为世界提供了美好，并且受到了赞赏。

## 艺术
## The Arts

天竺葵的花瓣不仅可以吃，而且缤纷多彩。我们烘焙一个蛋糕吧，用天竺葵的花瓣装点蛋糕！把花瓣放在蛋糕酥皮上，做成有趣的图案。比如，你可以从做一个笑脸开始。然后试着做更复杂的图案，或许可以做一朵五彩的花。使用天然的材料来赞美自然，然后，你就可以享用一个漂亮、味美的蛋糕了。

# TEACHER AND PARENT GUIDE

## 思维拓展
### Systems: Making the Connections

南非西开普地区具有独特的生物多样性。纳马夸兰高山硬叶灌木林是一个绝无仅有的植物王国和独特的生态系统，以拥有种类繁多的开花植物为傲，如今却因为外来生物（特别是桉树）的引进受到压力。按照重商主义的逻辑，桉树生长快，能为建筑和燃料提供充足的木材，因而被引进到这里。但是，桉树的引进给非洲南端脆弱的生物圈带来了严重破坏。严峻的现实是这种树的耗水量超过了生态系统的供给能力，剥夺了本地物种生存所必需的少量水分。桉树的种植在给当地环境带来压力的同时，也扼杀了与自然的生物多样性密切相关的真正的经济潜力。首先，这个地区独特的植物生物多样性只剩下了一种功能，即树木和花卉的出口。这种状况已经持续了三个世纪。其次，土著人的文化传统，包括科伊桑语和班图语，提供了数千年来积累下来的关于纳马夸兰高山硬叶灌木林中动植物知识的宝库。植物为人类提供了食物和药材，但是人类将植物的贡献降低到仅仅是好看的花朵。这使植物不能充分发挥潜力。南非西开普省失业率很高，虽然有如何利用植物的知识，但是迄今为止这些知识还没被转化成任何经济利益。

## 动手能力
### Capacity to Implement

我们需要了解植物对人类作出的特殊的贡献。查阅百科全书，列出天竺葵属植物的用途清单，通过网络搜索进行验证。这份清单将会很长。在获得总体印象后，识别出所有经过现代科学检验的医疗应用。"经过科学验证"是指，这些应用已经完成安慰剂对照研究。在这种研究中，研究者给一组被试用一种中性物质（安慰剂），把要测试的疗法实施在另一组被试身上。对比两组的测试结果，看看被测试的疗法是否比安慰剂更有效。在你的清单中有多少种天竺葵具有所声称的效果呢？检查一下那些已经经过科学验证的天竺葵属植物，是否能在当地商店里买到它们的提取物。或许，这里面有你意想不到的机遇！

# 教师与家长指南

故事灵感来自

## 特瑞达·普雷克尔
### Truida Prekel

特瑞达·普雷克尔生于南非比勒陀利亚。她在南非大学获得商务领导力硕士学位，并从事妇女管理潜能领域的开创性研究。她在鼓励妇女在职业领域承担领导责任，并促进管理的创新和改革。她的志愿工作涵盖了广泛的领域，从为社会变革而进行的音乐教育到社区安全和扫盲教育。特瑞达在勒娜特·库切的经典著作《自然的盛宴》的出版工作中起了关键作用，这本书歌颂了早期人类的饮食文化传统以及曾经与自然环境和谐相处的科伊桑人。它是同类书中第一本，而且是用石头纸印刷的。

更多资讯

https://www.teachengineering.org/view_lesson.php?url=collection/cub_/lessons/cub_catapult/cub_catapult_lesson01.xml

http://www.entente-florale.eu/results-2015/

http://www.livescience.com/14635-impression-smell-thoughts-behavior-flowers.html

### 图书在版编目（CIP）数据

华丽的天竺葵：汉英对照 /（比）冈特·鲍利著；
（哥伦）凯瑟琳娜·巴赫绘；何家振译. —— 上海：学林
出版社，2016.6
（冈特生态童书. 第三辑）
ISBN 978-7-5486-1061-8

Ⅰ. ①华… Ⅱ. ①冈… ②凯… ③何… Ⅲ. ①生态环
境－环境保护－儿童读物－汉、英 Ⅳ. ① X171.1-49

中国版本图书馆 CIP 数据核字（2016）第 125789 号

―――――――――――――――――――――――――――

ⓒ 2015 Gunter Pauli
著作权合同登记号 图字 09-2016-309 号

### 冈特生态童书
#### 华丽的天竺葵

| | | |
|---|---|---|
| 作　　者—— | 冈特·鲍利 | |
| 译　　者—— | 何家振 | |
| 策　　划—— | 匡志强 | |
| 责任编辑—— | 匡志强　蔡雪奇 | |
| 装帧设计—— | 魏　来 | |
| 出　　版—— | 上海世纪出版股份有限公司　学林出版社 | |
| | 地　址：上海钦州南路81号　电话/传真：021-64515005 | |
| | 网址：www.xuelinpress.com | |
| 发　　行—— | 上海世纪出版股份有限公司发行中心 | |
| | （上海福建中路193号　网址：www.ewen.co） | |
| 印　　刷—— | 上海丽佳制版印刷有限公司 | |
| 开　　本—— | 710×1020　1/16 | |
| 印　　张—— | 2 | |
| 字　　数—— | 5万 | |
| 版　　次—— | 2016年6月第1版 | |
| | 2016年6月第1次印刷 | |
| 书　　号—— | ISBN 978-7-5486-1061-8/G·396 | |
| 定　　价—— | 10.00元 | |

（如发生印刷、装订质量问题，读者可向工厂调换）

# Health 79

# 蚊子的叮咬

## Mosquito Bites

**Gunter Pauli**

冈特·鲍利 著
凯瑟琳娜·巴赫 绘
何家振 译

## 丛书编委会

主　任：贾　峰
副主任：何家振　郑立明
委　员：牛玲娟　李原原　李曙东　吴建民　彭　勇
　　　　冯　缨　靳增江

## 丛书出版委员会

主　任：段学俭
副主任：匡志强　张　蓉
成　员：叶　刚　李晓梅　魏　来　徐雅清　田振军
　　　　蔡雩奇　程　洋

特别感谢以下热心人士对译稿润色工作的支持：

姜竹青　韩　笑　贾　芳　刘　晓　张黎立　刘之杰
高　青　周依奇　彭　江　于函玉　于　哲　单　威
姚爱静　刘　洋　高　艳　孙笑非　郑莉霞　周　蕊

# 目录

| | |
|---|---|
| 蚊子的叮咬 | 4 |
| 你知道吗？ | 22 |
| 想一想 | 26 |
| 自己动手！ | 27 |
| 学科知识 | 28 |
| 情感智慧 | 29 |
| 艺术 | 29 |
| 思维拓展 | 30 |
| 动手能力 | 30 |
| 故事灵感来自 | 31 |

# Contents

| | |
|---|---|
| Mosquito Bites | 4 |
| Did you know? | 22 |
| Think about it | 26 |
| Do it yourself! | 27 |
| Academic Knowledge | 28 |
| Emotional Intelligence | 29 |
| The Arts | 29 |
| Systems: Making the Connections | 30 |
| Capacity to Implement | 30 |
| This fable is inspired by | 31 |

一匹斑马在水潭旁喝水，这时一只猴子停下来哀叹道："我昨晚睡不着，那些蚊子简直太讨厌啦！我想在白天打个盹都不行，它们嗡嗡的声音不停地烦我！"

"真有意思！"斑马说，"蚊子从来不来烦我。"

A zebra is taking a long drink at the waterhole when a monkey stops by and laments, "I couldn't sleep last night, you know. Those mosquitoes were simply impossible! And I cannot even catch a nap during the day either. Their buzzing irritates me endlessly!"

"Interesting," says the zebra, "I have no mosquito problems."

那些蚊子简直太过厌啦!

Those mosquitoes were simply impossible!

# 黑白相间的斑纹

Black and white stripes

"那是因为你的皮肤很厚!"
"不,不是那样。我黑白相间的斑纹能驱赶蚊子。"
"我知道你黑白相间的条纹能让你凉快,但我还是头一回听说它能驱赶蚊子。"

"It's because your skin is so thick!"
"No, it's not that. My black and white stripes keepthe mosquitoes away."
"I know your black and white stripes cool you down, but that it keeps mosquitoes away, now that is news to me."

"的确是这样,我身上有阵阵微风吹过,蚊子没法停在那里。"

"太聪明了!我也希望我有办法让蚊子不叮我。蚊子简直一无是处。"

"自然界里,任何事物都有存在的理由。"

"Sure, too many micro gusts of wind blow over my body, so they cannot land."

"That's smart! I wish I had a way of keeping mosquitoes away. They are good for nothing."

"Everything in nature has a reason to exist."

任何事物都有存在的理由

Everything has a reason to exist

老兄，在蚊子吸过我的血之后，我的确感到痒！

Boy, does it itch after they sucked my blood!

"是的,我知道。但我需要绞尽脑汁才能想到蚊子在这个世界上究竟有什么益处。我甚至在想,也许蚊子能提供微按摩。"

"天晓得。当蚊子叮你的时候你觉得疼吗?"

"不,一点都不疼。不过,老兄,在蚊子吸过我的血之后,我的确感到痒。"

"Yes, I know. But I've been racking my brain to try to imagine what good mosquitoes could do in the world. Once I thought perhaps mosquitoes offer tiny massages."

"Who knows? But have you ever felt pain when a mosquito stings you?"

"No, nothing. But, boy, does it itch after they sucked my blood!"

"你胳膊上打过针吗?"
"当然打过啊,现在根本没法逃过接种疫苗,不管你喜不喜欢。"
"你害怕打针吗?"
"害怕?我一看到针就晕了。只是想到打针,就会让我感到虚弱。"

"And have you ever had an injection in your arm?"
"Sure, there is no way to escape vaccinations these days, whether you like it or not."
"Are you scared of needles?"
"Scared? I nearly faint when I see one. Just the thought of it makes me weak."

你害怕打针吗?

Are you scared of needles?

两名护士不得不摁住我

Two nurses had to hold me down

"打针时你觉得痛吗?"

"哦,是的,很痛。有一次我还跳了起来,结果不小心被针扎破了,两名护士不得不摁住我。我妈妈觉得很难为情。"

"你想过吗,为什么蚊子咬你的时候你感觉不到,但是打针的时候却觉得疼呢?"

"And does it hurt when you get an injection?"

"Oh yes, it hurts. Once I jumped up and was accidentally scratched by a needle. Two nurses had to hold me down. My mum was so embarrassed."

"Have you ever wondered why you do not feel it when a mosquito bites you, but you do feel pain when a needle goes into your arm?"

"因为注射针更粗?"
"最细的针头几乎和蚊子的喙一样细。"
"是不是蚊子用了止疼药,让我感觉不到痛?"
"不,那是由蚊子喙的形状决定的。这完全是几何结构的问题。"

"The needle is thicker?"
"The finest needles are nearly the same size as a mosquito's proboscis."
"Mosquitoes use a painkiller so I feel nothing?"
"No, it's because of the shape of the proboscis. So it's all about geometry."

# 这完全是几何结构的问题

It's all about geometry

圆锥形的，而且还有锯齿状的表面

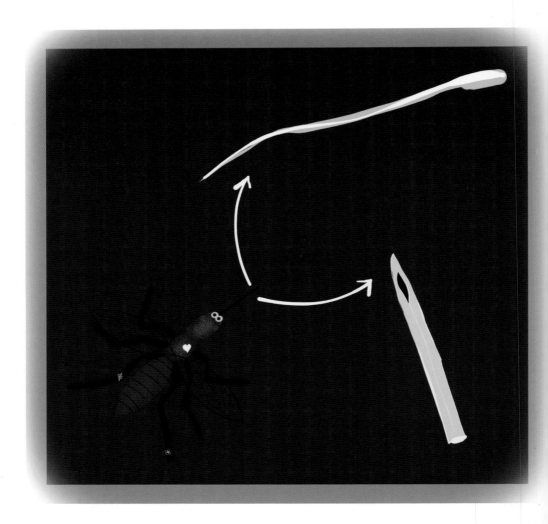

shape of a cone and surface like a saw

"你是想显摆你很聪明吗？你想聊聊数学之类的？我在数学考试时可痛苦了。"

"告诉我，蚊子的喙是什么形状的？"

"你是说它的长鼻子吗？让我想想……圆柱形？"

"不对，喙是圆锥形的，而且还有锯齿状的表面。"

"You want to be smart and talk about mathematics and so on? You know I had a hard time passing those exams."

"Tell me, what shape is a mosquito's proboscis?"

"You mean its long snout? Let me think … a cylinder?"

"No, it has the shape of a cone and has a surface like a saw."

"这是为了扎进我的皮肤？"

"不，是为了确保你什么都感觉不到！"

"为什么医生不把注射器做成那样呢？这样我妈妈再也不会感到难堪了！"

……这仅仅是开始！……

"Cutting into me?"
"No, making sure you feel nothing!"
"So why do the doctors not make use of syringes like that? My mum will never have to be embarrassed again!"
... AND IT HAS ONLY JUST BEGUN!...

……这仅仅是开始!……

…AND IT HAS ONLY JUST BEGUN!…

# Did You Know?
## 你知道吗？

Zebras' black and white stripes create temperature differences and wind on their hides, keeping them cool. Wind blowing over their skins prevents mosquitoes from landing.

斑马黑白相间的斑纹，使其皮肤表面有温差并产生了空气流动，让它们保持凉爽。在皮肤表面吹过的风还可以阻止蚊子叮在它们身上。

Mosquitoes do not sting or bite, they probe by twisting and bending their mouth parts (proboscis) in search of a blood vessel in your skin. A mosquito does not have a sharp and rigid 'needle'; its 'snout' rather operates like a tentacle with a conical end.

蚊子并不叮或者咬，而是不断扭动它们的喙来探寻你身上的血管。蚊子没有坚硬锋利的针头；它们的喙更像一个带圆锥尖的触须。

蚊子的喙尖是圆锥体形的，这就是为什么它刺进你的皮肤时，你什么都感觉不到。

The shape of the top of the mosquito's proboscis is conical, which explains why you do not feel anything when the mosquito probes.

"蚊子"这个词来自西班牙语，意思是"小飞虫"。蚊子传播疟疾、黄热病、登革热。蚊子有冬眠期或者滞育期，在太冷或者湿度太低时停止发育。

The word 'mosquito' is borrowed from Spanish, which means 'little fly'. Mosquitoes transmit malaria, yellow fever, and dengue fever. Mosquitoes hibernate, or diapause, delaying their development when the weather is too cold and the humidity too low.

Male mosquitoes live five to seven days, feeding on nectar and sugar, and pollinating flowers. The female searches for a full blood meal, which is needed to develop her eggs. Once the eggs have been laid, she starts searching for more blood.

雄性蚊子能活5到7天，以花蜜和糖为食，给花儿授粉。雌性蚊子为了蚊卵的发育必须吸食充足的血液。产卵后，它又开始寻找更多的血液。

Mosquitoes are the deadliest insects on Earth, especially as malaria vectors. Still, as aquatic insects, mosquito larvae play a key role in the food chain. Mosquito larvae are nutrient-packed snacks for fish. As adults, they represent a considerable biomass of food for birds.

蚊子是地球上最致命的昆虫，尤其是当它携带了疟疾病菌时。此外，作为水生昆虫，蚊子的幼虫是食物链中重要的一环。蚊子幼虫是鱼类的营养快餐。成年蚊子为鸟类提供了大量的生物质食物。

The estimated ratio of insects to humans is 200 million to one. There are 160 million insects per hectare of land.

据估计，昆虫与人类数量的比例是 2 亿比 1。1 公顷土地上有 1.6 亿只蚊子。

A million people die worldwide per year due to unclean syringes. The disposable syringe is only a temporary solution. Needle-free injectors are a much more sustainable solution but rely on power.

全球每年有 100 万人死于不干净的注射。一次性注射器只是临时的解决方案，无针注射才是更长久的解决方案，但是依赖于电力。

# Think About It
# 想一想

Imagine insects are the landowners and we are the tenants. Who has the greatest chance of survival?

想象一下，如果昆虫是地主，人类是佃农。谁活下来的机会最大？

你愿意了解更多蚊子在生态系统中的积极用途，还是愿意学习如何消灭蚊子？

Would you like to learn more about the positive role of mosquitoes in ecosystems, or would you rather want to learn how to kill them?

If the shape of a needle eliminates pain, would less people have a phobia for needles?

如果注射器针头的形状能够消除疼痛，恐惧打针的人会减少吗？

你相信自然界里的一切都有其存在的理由吗？

Do you think that everything in nature has a reason to exist?

# Do It Yourself!
# 自己动手！

You are in charge of an immunisation campaign. Children are scared of needles; even some of their parents are scared. How do you overcome this fear? Would you impose strong rules and control, talk about the benefits of getting the vaccine, have a smiling team to accompany doctors, buy extra thin needles that give less painful injections, have needle-free injection systems, tell funny stories while injecting to distract the patients, or play noisy TV cartoons? Discuss your ideas with your friends and come up with some strategies. Perhaps you have different strategies you would like to share with us, ones for children of different ages or even a special one for adults who are scared of needles.

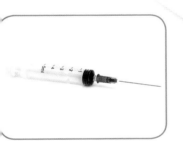

假设你在负责一项免疫接种活动。孩子们害怕打针；甚至有些父母也害怕。你如何帮助他们克服这种恐惧？强行控制？宣传接种疫苗的好处？组建微笑团队配合医生？购买超细针头以减轻注射时的疼痛？使用无针注射器？在打针时讲一些搞笑的故事分散孩子的注意力？或者播放喧闹的电视卡通？与你的朋友讨论，想出一些办法。也许你可以和我们分享一些不同的办法，包括适合不同年龄段孩子甚至是专门针对害怕打针的成年人的办法。

# TEACHER AND PARENT GUIDE

## 学科知识
### Academic Knowledge

| | |
|---|---|
| 生物学 | 一次按摩可以减轻焦虑，降低血压和心率，持续性的按摩治疗可以减轻特质性焦虑、抑郁和疼痛；蚊子喙可以作为探测血管的触角；已知的蚊子有3500种；引进食蚊鱼的生物控制方法已在南美和黑海附近的俄罗斯南部成功减少疟疾，但是在另外一些地方，食蚊鱼对当地鱼类造成了损害；蚊子能通过翅膀震动的声音辨识其家庭成员；蚊子的耳朵能听到并分辨不同的嗡嗡声，并且能够通过变声传达特殊信息；蚊子在静止的水中产卵，包括湖水和肮脏的死水。 |
| 化 学 | 按摩有助于释放身体里的化学物质；蚊子的唾液会刺激人的皮肤；蚊子唾液是一种抗阻剂，能防止喙在刺入皮肤时被阻塞；二氧化碳、辛醇、汗水以及人类身体发出的红外线对蚊子有吸引力。 |
| 物 理 | 利用白色白天反射热量，夜晚保存热量；利用黑色白天吸收热量，夜晚散发热量；黎明和傍晚，蚊子会在无风的环境里叮咬人；叮意味着注射毒液，咬意味着吸血；雌性蚊子体型更大一些，因此它们的翅膀扇动得更慢一些。 |
| 工程学 | 无针注射器通过喷嘴喷出高速液体射穿皮肤，从而避免了因不正确的针头消毒方法带来的问题；蚊子通过化学、视觉和热导感受器发现猎物。 |
| 经济学 | 治疗疟疾的成本使非洲经济增长率下降了约1.3%。 |
| 伦理学 | 自然界没有"好"和"坏"，万物皆有存在的理由；人类决定消灭某些物种（例如蚊子）是一种自大的表现；引进非本地物种以解决某一问题往往会引发很多新的问题。 |
| 历 史 | 世界上最古老的蚊子是在7900万年前的加拿大琥珀中发现的。 |
| 地 理 | 除了南极和冰岛外，世界各地都有蚊子。 |
| 数 学 | 数学模型有助于预测流行病爆发的趋势，评估蚊子控制技术的实效。 |
| 生活方式 | 每年因注射针刺伤而意外死亡的人数超过100万；在影片《侏罗纪公园》中，科学家通过抽取200万年前曾经吸食过恐龙血的蚊子身上的基因，克隆出恐龙。 |
| 社会学 | 共同威胁（如蚊子）意识动员人们团结起来应对威胁；会嗡嗡叫的蚊子不咬人的说法是对的，因为只要它在发出声响，它一定是在飞行，而在飞行中的蚊子是不可能咬人的。 |
| 心理学 | 恐惧症是对一种真实威胁的非理性恐惧反应。 |
| 系统论 | 蚊子对人的健康是有害的，但同时，它们的生物量是巨大的，它们在生态系统中的作用是明确的，它们为鱼类和鸟类提供营养；一物种对另一物种的害处，并不是消灭该物种的理由，因为那样将打破生态平衡，结果将事与愿违。 |

# 教师与家长指南

## 情感智慧
### Emotional Intelligence

**斑马**

斑马很放松。他知道自己防御蚊子的机理,并且很乐意分享他的知识,尽管这并不能解决猴子的烦恼。斑马站在哲学立场上,认为蚊子是一种讨厌的东西。他相信万物皆有存在的理由。斑马通过一连串的问话,了解到猴子害怕蚊子和打针。这显示了斑马的态度和求知欲,表明他的思维很清晰。他有一个高尚的目标:消灭痛苦。

**猴子**

猴子很紧张,睡不着觉。他渴望学习,但是缺乏知识、洞察力和想象力。他的思路不畅,想象不出蚊子对生态系统的益处。斑马的问题揭示了猴子恐惧打针的事实。猴子清醒而诚实,说出了他对蚊子的厌恶。他坦承希望找到克服非理性恐惧的办法,这并不丢面子。恐惧使他和他的家人感到难堪。这些对话使猴子更加自信,并在猴子和斑马之间建立了同理心。

## 艺术
### The Arts

蚊子具有非常与众不同的嗡鸣声。我们可以模仿它。我们可以开一个蚊子嗡鸣声的音乐会,除了我们的嘴巴,不使用任何乐器。你知道一首你和你的小伙伴都能嗡嗡哼唱的歌吗?

# TEACHER AND PARENT GUIDE

## 思维拓展
### Systems: Making the Connections

全世界每年有2.5亿人感染疟疾，100万人因疟疾死亡。因此，蚊子给人类带来沉重的医疗和财政负担。很多人会不假思索地认为应该消灭蚊子。然而，这种想法会带来一些问题。首先，只有10%的蚊子是咬人或者烦人的。这种昆虫在地球存在了上亿年，已经成为生态系统的一部分。人们并没有充分认识到蚊子在生态系统中扮演的角色以及它们作出的贡献。没有蚊子，蚊子的天敌就没有食物，某些植物就失去了为其授粉的昆虫。消灭蚊子将在北极地区造成最大的生态差异。在一年中很短的一段时间内，北极地区的蚊子数量异常多，形成一团团厚厚的蚊子云。它们是食物源，如果没有蚊子，候鸟的数量将减少50%。北美驯鹿迎着风逃离蚊子的烦扰，而逃离的路径决定了北美驯鹿的迁徙路线。如果没有蚊子的幼虫，数百种鱼类将不得不改变它们的饮食，受影响的可不仅仅是灭蚊高手食蚊鱼（它们常被引进稻田或池塘里以灭蚊）。在安第斯山脉的热带地区，蚊子是可可树的传粉者。如果没有蚊子，我们将会为吃巧克力付更多的钱。消灭一个物种也许能暂时缓解人类的痛苦，但此物种将很快被其他物种取代，而这些新物种，很有可能也会给人类带来多种疾病。人类无意间将一些的有益物种逼到濒临灭绝的边缘，比如金枪鱼、珊瑚，他们现在又想消灭蚊子。但是蚊子在自然界占据着至关重要的地位。我们必须认识到，如果蚊子被成功消灭，生态系统可能只是会遇到一些小问题，然后继续发展，但我们又会遇到其他东西，可能更好，也可能更坏。

## 动手能力
### Capacity to Implement

发明一个捕蚊器。目标是让蚊子不能逃走，但也不要杀死它们。只是捉到它们，然后用捉到的蚊子喂鱼。你要做的第一件事，是要想一想什么东西可以吸引蚊子：什么食物、气味和音乐。接着，你要找到什么能驱赶它们，换句话说，怎样做会捉不到蚊子。列出一个吸引蚊子的东西的清单，并找到现成的、低成本的东西。假如你找到了一种行得通的方式，你将很快成为一名重要的社会企业家，因为每个人都想在家里有一个捕蚊器。

故事灵感来自

# 冈野雅行
## Masayuki Okano

　　冈野雅行在 16 岁时从初中退学，加入了他父亲的公司，制造模具和压模。作为匠人的父亲认为，儿子是一位变不可能为可能的梦想家。冈野雅行发明了有更大开口的易拉罐。他相信工作上的成功不仅与能力有关，也与良好的人际关系有关。他后来又发明了无痛胰岛素注射针，直径只有接种疫苗针头的一半。这种针头直径只有 0.2 毫米，针尖是像蚊子的喙一样的圆锥形。虽然这种设计想象起来相当容易，但是如何将其规模化生产是个巨大的挑战。泰尔茂公司认为这是不可能的，但冈野雅行解决了这个问题。他的发明将数百万需要定期打针的糖尿病人从痛苦中解放出来。

更多资讯

　　https://www.youtube.com/watch?v=RLIYuXlUS3k

　　http://www.nature.com/news/2010/100721/full/466432a.html?s=news_rss

　　http://www.mosquitoreviews.com/mosquitoes-buzz-ears.html

　　http://www.jpo.go.jp/shoukai_e/soshiki_e/pdf/relay/3.pdf

图书在版编目（CIP）数据

蚊子的叮咬：汉英对照 /（比）冈特·鲍利著；
（哥伦）凯瑟琳娜·巴赫绘；何家振译. -- 上海：学林
出版社，2016.6
（冈特生态童书. 第三辑）
ISBN 978-7-5486-1060-1

Ⅰ. ①蚊… Ⅱ. ①冈… ②凯… ③何… Ⅲ. ①生态环
境－环境保护－儿童读物－汉、英 Ⅳ. ① X171.1-49

中国版本图书馆 CIP 数据核字 (2016) 第 125791 号

----

© 2015 Gunter Pauli
著作权合同登记号 图字 09-2016-309 号

## 冈特生态童书
### 蚊子的叮咬

| | | |
|---|---|---|
| 作　　者—— | 冈特·鲍利 | |
| 译　　者—— | 何家振 | |
| 策　　划—— | 匡志强 | |
| 责任编辑—— | 匡志强　蔡雩奇 | |
| 装帧设计—— | 魏　来 | |
| 出　　版—— | 上海世纪出版股份有限公司 学林出版社 | |
| | 地　址：上海钦州南路81号　　电　话/传　真：021-64515005 | |
| | 网　址：www.xuelinpress.com | |
| 发　　行—— | 上海世纪出版股份有限公司发行中心 | |
| | （上海福建中路193号 网址：www.ewen.co） | |
| 印　　刷—— | 上海丽佳制版印刷有限公司 | |
| 开　　本—— | 710×1020　1/16 | |
| 印　　张—— | 2 | |
| 字　　数—— | 5万 | |
| 版　　次—— | 2016年6月第1版 | |
| | 2016年6月第1次印刷 | |
| 书　　号—— | ISBN 978-7-5486-1060-1/G·395 | |
| 定　　价—— | 10.00元 | |

（如发生印刷、装订质量问题，读者可向工厂调换）

# Health 74

# 你真聪明
## How Smart Are You

**Gunter Pauli**

冈特·鲍利 著

凯瑟琳娜·巴赫 绘
何家振 译

学林出版社
www.xuelinpress.com

## 丛书编委会

主　　任：贾　峰

副主任：何家振　郑立明

委　　员：牛玲娟　李原原　李曙东　吴建民　彭　勇
　　　　　冯　缨　靳增江

## 丛书出版委员会

主　　任：段学俭

副主任：匡志强　张　蓉

成　　员：叶　刚　李晓梅　魏　来　徐雅清　田振军
　　　　　蔡雩奇　程　洋

特别感谢以下热心人士对译稿润色工作的支持：

姜竹青　韩　笑　贾　芳　刘　晓　张黎立　刘之杰
高　青　周依奇　彭　江　于函玉　于　哲　单　威
姚爱静　刘　洋　高　艳　孙笑非　郑莉霞　周　蕊

# 目录

| | |
|---|---|
| 你真聪明 | 4 |
| 你知道吗? | 22 |
| 想一想 | 26 |
| 自己动手! | 27 |
| 学科知识 | 28 |
| 情感智慧 | 29 |
| 艺术 | 29 |
| 思维拓展 | 30 |
| 动手能力 | 30 |
| 故事灵感来自 | 31 |

# Contents

| | |
|---|---|
| How Smart Are You | 4 |
| Did you know? | 22 |
| Think about it | 26 |
| Do it yourself! | 27 |
| Academic Knowledge | 28 |
| Emotional Intelligence | 29 |
| The Arts | 29 |
| Systems: Making the Connections | 30 |
| Capacity to Implement | 30 |
| This fable is inspired by | 31 |

一只章鱼正在抱怨有些人喜欢活吃他们。一只牡蛎坐在一旁，倾听章鱼冗长枯燥的抱怨。

"你以为只有你们章鱼会被活吃吗？"牡蛎问道，"我们也经常被人活吃呢。"

An octopus is complaining that some people like to eat him alive. An oyster sitting nearby listens to this litany of complaints from the octopus.

"So you think you are the only one eaten alive?" asks the oyster. "I am also devoured by people while I am still kicking."

一只章鱼正在抱怨

An octopus is complaining

你真的有很多腿啊!

you do have an awful lot of legs!

"我不知道,为什么把我杀死,而且在我的肢体还在蠕动的时候咀嚼它,会给人们带来快乐。"

"你真的有很多腿啊!"

"I don't know why it gives people joy to slaughter me and then start chewing on my limbs while I am still moving them," complains the octopus.

"You do have an awful lot of legs!"

"我原谅你把我的胳膊说成'腿'。我其实是有四对胳膊,就是一共八条胳膊。"

"你被认为是这附近最聪明的动物之一!"

"我并不认为我有那么聪明,但是如果有只螃蟹藏在箱子里,人们应该知道我会钻进那个箱子。如果我知道船上装满了螃蟹,我会跳上那条船——当然了,还会吃掉那些螃蟹。"

"You are forgiven for thinking that I have legs. I really have four pairs of arms. So that makes it eight arms in total."

"And you are supposed to be one of the smartest animals around!"

"I don't think I'm that smart, but if there is a crab in a box then people had better know that I will get inside that box. I will even jump on a boat when I know it's full of crabs – and eat them, of course."

我其实是有四对胳膊

I really have four pairs of arms

如果你被困在箱子里怎么办?

And what if you are trapped in a box?

"如果你被困在箱子里怎么办？"

"只要我能发现一个2厘米的洞，我就能钻出来！"

"不可能，你太大了！如果你试图钻出来，你的背部会被挤断的。"

"And what if you are trapped in a box?"
"If I can find a 2 centimetre hole, then I will get out!"
"Impossible, you are way too big! And if you ever try you will break your back."

"我根本没有脊骨——这就是为什么我会如此柔软。"

"你不像我一样有很硬的外壳。那你是咬个洞钻出来吗？"

"虽然我的舌头上长满了牙齿，但我不是靠它出来的。只要能将我的胳膊一条一条挤过去，我就可以从任何空隙钻出来。"

"可是，你的头怎么办呢？"

"I do not have a backbone – that's what makes me so flexible."

"You do not have a shell like me either. So will you bite your way through it?"

"My tongue is full of teeth, but I do not need them for that. I can squeeze through any space provided I can get one arm through after the other."

"And what about your head?"

我的舌头长满了牙齿！

My tongue is full of teeth!

我的头没有问题!

My head is no problem!

"我的头没有问题,别忘了我没有脊椎、没有颅骨。我得确保我的三颗心脏折叠好,只要它们能通过就没问题了。"

"你一定是在开玩笑吧,没人有三颗心脏。"

"My head is no problem; remember I have no back and no skull. I must simply ensure that my three hearts can be folded so that they get through."

"You've got to be joking, no one has three hearts."

"你为什么不相信呢？世界之大，无奇不有，总会有些例外。我需要一个额外的心脏给我的八条胳膊供血。"

"一颗心脏专门为你的胳膊供血？那另外两颗心脏是干吗的？"

"我的双鳃。"

"对不起，我只是一只牡蛎——我不明白，为什么你的两个鳃需要两个心脏？"

"Why can't you believe this? Nature is full of surprises, and there are always exceptions. I need an extra heart to pump blood through my eight arms."

"An extra one just for your arms? And what are the other two hearts used for?"

"My gills."

"Sorry, but I am just an oyster – why do you need two hearts for two gills?"

世界之大，无奇不有！

Nature is full of surprises!

我的鳃就是我的肺

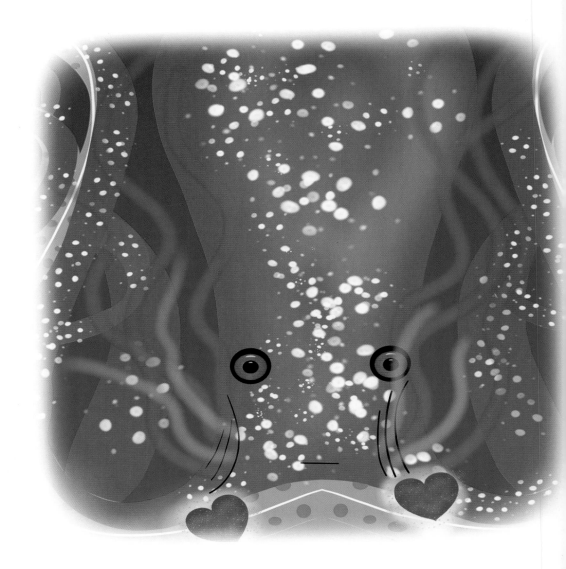

My gills are my lungs

"我的鳃就是我的肺。我从水中吸收氧气，再把二氧化碳排进水中。这需要很多能量，特别是当我在4000米深的海底游动时，那里的水中几乎没有氧气。"

"在水中呼吸一点也不稀奇，我们都是在水里呼吸的。但是，如果你断掉一条胳膊，真的还会再长出来吗？"

"My gills are my lungs. I take oxygen out of the water, and pump carbon dioxide back into the water. This requires a lot of energy, especially when you are swimming 4 000 metres deep, where there is hardly any oxygen in the water."

"Breathing in water is nothing special, we all do that. But is it true that if you rip off one of your arms it will grow again?"

"当然,而且在遇到危险时,我还会变颜色呢。如果需要,我们还能改变水的颜色,以便逃跑。"

"难怪人们那么喜欢活吃你呢!"

"人们最好明白,即使将我切碎,我仍然能钻回他们的喉咙。他们甚至会因此窒息!我已经决心要反抗到底。"

……这仅仅是开始!……

"Sure, and when there is danger around, I will change colour. If we need to, we can change the colour of the water too, allowing us to escape."

"You sure make these people run for the pleasure of eating you alive!"

"People had better realise that even if they cut me up, I will crawl back up their throat. They may even choke! I am ready to fight until the very bitter end."

... AND IT HAS ONLY JUST BEGUN!...

……这仅仅是开始!……

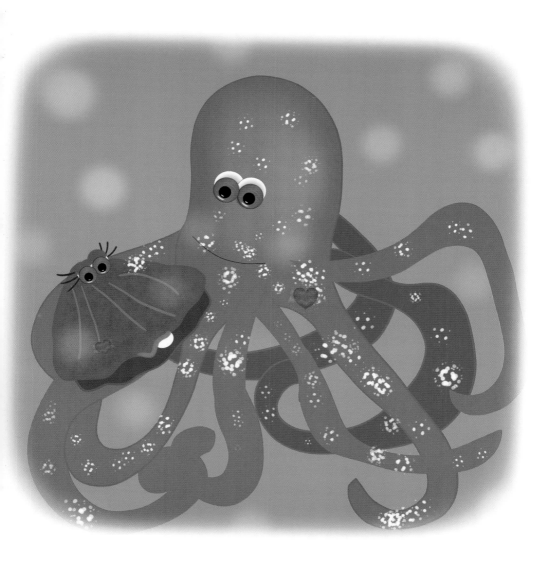

...AND IT HAS ONLY JUST BEGUN!...

# Did You Know?
## 你知道吗？

有289种不同的章鱼，它们全都居住在海水之中。最大的章鱼重300千克，而最小的章鱼只有1克重。它们都能喷出毒汁。

There are 289 different types of octopus, all living in salt water. The biggest one weighs up to 300 kg and the smallest only weighs one gram. All of them have toxic ink.

因为章鱼没有颅骨，它的牙不是长在下巴上，而是长在舌头上。章鱼不吃食物也能生存6个月。

The teeth of the octopus are not on its jaw, since it has no skull. They are located on its tongue. An octopus can remain without food for 6 months.

Octopuses' favourite food is crabs. They crush it to pieces in their beaks, which are made of keratin – the same material as human fingernails.

章鱼最喜欢的食物是螃蟹，它们用腭把螃蟹碾碎。章鱼的腭是角质的，成分类似于人类的指甲。

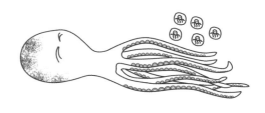

The eyes of an octopus focus like a camera, moving the lens in and out. They have no blind spot, and have polarised vision, meaning that they can clearly see jellyfish, which we can hardly see.

章鱼眼睛就像一个照相机镜头，能够伸缩调焦。章鱼眼睛没有盲点，有极好的视力，能够清楚地看到人类很难看到的水母。

Octopuses have excellent memory, and their brain has folds, indicating its complexity. The octopus can unscrew jars to get to the crabs inside.

章鱼有很强的记忆力，它们的大脑具有沟回，这意味着它们的大脑很复杂。章鱼能够拧开坛子的盖子，去抓里面的螃蟹。

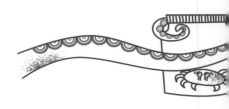

The octopus is a very efficient feeder. Multiply its weight by three and then you know how much it has eaten during its lifetime.

章鱼是非常高效的吃货。吃饱时其体重会增加三倍。由此你可以想象，章鱼的一生吃掉了多少东西。

章鱼的生命周期很短,寿命最长的章鱼只能活5年。章鱼生育后就停止进食,直到饿死,在它的蛋孵出后不久就死掉了。

The octopus has a short life span, with a maximum of 5 years. The octopus stops eating after breeding and starves to death, dying shortly after its eggs have hatched.

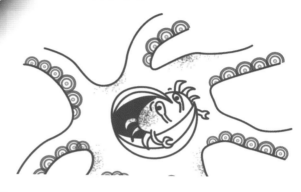

章鱼保罗准确地预测了2010年南非足球世界杯赛中德国队全部7场比赛的成绩,并且准确预测出西班牙队是冠军。在那之后不久他死了。

Paul the Octopus correctly predicted the winner of all 7 German matches during the 2010 Football World Cup in South Africa, and also correctly predicted Spain as the overall winner. He died shortly after that.

# Think About It 想一想

Can you imagine the feeling of a piece of an octopus's arm trying to crawl up someone's throat after it had been eaten alive?

你能想象一只章鱼被活吃后，他的一段胳膊爬过人的喉咙的那种感觉吗？

如果你的身体或头颅没有骨头，它们会是什么样子？

What would your body and head look like without bones or a skull?

What other surprises can you imagine from Nature that has, through evolution, developed an animal with eight arms and three hearts?

通过自然进化，一种动物有八条胳膊和三个心脏，你能想象还有更令人惊奇的事情吗？

你喜欢吃多新鲜的食物：刚出炉的还是活鲜的？

How fresh do you want to eat food: oven fresh or living fresh?

# Do It Yourself! 自己动手!

Start a search on the internet and look for cultures and traditions around the world where live animals are eaten. The goal is not to accuse or to judge, but to understand that eating live octopus is not an isolated case and that many culinary exploits are based on eating live animals.

在网上发起一个调查,看看世界上哪种文化传统里有活吃动物的习俗。目的不是起诉或审判,而是要理解生吃章鱼并不是孤立的案例,很多烹饪上的开拓是建立在生吃动物的基础上的。

# TEACHER AND PARENT GUIDE

## 学科知识
### Academic Knowledge

| | |
|---|---|
| 生物学 | 在自然界,有规则就有例外;章鱼的眼睛与人的眼睛相似,有角膜、虹膜、晶状体、玻璃体液和视网膜;章鱼是食肉动物,有时候甚至会吃它们的同类;再生能力:海星的肢体切段能够再生一个新海星,但是章鱼的肢体切段只能再生一个新肢体;章鱼有5亿个大脑神经细胞,而人类有1千亿个大脑神经细胞;章鱼并不从父母那里学习知识,小章鱼出生后它们的父母就死了;章鱼的胳膊是兼具嗅觉和触觉功能的器官,它们可以用胳膊闻气味。 |
| 化学 | 章鱼的墨汁是由黑色素组成的,与头发和皮肤中的黑色是同样的色素;章鱼的食谱是低热量的海鲜,不但富含钙、钾、磷、硒等元素,而且还含有Ω3脂肪酸;章鱼有能折射光线的色素,能使章鱼具有彩虹般的光彩;有些种类的章鱼具有由神经系统控制的发光器,发出光亮的持续时间从不到一秒到几分钟;由于章鱼血液里有一种叫血蓝蛋白的含有铜元素的色素,章鱼的血液是蓝色的。 |
| 物理 | 章鱼有四对胳膊,需要提供额外的运送血液的压力,章鱼的第三个心脏专职负责向全身提供血液的常规功能;章鱼的幼体漂浮在海面上,捕食它们发现的任何食物,等它长得足够重,就沉入海底继续生长;章鱼不仅能通过色素模拟颜色,而且还能根据环境变化身体形状,把自己隐藏起来,比如可以变得像一块被海藻覆盖的石头。 |
| 工程学 | 通过氧化作用将受污染的水净化成灌溉用水或饮用水。 |
| 经济学 | 鲜鱼和甲壳纲动物市场是价值1250亿美元的全球性大市场,而藻类市场只有20亿美元,但鱼类市场面临压力,而藻类市场仍在增长。 |
| 伦理学 | 什么论据能够支持人类活吃鱼类的欲望?如果我们反对活吃动物,那么我们也应该反对生吃蔬菜和水果吗? |
| 生活方式 | 韩式活章鱼(一种受欢迎的韩国菜)是把活章鱼切成小块,上菜的时候,章鱼的爪子仍在扭动;日本的活鱼料理是一种由活的章鱼、虾、龙虾切成的生鱼片;在西方文化里,牡蛎是生吃的,比利时人过去常喝杯子里有条活鱼的啤酒。 |
| 社会学 | 活吃动物的传统在亚洲很广泛,但是在欧洲,由于动物权利意识的出现,这种传统不再盛行。 |
| 心理学 | 多数人不仅吃动物肉,而且也关心动物权利,经受着"食肉悖论"的心理挣扎。 |
| 系统论 | 认识到我们被有意识的生命所包围,不只是人类才有知觉。 |

# 教师与家长指南

## 情感智慧
### Emotional Intelligence

**牡 蛎**　　牡蛎没有强化章鱼作为受害者的感觉；相反她证实了其他生物也正遭遇完全相同的境遇。当章鱼坚持强调可怕的未来时，牡蛎一开始嘲笑章鱼有很多胳膊，但紧接着就表达了尊重，最终发展为对其独特性的钦佩。这导致了一场对话。牡蛎对这位朋友越来越感兴趣，对他不断揭示的事实惊讶不已，章鱼以一个关键人物的形象出现，如同科幻电影里的主角一样。牡蛎最终把章鱼视为一个罗宾汉似的英雄，能够与捕猎者一较高下。

**章 鱼**　　章鱼一开始充当了受害者的角色，抱怨所遭受的各种虐待。牡蛎直面章鱼，显露出某种现实主义态度。但听了牡蛎赞美其"腿"的笑话，章鱼并不受用，他平静地纠正牡蛎，但也维护了牡蛎的自尊。当牡蛎奉承章鱼时，章鱼的第一反应是少谈自己的智慧，而是更多地展示各种实用本领。他通过一系列令人惊奇的信息说服牡蛎。他明确清晰地述说他的每一个独有的特征。当牡蛎对章鱼的胳膊再生能力感到好奇时，他不以为然，而是强调了他变化颜色和形体的能力，他把这视为自己最惊人的特征。最终他以轻蔑的口气，完成了从受害者角色到拥有优势地位的角色的转换。

## 艺术
### The Arts

　　画一幅"章鱼风格"的油画。想象你是一只章鱼，你可以随意在白色帆布上喷洒黑色墨水。你将怎样喷墨呢？是用一个细小的高压水针画，还是直接在帆布上泼墨？画几幅水墨画，心理分析师能够从你的画风里了解到你的一些重要的性格特质。当你选择做何种类型的喷墨画家时，你也传递出了你是怎样的人。

# TEACHER AND PARENT GUIDE

## 思维拓展
### Systems: Making the Connections

　　当学习物理学时，我们发现所有定律几乎都没有例外；学习化学时，我们发现很多情况下，化学反应取决于温度、气压和使用的催化剂；在学习生物学时，我们又发现似乎每件事物都是例外，没有任何事物是完全相同的。不知何故，自然界已经从原始生命、从几个简单的细胞进化为一连串独特的生态系统，有上亿种不同形态的生命形式。生物多样性是神奇的，而其细节的发现，唤起了人们对与我们共享这颗星球的所有生命形态深深的敬意和钦佩。然而，人类似乎经常是无知的，没有觉察到这些难以置信的、好像直接来自科幻小说的生命力量。基于这样的背景，活吃动物有违人类不同于其他一切生命形式的基本价值观。对所有生命形态的赞美和尊重意识正在不断提高，同时却有很多人认为动物必须服从于人类。但这就赋予我们虐待动物的特权吗？也许虐待有轻有重，但必须确保所有生物都能有尊严地生存和繁衍。只有每种生物相互尊重和欣赏彼此在这颗星球上的独特性，万物才能生生不息地循环。

## 动手能力
### Capacity to Implement

　　生命的关键是保持正能量。有太多的理由让人觉得自己是受害者，哀叹自己的不幸。创造新的机会和学习新的知识是很重要的，甚至包括从我们不喜欢的或者让我们感到痛苦的事件中学习。我们有与时俱进的能力，但任何时候，我们都要尊重差异和传统。或许此时，你应该停下来仔细看看你的饮食，看清你在吃什么。你有过"食肉悖论"吗？在你内心，是怎样调节对动物的爱与为获取动物蛋白而吃动物肉的矛盾的？你可以先思考一下这些问题，归纳出正面和反面的理由，然后再与朋友进行辩论，分享你的观点。不管结论是什么，就"食肉悖论"挑战你父母，找到一种你良心能够接受，而且吃得舒服的饮食结构。

# 教师与家长指南

故事灵感来自

## 查尔斯·范·德·阿埃让
## Charles van der Haegen

查尔斯是十个孩子的父亲，他的妻子为亲生和收养的子女的美好未来奉献了自己的全部身心，这一行为对查尔斯的思想和工作产生了很大影响。他毕业于商务工程专业，拥有MBA学位。后来，他进入了医疗器械产业。1982年，作为总经理的他被要求关闭仪器仪表公司时，他拒绝了，并把公司转变为比利时首家机器人公司。他从商务经理人转变为投资企业家，带着没有光明前途的公司进入一个竞争激烈的行业。他的职业生涯中不仅有成功，而且也有失败。正是他顽强的求存精神，让他在困难重重的黑暗中看到光明。查尔斯在商务和社会活动领域贡献着他丰富的经验，特别是作为传递者，在欧洲推行蓝色经济项目。

更多资讯

www.orma.com/sea-life/octopus-facts

www.schweisfurth.de/english-version.html

图书在版编目（CIP）数据

你真聪明：汉英对照 /（比）冈特·鲍利著；
（哥伦）凯瑟琳娜·巴赫绘；何家振译. —— 上海：学林
出版社，2016.6
（冈特生态童书. 第三辑）
ISBN 978-7-5486-1059-5

Ⅰ. ①你… Ⅱ. ①冈… ②凯… ③何… Ⅲ. ①生态环
境－环境保护－儿童读物－汉、英 Ⅳ. ① X171.1-49

中国版本图书馆 CIP 数据核字 (2016) 第 125792 号

————————————————————————

© 2015 Gunter Pauli
著作权合同登记号 图字 09-2016-309 号

## 冈特生态童书
### 你真聪明

| | |
|---|---|
| 作　　者—— | 冈特·鲍利 |
| 译　　者—— | 何家振 |
| 策　　划—— | 匡志强 |
| 责任编辑—— | 匡志强　蔡雪奇 |
| 装帧设计—— | 魏　来 |
| 出　　版—— | 上海世纪出版股份有限公司 学林出版社 |
| | 地　址：上海钦州南路81号　电话/传真：021-64515005 |
| | 网址：www.xuelinpress.com |
| 发　　行—— | 上海世纪出版股份有限公司发行中心 |
| | （上海福建中路193号 网址：www.ewen.co） |
| 印　　刷—— | 上海丽佳制版印刷有限公司 |
| 开　　本—— | 710×1020　1/16 |
| 印　　张—— | 2 |
| 字　　数—— | 5万 |
| 版　　次—— | 2016年6月第1版 |
| | 2016年6月第1次印刷 |
| 书　　号—— | ISBN 978-7-5486-1059-5/G·394 |
| 定　　价—— | 10.00元 |

（如发生印刷、装订质量问题，读者可向工厂调换）

# Health 99

# 根与芽
## Roots and Shoots

Gunter Pauli

冈特·鲍利 著
凯瑟琳娜·巴赫 绘
何家振 译

学林出版社
www.xuelinpress.com

## 丛书编委会

主　任：贾　峰
副主任：何家振　郑立明
委　员：牛玲娟　李原原　李曙东　吴建民　彭　勇
　　　　冯　缨　靳增江

## 丛书出版委员会

主　任：段学俭
副主任：匡志强　张　蓉
成　员：叶　刚　李晓梅　魏　来　徐雅清　田振军
　　　　蔡雩奇　程　洋

**特别感谢以下热心人士对译稿润色工作的支持：**

姜竹青　韩　笑　贾　芳　刘　晓　张黎立　刘之杰
高　青　周依奇　彭　江　于函玉　于　哲　单　威
姚爱静　刘　洋　高　艳　孙笑非　郑莉霞　周　蕊

# 目录

| | |
|---|---|
| 根与芽 | 4 |
| 你知道吗？ | 22 |
| 想一想 | 26 |
| 自己动手！ | 27 |
| 学科知识 | 28 |
| 情感智慧 | 29 |
| 艺术 | 29 |
| 思维拓展 | 30 |
| 动手能力 | 30 |
| 故事灵感来自 | 31 |

# Contents

| | |
|---|---|
| Roots and Shoots | 4 |
| Did you know? | 22 |
| Think about it | 26 |
| Do it yourself! | 27 |
| Academic Knowledge | 28 |
| Emotional Intelligence | 29 |
| The Arts | 29 |
| Systems: Making the Connections | 30 |
| Capacity to Implement | 30 |
| This fable is inspired by | 31 |

一些细菌被困在牙齿的龋洞里,因为上面被金属盖盖住了,他们出不来。

一个细菌说:"看看这个可怕的、黑乎乎的地方,我们被困在这里了!我们只不过是想吃那些粘在牙齿上的甜食而已。"

Some bacteria are trapped in a tooth cavity under a metal cap and cannot get out.

One says, "Look at this scary, dark place where we've ended up! And all we wanted to do was to eat the sweet food rests sticking to this tooth."

一些细菌被困在牙齿的龋洞里

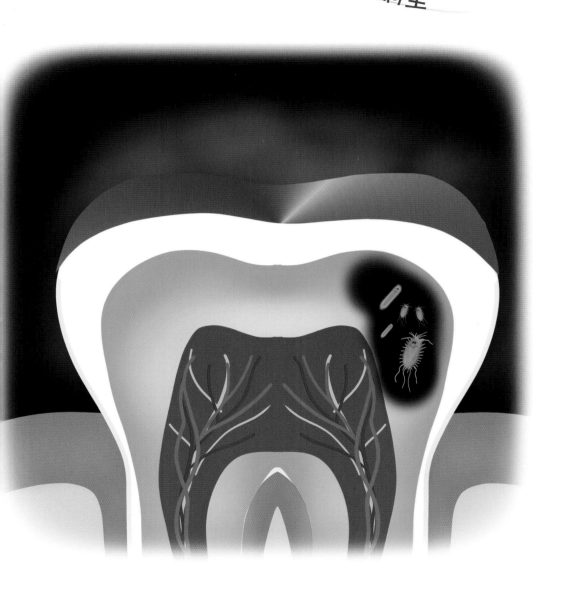

Some bacteria are trapped in a tooth cavity

这里除了骨头，什么都没有

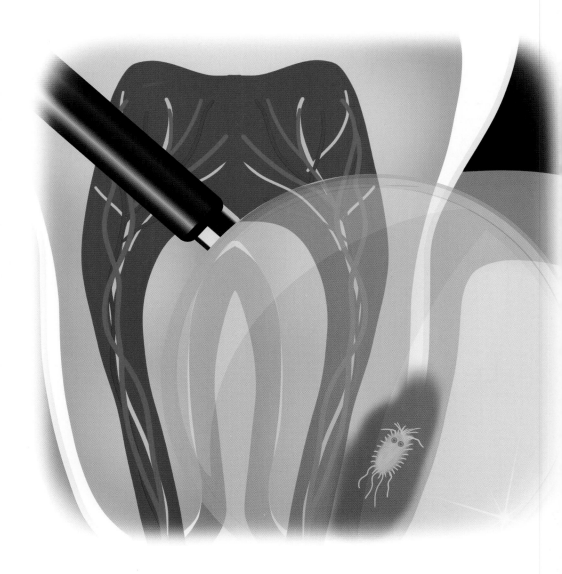

Nothing but bone around here

"现在我们被困在牙齿里这么深的地方。这里好像除了骨头什么都没有。"另一个细菌说。

"那一定是因为我们太贪吃了,想把能找到的东西都吃掉。现在可好,只剩下骨头和金属这些硬邦邦的东西了。"

"Now we are stuck deep inside this tooth. It feels like there's nothing but bone around here," says another.

"We must've been too greedy trying to eat everything we could find. And now all that's left is this hard stuff – bone and metal."

"你总是起劲儿地挖,把所有龋洞里的糖渣挖得一点儿不剩!"他的朋友指责道。"我们一直挖呀挖,到处打洞,几乎所有牙齿都被我们打通了。"

"嗯,这些洞被称为龋洞。我们通常都会把洞打得这么深。你大概不会相信,能做到早餐前刷刷牙,把我们踢出来的人真不多。"

"You're the one that was so keen to get to the last bit of sugar out of every hole!" his friend accuses him. "We've been digging and digging through everything until there was hardly any tooth left."

"Well, these holes are called cavities. It's normal for us to tunnel our way through here. You won't believe how few people brush their teeth early in the morning to kick us out before they have breakfast."

刷牙的人真不多

How few people brush their teeth

我喜欢这些吃剩的饭菜

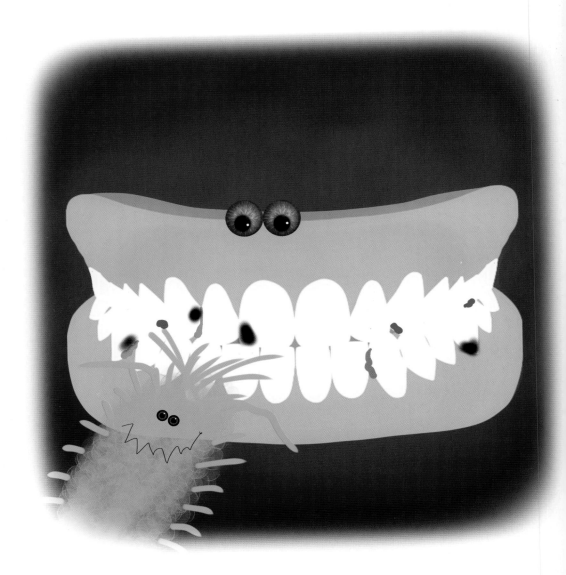

I love these left-overs

"那有什么不好呢?你不想吃他们吃剩的饭菜吗?"

"我喜欢这些吃剩的饭菜。不过,这里的细菌太多了,我们得快点儿吃。"

"So what's wrong with that? Don't you want to eat all their left-overs?"

"I love these left-overs. But as there are so many of us around here, we'd better get everything as quickly as we can."

"我知道这里的细菌很多。"他的朋友答道,"我听说,一个不注意口腔卫生的人嘴里的细菌数量比地球上的总人口还多。"

"哈哈,这就是不注意口腔卫生的人口臭严重的原因!因为我们在他的舌头、牙龈和牙齿上举办晚会呢。"

"I know that there are a lot of us," replies his friend. "I've been told that there are more bacteria in the mouth of a person who neglects his teeth than there are people on earth."

"Ha-ha, that must be why this guy has such bad breath! It's because we're having a party on his tongue, gums, and teeth."

在他的舌头、牙龈和牙齿上举办晚会

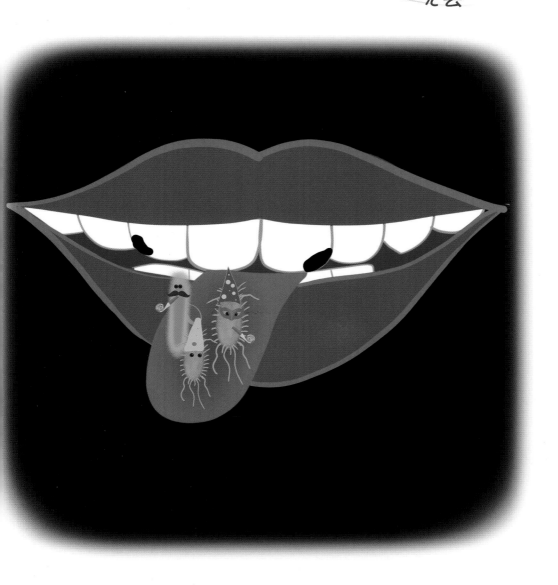

Having a party on his tongue, gums and teeth

坐在牙根管里

Sitting in a root canal

"但是自从我们被逼到这个洞底,快乐的时光好像彻底结束了。"他的朋友回应道。

"我觉得我们正坐在一个牙根管里,以前这里可是牙神经待的地方。或许这是一个移植管,我分辨不出来。我能肯定的是这里根本没有糖。"

"But party time seems to be over for us since we've been pushed deep down this hole," replies his friend.

"I think we're sitting in a root canal, where the nerve used to be. Or maybe it is an implant. I cannot tell. What I do know is that there's certainly no sugar down here."

"我想,牙主人的牙被我们吃掉了这么多,他一定疼得厉害,所以牙医不得不把他的牙神经抽出来。"

"我们肯定就是这样被逼到这里来的。"

"牙医确实想用他的刺激性化学药品杀死我们,但是你知道的,人们越想杀死我们,我们中活下来的细菌就会变得越强。"

"You know, I think we ate so much of our host's tooth away that it started hurting so badly that the dentist had to take the nerve out."

"And that must be how we got pushed in."

"The dentist did try to kill us with his harsh chemicals but, as you know, the more people try to kill us bacteria, the stronger those that survive will become."

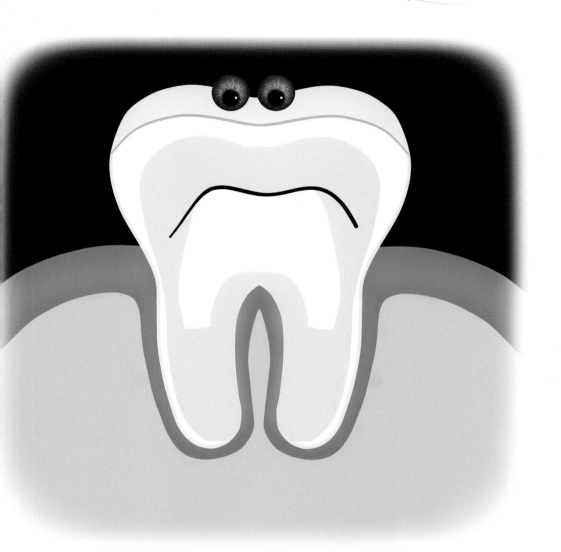

最好改变我们的饮食习惯

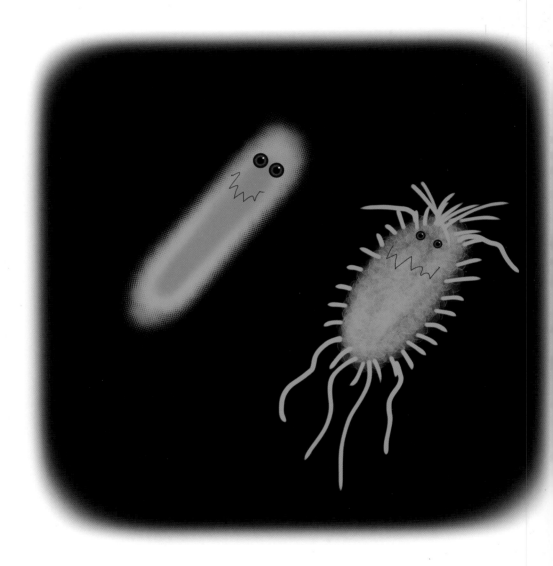

Better change our diet

"那么，你建议我们现在做点什么呢？"
"我想，我们应该改变我们的饮食习惯。"
"有什么用呢？在这里，我们无路可逃。上方的金属盖挡住了我们的去路，而且我们也没法穿过牙釉质或者牙根。"

"So what do you suggest we do now?"
"I think that we'd better change our diet."
"To what? We have no choice down here. And this metal cap above us is blocking the way out. And we cannot get out through the enamel or the roots either."

"因此我想,我们唯一的选择就是留在这里,吃掉牙齿的骨质部分,然后从牙齿的另一面逃出去。"

"那好,现在就开始享受我们的盛宴吧!"

……这仅仅是开始!……

"Then I suppose the only option we have left is to eat away at the bone and hope we can one day get out on the other side."

"And start feasting again!"

... AND IT HAS ONLY JUST BEGUN!...

……这仅仅是开始！……

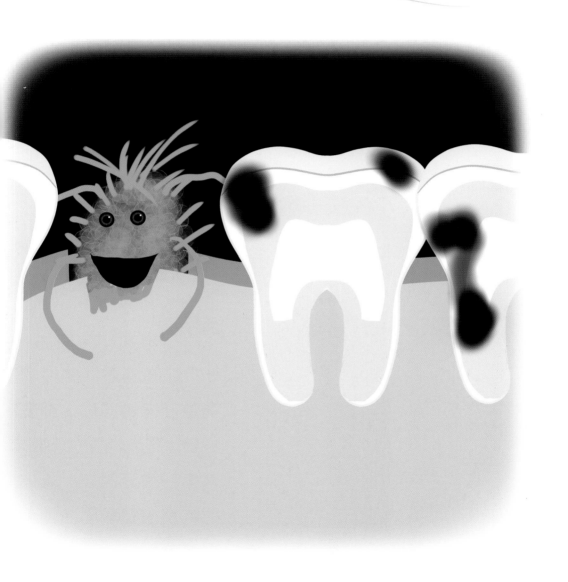

...AND IT HAS ONLY JUST BEGUN!...

# Did You Know?

## 你知道吗?

When simple sugars and carbohydrates are not cleaned off teeth, bacteria eat them and produce acids. When many bacteria feed and multiply, they form plaque. If this sticky film is not removed when it is soft, it will get hard and be tough to get rid of.

如果单糖或碳水化合物残留在牙齿上,细菌就会吃它们,并产生酸性物质,当很多细菌都来吃并开始繁殖时,牙菌斑就形成了。如果人们没有趁粘性菌膜还软的时候将其清除,它就会变硬,且很难清除。

The acids made by bacteria remove the tooth's enamel, reaching the inner layer called dentin. When bacteria reach beyond this, they devour the inner tooth pulp. The body responds to this by sending white blood cells and this may result in a tooth abscess.

细菌制造的酸性物质腐蚀了牙釉质后,细菌就会到达牙齿内层的牙质层,然后吞食牙髓质。人体对此的反应是输送白血球,这可能会导致牙周脓肿。

During a root canal treatment, the infected pulp is removed, along with the nerve. The inside is cleaned and disinfected, and then filled with a filling and sealed with a crown.

在牙根管的治疗过程中,被感染的牙髓质与牙神经一起被移除。医生在牙根管内部进行清洁并消毒,然后填上填充物,并用金属盖封上。

There are an estimated 20 billion bacteria (from 500 to 1 000 different families of bacteria) in our mouths. Some reproduce every five hours. Most of the bacteria are beneficial and aid in preventing disease.

据估计,我们的口腔里住着 200 亿个细菌(它们来自 500 到 1000 个不同的细菌家族)。有些细菌每五小时繁殖一次。大多数细菌是有益的,有助于预防疾病。

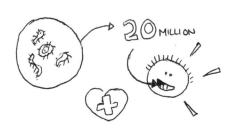

Oral bacteria are the first line of defence of our immune system. Without these good bacteria, our body would be overloaded with airborne and saliva-transferred germs. These good bacteria also control the growth of fungi.

口腔里的有益细菌是我们免疫系统的第一道防线。如果没有它们，大量的细菌将通过空气和唾液的传播进入我们体内，人体将无法承受。有益菌还能抑制真菌的增长。

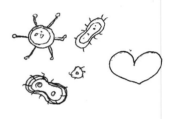

Saliva in the mouth contains the calcium ions needed to repair the minimal damage caused by bacterial acids. The protein in saliva binds bacteria, preventing it from sticking to any surface, and it is then washed away by the saliva and swallowed.

细菌产生的酸性物质会导致微创伤，唾液里的钙离子能修复这些创伤。唾液里的蛋白质能抑制细菌，防止其沾在口腔内部，然后唾液会将其冲走并吞掉。

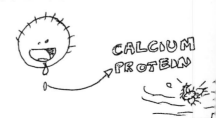

Brushing and flossing teeth prevent, or at least reduce, infection and disease. However, reducing the intake of sugars and carbohydrates in our diet has a greater influence on oral health.

刷牙或者用牙线洁牙,能够防止或减少感染和疾病。但是,减少糖和碳水化合物的摄入更为重要。

It is best to brush your teeth in the morning before breakfast, as the greatest amount of plaque is present in our mouths when we wake up. It is also easier to remove it then.

每天早餐前最好刷刷牙,因为在刚醒的时候牙菌斑最多,而且最容易清除。

# Think About It
# 想一想

Do you think you can ever get rid of all the bacteria in your mouth?

你认为你能清除口腔里所有的细菌吗?

如果你深陷一个大坑中,而且被封在里面,你是否准备为了生存而改变你的饮食习惯?

If you were trapped in a big hole that is sealed off, would you be prepared to change your diet to survive?

When there is a lot of food available, would you eat more than you usually do?

当有很多食物任你享用时,你是否比平常吃得更多呢?

你认为是细菌太贪吃,理应被困在牙齿深处,还是牙主人吃的食物有问题呢?

Do you think the bacteria were greedy and therefore to blame for their predicament or did the person eat the wrong food?

Do you have your microscope ready? Ask a parent or teacher to assist you in scraping some plaque off your teeth with a metal scraper. Put it on a glass slide and have a look through the microscope at the live bacteria moving around. Is the sight of these micro-organisms, fresh off your teeth, motivating you to take better oral care?

准备好你的显微镜了吗?请一位家长或老师用金属刮刀帮你刮掉牙齿上的牙菌斑。将牙菌斑放在玻璃片上,通过显微镜观察活动的细菌。观察这些刚从你牙齿上清除下来的微生物,会促使你做好口腔护理吗?

# TEACHER AND PARENT GUIDE

## 学科知识
### Academic Knowledge

| | |
|---|---|
| 生物学 | 口腔里最常见的两种有害菌是变异链球菌和牙龈卟啉单胞菌；细菌通过二分裂法生长；因为没有核，细菌会快速地发生突变；不好的气味来自舌头上的小裂隙，导致口臭的细菌在那里进行繁殖；多数产生口臭的细菌是厌氧菌，它们在人们闭上嘴睡觉的时候生长得最快。 |
| 化 学 | 溶菌酶是人唾液中的抗菌物质，它能杀死细菌，或者至少可以抑制其活性；唾液中含有磷酸盐和钙离子，能够帮助修复由细菌酸引起的牙齿微损伤；我们唾液中的维生素K可以防止酸性物质的形成，人们在发酵食物（如德国泡菜）中发现了维生素K，蛋黄和食草动物的制品中也有维生素K；唾液酸性越强，龋齿就越多。 |
| 物 理 | 人在一天中通过咀嚼食物会产生15 000次电脉冲，可使身体保持活力，因为通过神经网络，每颗牙齿都与特定器官相连。 |
| 工程学 | 牙科医生不断探索贵金属合金和新瓷材料的结合，以提高填充材料和移植物的强度和弹力。 |
| 经济学 | 口腔卫生不良的代价很大，它会增加患心脏病、中风、痴呆、呼吸系统疾病、糖尿病和癌症的风险。 |
| 伦理学 | 贪婪对你和他人的生活会造成很大的影响；人类倾向于摧毁他们不理解的东西，结果导致有益菌被杀死。 |
| 历 史 | 荷兰科学家安东尼·范·列文虎克是第一位用显微镜观察细菌的人，当时他在研究牙菌斑；3500年前人类就开始使用牙刷了；牙刷的工业化生产始于15世纪的中国；在19世纪，家庭制作的牙膏中含有木炭粉。 |
| 地 理 | 波兰儿童12岁之前平均掉4颗恒牙，而相同年龄的德国和英国儿童掉牙不超过一颗。 |
| 数 学 | 口腔中的细菌在很短的时间内以指数方式增长。 |
| 生活方式 | 除了有规律地刷牙，改变饮食也是对牙齿很好的保护：少吃单糖和碳水化合物；牙刷已经成为一次性物品；牙刷毛是由尼龙制成的，降解非常慢，半衰期长达数百年。 |
| 社会学 | 总体而言，中等收入家庭成员的牙齿健康最糟糕，而贫困家庭和富裕家庭稍好一些。 |
| 心理学 | 蛀牙的发展与消极情绪、压力之间的关系。 |
| 系统论 | 少吃精加工的糖和精白面粉可以降低蛀牙的概率；免疫系统的第一道防线是口腔里的有益菌和唾液。 |

# 教师与家长指南

## 情感智慧
## Emotional Intelligence

细菌

细菌肆无忌惮，充满自信。它们认识到其生存环境已经发生了根本变化。它们质疑自身的行为，并且想知道，自己忘乎所以地吃掉所有能吃的东西，是否太贪婪了。细菌很清楚人们知道如何摆脱它们，但是多数人懒得去做。细菌们明白发生了什么，并决心生存下去，即使这意味着要改变饮食习惯。它们怀抱希望，相信很快就能恢复原有的习惯，继续它们数百万年来的生活方式。

## 艺术
## The Arts

如何处理旧牙刷？很多旧牙刷进了垃圾填埋场。让我们收集旧牙刷，并将它们变成油画刷！我们用牙刷画幅画吧。就拿旧牙刷作为你的画笔，蘸上油画颜料，在厚纸板或者帆布上画吧。这些旧牙刷可能没剩多少毛可以刷牙了，但是用来当油画刷已经足够了。这给你提供了一个利用旧物表达自我的有趣方式。

# TEACHER AND PARENT GUIDE

## 思维拓展
### Systems: Making the Connections

　　口腔是人体与外部世界之间的第一道主要屏障，因此是人类免疫系统的第一道防线。如果没有口腔提供的独特的保护，数十亿通过空气传播的细菌和真菌将进入人体。作为防御机制，免疫系统与其说是一个杀菌机器，不如说是一个不断寻求平衡和排除的过程。单糖和碳水化合物摄入过多的现代饮食习惯，增加了免疫系统的负担。唾液中的钙离子自然修复系统，无法负担过量"细菌食物"的负荷。过多的不健康食品引来了对我们的健康存在潜在威胁的细菌。为了保持最好的口腔卫生状态，必须经常刷牙并用牙线清洁牙齿。理想的解决方案是大量减少糖和碳水化合物的摄入。除了作为阻挡细菌入侵的屏障，口腔和牙齿还经过神经系统与人体其他器官相连。咀嚼本身就能激活和刺激身体，掉一颗牙齿就会失去部分刺激功能。金属的引入，从人造牙冠、填充物到人工植牙，都会阻碍人体的能量流动。而这种能量流动对保持身体和免疫系统处于激活状态并持续发挥功能是非常必要的。如果人们长期不健康饮食，细菌也会改变它们的饮食，从而导致失衡，摧毁健康系统的根基。如果链球菌过多、真菌失衡、骨质被细菌吃掉，牙齿就会失去刺激身体其他部位的能力。重金属从口腔渗透体内，使身体变得不堪重负，甚至中毒，需要专业手段才能解毒。金属等外来物质应该从口腔里清除，否则它会持续释放微量金属，增加免疫系统的负担，甚至导致免疫系统崩溃。从改变饮食习惯开始，清除各种来源的重金属，保持口腔卫生，恢复牙齿对各器官的刺激功能，这才是通往健康和幸福真正的系统方法，远比刷牙和用牙线洁牙更有意义。

## 动手能力
### Capacity to Implement

　　组织一次刷牙音乐会。世界上第一次刷牙音乐会是在尼日利亚举办的，30万儿童同时开始刷牙。你可以在学校或家里举行一次这样的活动。确保每个人都有一把好牙刷，一把很耐用的牙刷。放点音乐，跟着节奏刷掉牙菌斑。当所有人做好刷牙的准备后，喊："预备，开始！"仔细地将牙刷上下（垂直）移动以确保不会损害牙龈。只有当牙刷放在你牙齿的表面上时，才能左右（水平）移动。你可以以牙线洁牙结束你的刷牙音乐会。

故事灵感来自

# 托马斯·拉乌医生
## Dr Thomas Rau

托马斯·拉乌是一位医师，他毕业于伯尔尼大学。在一家康复中心工作时，他发现尽管他按照学校所教的知识尽心地给病人治疗，但是病人的病情并没有好转。于是拉乌医生转而成为印度草药按摩和中医的终生学习者。从那以后，他成为整体论医学的拥护者，并阐述了一个理论：健康的恢复建立在中西医结合的基础上，包括解毒、营养、消化和提高免疫力。他遵循生物医学的原则，把口腔卫生和无金属牙科治疗融入他的医疗实践。拉乌医生现在是瑞士帕拉塞尔苏斯诊所的医疗总监。

更多资讯

http://www.drrausway.com/

## 图书在版编目（CIP）数据

根与芽：汉英对照 /（比）冈特·鲍利著；（哥伦）凯瑟琳娜·巴赫绘；何家振译. -- 上海：学林出版社，2016.6

（冈特生态童书. 第三辑）

ISBN 978-7-5486-1062-5

Ⅰ. ①根… Ⅱ. ①冈… ②凯… ③何… Ⅲ. ①生态环境－环境保护－儿童读物－汉、英 Ⅳ. ① X171.1-49

中国版本图书馆 CIP 数据核字 (2016) 第 125788 号

---

ⓒ 2015 Gunter Pauli

著作权合同登记号 图字 09-2016-309 号

### 冈特生态童书

#### 根与芽

| | |
|---|---|
| 作　　者—— | 冈特·鲍利 |
| 译　　者—— | 何家振 |
| 策　　划—— | 匡志强 |
| 责任编辑—— | 匡志强　蔡雪奇 |
| 装帧设计—— | 魏　来 |
| 出　　版—— | 上海世纪出版股份有限公司 学林出版社 |
| | 地　址：上海钦州南路81号　电话／传真：021-64515005 |
| | 网址：www.xuelinpress.com |
| 发　　行—— | 上海世纪出版股份有限公司发行中心 |
| | （上海福建中路193号　网址：www.ewen.co） |
| 印　　刷—— | 上海丽佳制版印刷有限公司 |
| 开　　本—— | 710×1020　1/16 |
| 印　　张—— | 2 |
| 字　　数—— | 5万 |
| 版　　次—— | 2016年6月第1版 |
| | 2016年6月第1次印刷 |
| 书　　号—— | ISBN 978-7-5486-1062-5/G · 397 |
| 定　　价—— | 10.00元 |

（如发生印刷、装订质量问题，读者可向工厂调换）